Second Edition

Real and Functional Analysis

PART B
FUNCTIONAL ANALYSIS

MATHEMATICAL CONCEPTS AND METHODS IN SCIENCE AND ENGINEERING

Series Editor: Angelo Miele
Mechanical Engineering and Mathematical Sciences
Rice University

Recent volumes in this series:

22 **APPLICATIONS OF FUNCTIONAL ANALYSIS IN ENGINEERING** • *J. L. Nowinski*

23 **APPLIED PROBABILITY** • *Frank A. Haight*

24 **THE CALCULUS OF VARIATIONS AND OPTIMAL CONTROL: An Introduction** • *George Leitmann*

25 **CONTROL, IDENTIFICATION, AND INPUT OPTIMIZATION** • *Robert Kalaba and Karl Spingarn*

26 **PROBLEMS AND METHODS OF OPTIMAL STRUCTURAL DESIGN** • *N. V. Banichuk*

27 **REAL AND FUNCTIONAL ANALYSIS, Second Edition Part A: Real Analysis** • *A. Mukherjea and K. Pothoven*

28 **REAL AND FUNCTIONAL ANALYSIS, Second Edition Part B: Functional Analysis** • *A. Mukherjea and K. Pothoven*

29 **AN INTRODUCTION TO PROBABILITY THEORY WITH STATISTICAL APPLICATIONS** • *Michael A. Golberg*

30 **MULTIPLE-CRITERIA DECISION MAKING: Concepts, Techniques, and Extensions** • *Po-Lung Yu*

31 **NUMERICAL DERIVATIVES AND NONLINEAR ANALYSIS** • *Harriet Kagiwada, Robert Kalaba, Nima Rasakhoo, and Karl Spingarn*

32 **PRINCIPLES OF ENGINEERING MECHANICS Volume 1: Kinematics—The Geometry of Motion** • *Millard F. Beatty, Jr.*

33 **PRINCIPLES OF ENGINEERING MECHANICS Volume 2: Dynamics—The Analysis of Motion** • *Millard F. Beatty, Jr.*

34 **STRUCTURAL OPTIMIZATION Volume 1: Optimality Criteria** • *Edited by M. Save and W. Prager*

A Continuation Order Plan in available for this series. A continuation order will bring delivery of each new volume immediately upon publication. Volumes are billed only upon actual shipment. For further information please contact the publisher.

Second Edition

Real and Functional Analysis

PART B
FUNCTIONAL ANALYSIS

A. Mukherjea and K. Pothoven

University of South Florida
Tampa, Florida

Plenum Press · New York and London

Library of Congress Cataloging in Publication Data

Mukherjea, Arunava, 1941–
Real and functional analysis.

(Mathematical concepts and methods in science and engineering; 27–28)
Includes bibliographical references and indexes.
Contents: pt. A. Real analysis—pt. B. Functional analysis.
1. Functions of real variables. 2. Functional analysis. I. Pothoven, K. II. Title. III. Series.
QA331.5.M84 1984 515.8 84-8363
ISBN 0-306-41558-5 (v. 2)

A Division of Plenum Publishing Corporation
233 Spring Street, New York, N.Y. 10013

Printed in the United States of America

Preface to the Second Edition

The second edition is composed of two volumes. The first volume, Part A, is entitled *Real Analysis* and contains Chapters 1, 2, 3, 4, parts of Chapter 5, and the first five sections of Chapter 7 of the first edition, together with additional material added to each of these chapters. The second volume, Part B, is entitled *Functional Analysis* and contains Chapters 5 and 6 and the Appendix of the first edition together with additional topics in functional analysis including a new section on topological vector spaces, a complete chapter on spectral theory, and an appendix on invariant subspaces. Included in this edition are many new problems, new proofs of theorems, and additional material. Our goal has been, as before, to present the essentials of analysis as well as to include in the book many interesting, useful, and relevant results (usually not available in other books) so that the book can be useful as a reference for the student of analysis.

As in the first edition, certain portions of the text designated by (•) can be omitted. In this volume, problems that are designated by (×) are an integral part of the text and should be worked by the student. Problems that are difficult are starred (⋆).

We are again grateful to friends and colleagues who have pointed out errors in the first edition and given suggestions for improving the text. Particularly, we thank Professors G. Högnäs, R. A. Johnson, and B. Schreiber.

A. Mukherjea
Tampa, Florida K. Pothoven

Contents

6

Banach Spaces

Integral equations occur in a natural way in numerous physical problems and have attracted the attention of many mathematicians including Volterra, Fredholm, Hilbert, Schmidt, F. Riesz, and others. The works of Volterra and Fredholm on integral equations emphasized the usefulness of the techniques of the integral operators. Soon it was realized that many problems in analysis could be attacked with greater ease if placed under a suitably chosen axiomatic framework. Axioms closely related to those of a Banach space were introduced by Bennett.[†] Using the axioms of a Banach space, F. Riesz[‡] extended much of the Fredholm theory of integral equations. In 1922 using similar sets of axioms for such spaces, Banach,[§] Wiener,[‖] and Hahn[¶] all independently published papers. But it was Banach who continued making extensive and fundamental contributions in the development of the theory of these spaces, now well known as Banach spaces. Banach-space techniques are widely known now and applied in numerous physical and abstract problems. For example, using the Hahn–Banach Theorem (asserting the existence of nontrivial continuous linear real-valued functions on Banach spaces), one can show the existence of a translation-invariant, finitely additive measure on the class of all bounded subsets of the reals such that the measure of an interval is its length.

The purpose of this chapter is to present some of the basic properties and principles of Banach spaces. In Section 6.1 we introduce the basic

[†] A. A. Bennett, *Proc. Nat. Acad. Sci. U.S.A.* **2**, 592–598 (1916).
[‡] F. Riesz, *Acta Math.* **41**, 71–98 (1918).
[§] S. Banach, *Fund. Math.* **3**, 133–181 (1922).
[‖] N. Wiener, *Bull. Soc. Math. France* **50**, 119–134 (1922).
[¶] H. Hahn, *Monatsh. Math. Phys.* **32**, 3–88 (1922).

concepts and definitions. In Sections 6.2 and 6.3 we present what are acknowledged as the four most important theorems in Banach spaces—the Hahn–Banach Theorem, the Open Mapping Theorem, the Closed Graph Theorem, and the Principle of Uniform Boundedness. In Section 6.4 we introduce the reflexive spaces, derive representation theorems for the duals of various important Banach spaces, and present in detail the interplay between reflexivity and weak topology. In Section 6.5 we introduce compact operators, present the classical Fredholm alternative theory, and discuss spectral concepts for such operators. In Section 6.6, topological vector spaces, locally convex spaces, and the Krein–Milman theorem are introduced. We present also Liapounoff's convexity theorem, as an application of the Krein–Milman theorem, in this section. The final section contains the Kakutani fixed point theorem and its application showing the existence of an invariant measure on a compact Hausdorff topological group.

For the sake of completeness and continuity, a slight overlap between parts of Sections 3.5 and 4.3 in Part A and parts of Sections 6.1 and 6.2 in this chapter has been unavoidable. The reader who is already familiar with those sections in Part A may, of course, skip this material in this chapter.

Throughout this chapter, F will denote either the real numbers R or the complex numbers C.

6.1. Basic Concepts and Definitions

We begin with several fundamental definitions.

Definition 6.1. A *linear space* X over a field F is an Abelian group under addition $(+)$, together with a scalar multiplication from $F \times X$ into X such that

(i) $\alpha(x + y) = \alpha x + \alpha y$,
(ii) $(\alpha + \beta)x = \alpha x + \beta x$,
(iii) $(\alpha\beta)x = \alpha(\beta x)$,
(iv) $1x = x$,

for all $\alpha, \beta \in F$, x and $y \in X$. (Here 1 denotes the multiplicative identity in F and 0 will denote the additive identity in X.)

Definition 6.2. A linear space X over a field F is called a *normed linear space* if to each $x \in X$ is associated a nonnegative real number $\| x \|$, called the norm of x such that

(i) $\| x \| = 0,$ if and only if $x = 0$;
(ii) $\| \alpha x \| = | \alpha | \; \| x \|,$ for all $\alpha \in F$;
(iii) $\| x + y \| \leq \| x \| + \| y \|,$ for all $x, y \in X$.

Defining $d(x, y) = \| x - y \|$, it can easily be verified that d defines a metric in X. A normed linear space X is called a *Banach* space if it is complete in this metric. The topology induced by d will usually be referred to as the topology in X.

A linear space is also called a vector space. Let X be a vector space over a field F. (The elements of X are often called vectors and those of F scalars.) Let $A \subset X$. If the subset A is also a vector space over F with respect to the same operations, then A is called a linear subspace (or simply, a subspace). The subset A is said to span X if any element of X can be expressed as a finite linear combination of elements from A, that is, for each $x \in X$ there exist scalars $a_1, a_2, \ldots, a_n$ in F, and vectors $x_1, x_2, \ldots, x_n$ in X such that $x = a_1x_1 + a_2x_2 + \cdots + a_nx_n$. Note that in a vector space, for any given subset B there is always a subspace (denoted by $\langle B \rangle$) spanned by B, namely, the set of all finite linear combinations of elements from B. The subspace $\langle B \rangle$ is the smallest subspace of X containing B. In other words, if S is a subspace of X containing B, then S contains the subspace $\langle B \rangle$.

A finite subset $\{x_1, x_2, \ldots, x_n\}$ of a vector space X over F is called linearly independent if any relation of the form

$$a_1x_1 + a_2x_2 + \cdots + a_nx_n = 0, \qquad a_i \in F \; (i = 1, 2, \ldots, n)$$

implies that $a_1 = a_2 = \cdots = a_n = 0$. A subset (not necessarily finite) of a vector space is called linearly independent if every finite subset of it is linearly independent. Notice that if $(A_k)_{k \in K}$ is a family (possibly infinite) of linearly independent subsets in a vector space such that for k_1, k_2 in K $(k_1 \neq k_2)$, either $A_{k_1} \subset A_{k_2}$ or $A_{k_2} \subset A_{k_1}$, then $\cup\{A_k : k \in K\}$ is also a linearly independent subset. Thus, considering the partial order "inclusion" in the family $\mathscr{F}$ of all linearly independent subsets in a vector space X $(X \neq \{0\})$, it is clear that $\mathscr{F}$ is nonempty (since it contains all nonzero singleton subsets), and each nonempty totally (or linearly) ordered subset of $\mathscr{F}$ has an upper bound. It follows by using Zorn's lemma (Part A, Theorem 1.2) that X has a maximal linearly independent set H. Such a set is called a Hamel basis for X. Equivalently, a subset H of X is a Hamel basis for X if every vector x in X can be expressed uniquely as a finite linear combination of vectors from H. Note that any linearly independent subset of X is contained in a Hamel basis for X. (This is proven in the same manner as before using Zorn's lemma in the family of all linearly independent subsets of X which contain as a subset the given linearly independent set.) The cardinality of a Hamel basis for a vector space X is a number independent of the Hamel basis, and is called the dimension of X. (See Problem 6.1.28 and the hint therein.)

Now we consider some examples of normed linear spaces.

Examples

6.1. $R^n = \{(a_1, a_2, \ldots, a_n): a_i \in R,\ 1 \leq i \leq n\}$ is a Banach space over R if we define addition and scalar multiplication in the natural way and

$$\| (a_1, a_2, \ldots, a_n) \| = \sum_{i=1}^{n} | a_i |.$$

6.2. $C_1[0, 1]$, the usual linear space of complex-valued continuous functions over $[0, 1]$, is a Banach space over F if we define

$$\| f \| = \sup_{0 \leq x \leq 1} | f(x) |.$$

(See Problem 6.1.5.) This norm will be referred to as the uniform or "sup" norm.

6.3. The set of all polynomials on $[0, 1]$ as a subspace of $C[0, 1]$ in Example 6.2 is a normed linear space, but not a Banach space. The reason is that nonpolynomial continuous functions can be uniformly approximated by polynomials on $[0, 1]$, by Weierstrass' theorem.

6.4. Let $(X, \mathscr{A}, \mu)$ be a measure space. For real number $p \geq 1$, let L_p be the linear space of all scalar-valued measurable functions f such that $| f |^p$ is integrable. If functions that are equal a.e. are identified, then L_p becomes a Banach space with the norm

$$\| f \|_p = \left(\int | f |^p \, d\mu \right)^{1/p}.$$

These spaces were introduced in Section 3.5 as were the spaces L_∞. L_∞ is the space of all measurable functions that are bounded except possibly on a set of measure zero. Again, if we identify functions in L_∞ that are equal a.e., L_∞ is a Banach space with the norm

$$\| f \| = \text{ess sup}\, | f |,$$

where

$$\text{ess sup} \mid f \mid = \inf\{M: \mu\{x: \mid f(x) \mid > M\} = 0\}.$$

When the measure space X is the set of positive integers with each integer having measure 1, the L_p spaces are called l_p spaces. Thus $l_p (1 \leq p < \infty)$ is the set of all sequences $x = (x_n)_{n=1}^{\infty}$, $x_n \in F$ with $\sum_{n=1}^{\infty} \mid x_n \mid^p < \infty$. l_p, like L_p, is a Banach space under the norm

$$\| x \|_p = \left(\sum_{n=1}^{\infty} \mid x_n \mid^p \right)^{1/p}.$$

The l_∞ space is the set of all bounded sequences $x = (x_n)_{n=1}^{\infty}$, $x_n \in F$. Again, l_∞, like L_∞, is a Banach space under the norm $\| x \|_\infty = \sup_n \mid x_n \mid$.

Definition 6.3. Suppose X is a linear space over F, and $\| \cdot \|_1$ and $\| \cdot \|_2$ are two norms defined on it. Then these norms are called *equivalent* if there exist positive numbers a and b such that

$$a \cdot \| x \|_1 \leq \| x \|_2 \leq b \cdot \| x \|_1$$

for all $x \in X$. ∎

Equivalence of norms is clearly reflexive, symmetric, and transitive. In an infinite-dimensional normed linear space, there are always two norms that are not equivalent (Problem 6.1.12). But the situation is much nicer in the finite-dimensional case, as the following theorem shows.

Theorem 6.1. In a finite-dimensional linear space X, all norms are equivalent. ∎

Proof. Let $\{z_1, z_2, \ldots, z_n\}$ be a basis for X. Let $x \in X$ and $x = \sum_{i=1}^{n} x_i z_i$. We define: $\| x \| = \sup_{1 \leq i \leq n} \mid x_i \mid$. Then $\| \cdot \|$ defines a norm in X, and X is complete in this norm (see Problem 6.1.2). We show that any given norm $\| \cdot \|_1$ on X is equivalent to this sup-norm.

Let $x = \sum_{i=1}^{n} x_i z_i$ and $b = \sum_{i=1}^{n} \| z_i \|_1$. Then $b > 0$ and

$$\| x \|_1 \leq b \| x \| \text{ for all } x \in X.$$

To find a similar inequality in the other direction, we use induction on n. When $n = 1$,

$$\| x \|_1 = \mid x_1 \mid \| z_1 \|_1 = \| x \| \| z_1 \|_1.$$

Thus, the theorem is true for $n = 1$. Suppose that the theorem is true for all spaces of dimension less than n, $n > 1$. Let X_i be the subspace spanned by $\{z_1, \ldots, z_{i-1}, z_{i+1}, \ldots, z_n\}$. Then dim $X_i < n$. Since X_i is complete in $\| \cdot \|$ and since the norms $\| \cdot \|$ and $\| \cdot \|_1$ are equivalent in X_i by the induction hypothesis, the subspace X_i is also complete in $\| \cdot \|_1$ and is therefore a closed subspace of $(X, \| \cdot \|_1)$. Since $z_i \notin X_i$, there is a positive number d_i such that

$$\inf\{\| z_i + y \|_1 : y \in X_i\} > d_i.$$

This means that for all scalars $\alpha_j \in F$,

$$\left\| \sum_{j=1}^{n} \alpha_j z_j \right\|_1 \geq | \alpha_i | d_i.$$

We can repeat the process for each i, $1 \leq i \leq n$, and for each i find positive numbers $d_i > 0$. Then

$$\left\| \sum_{i=1}^{n} \alpha_i z_i \right\|_1 \geq \sup_{1 \leq i \leq n} | \alpha_i | \cdot \min_{1 \leq i \leq n} | d_i |,$$

which means that

$$a \| x \| \leq \| x \|_1, \qquad \text{for all } x \in X,$$

where $a = \min_{1 \leq i \leq n} | d_i | > 0$. ∎

Corollary 6.1. Every finite-dimensional normed linear space is complete. ∎

Corollary 6.2. Let $(X_1, \| \cdot \|_1)$ and $(X_2, \| \cdot \|_2)$ be any two finite-dimensional normed linear spaces of the same dimension over F. Then they are topologically isomorphic, i.e., there is a mapping from one onto the other, which is an algebraic isomorphism as well as a topological homeomorphism. ∎

The proof is an easy consequence of Theorem 6.1 and is left to the reader. Next, we present a result due to Riesz, often useful in proving various results in the theory of normed linear spaces, besides being of independent geometric interest.

Proposition 6.1. Let Y be a proper closed linear subspace of a normed linear space X over F. Let $0 < a < 1$. Then there exists some $x_a \in X$ such that $\| x_a \| = 1$ and $\inf_{y \in Y} \| x_a - y \| \geq a$. ∎

Proof. Let $x \in X - Y$ and $d = \inf_{y \in Y} \| x - y \|$. Then $d > 0$, since Y is closed. Now there exists $y_0 \in Y$ such that $0 < \| x - y_0 \| < d/a$. Let $x_a = [x - y_0]/\| x - y_0 \|$. Now the reader can easily check that x_a satisfies the requirements of the propositon. ∎

Note that in the above proposition, the proper subspace Y has to be necessarily *closed*. For instance, if X is $C[0,1]$ with the uniform† norm and Y is the subspace of all polynomials on $[0,1]$, then $\bar{Y} = X$ and therefore the proposition fails to work in this case.

Also, one cannot generally take $a = 1$ in Proposition 6.1. For example, let X be the real-valued continuous functions on $[0, 1]$ which vanish at 0, with the uniform norm, and $Y = \{f \in X: \int_0^1 f(x)\,dx = 0\}$. Then Y is a closed proper subspace of X. Suppose there exists $h \in X - Y$ such that $\inf_{f \in Y} \| h - f \| \geq 1$, where for $g \in X$, $\| g \| = \sup_{0 \leq x \leq 1} | g(x) |$. If $g \in X - Y$ and $a(g) = [\int_0^1 h(x)\,dx]/[\int_0^1 g(x)\,dx]$, then

$$\int_0^1 [h(x) - a(g) \cdot g(x)]\,dx = 0 \quad \text{or} \quad h - a(g) \cdot g \in Y;$$

therefore, $\| h - [h - a(g) \cdot g] \| = \| a(g) \cdot g \| \geq 1$. Let $g_n(x) = x^{1/n}$, $1 \leq n < \infty$. Then $g_n \in X - Y$ and hence $\| a(g_n) \cdot g_n \| \geq 1$. But $a(g_n) = [(n + 1)/n] \int_0^1 h(x)\,dx$. Hence, since $\| g_n \| = 1$, we have

$$\left| \int_0^1 h(x)\,dx \right| \geq \frac{n}{n+1},$$

for each positive integer n. This means that $| \int_0^1 h(x)\,dx | \geq 1$. But since $h(0) = 0$ and $\| h \| = 1$, $| \int_0^1 h(x)\,dx | < 1$, which is a contradiction.

Now we show how Proposition 6.1 can help us understand the notion of compactness in a normed linear space. We know that a set in R^n or in any finite-dimensional normed linear space is compact if and only if it is closed and bounded. But if X is an infinite-dimensional normed linear space, then by Proposition 6.1 we can find (x_n) with $\| x_n \| = 1$, $1 \leq n < \infty$ and for each n, $\| x_n - x_i \| \geq \frac{1}{2}$, for $1 \leq i < n$; clearly these x_n's cannot have a limit point and therefore the closed unit ball in X is not compact. Thus we have the following theorem.

Theorem 6.2. Let X be a normed linear space over F. Then its closed unit ball is compact if and only if X is finite dimensional. ∎

† The same as the "sup"-norm.

Definition 6.4. A series $\sum_{k=1}^{\infty} x_k$ in a normed linear space X over F is called *summable* if $\| \sum_{k=1}^{n} x_k - x \| \to 0$ as $n \to \infty$ for some $x \in X$. For a summable series, we write $\sum_{k=1}^{\infty} x_k = \lim_{n\to\infty} \sum_{k=1}^{n} x_k$. The series $\sum_{k=1}^{\infty} x_k$ is called absolutely summable if $\sum_{k=1}^{\infty} \| x_k \| < \infty$. ∎

We know that an absolutely summable series of real numbers is summable. This is a consequence of the completeness of the real numbers. In fact, we have the following theorem, which is often useful in establishing the completeness of a normed linear space.

Theorem 6.3. Every absolutely summable series in a normed linear space X is summable if and only if X is complete. ∎

Proof. For the "if" part, let X be complete and for each positive integer n let x_n be an element of X such that $\sum_{n=1}^{\infty} \| x_n \| < \infty$. Let $y_k = \sum_{n=1}^{k} x_n$. Then

$$\| y_{k+p} - y_k \| = \left\| \sum_{n=k+1}^{k+p} x_n \right\| \leq \sum_{n=k+1}^{k+p} \| x_n \|,$$

which converges to zero as $k \to \infty$. Hence the sequence $(y_k)_{k=1}^{\infty}$ is Cauchy in X. Therefore since X is complete, there exists $x \in X$ such that $x = \lim_{k\to\infty} \sum_{n=1}^{k} x_n$. This proves the "if" part.

For the "only if" part, suppose every absolutely summable series in X is summable. Let (x_n) be a Cauchy sequence in X. For each positive integer k, there is a positive integer n_k such that $\| x_n - x_m \| < 1/2^k$ for all n and m greater than or equal to n_k. We choose $n_{k+1} > n_k$. Let $y_1 = x_{n_1}$ and $y_{k+1} = x_{n_{k+1}} - x_{n_k}$, $k \geq 1$. Then $\sum_{k=1}^{\infty} \| y_k \| < \infty$. Therefore, there exists $y \in X$ such that

$$y = \lim_{m\to\infty} \sum_{k=1}^{m} y_k = \lim_{m\to\infty} x_{n_m}.$$

Since (x_n) is Cauchy, $\lim_{n\to\infty} x_n$ is also y. ∎

Finally, in this section we introduce the concept of quotient spaces. Quotient spaces are often useful in tackling certain problems in normed linear spaces, as will be seen in the later sections of this chapter. See also Problem 6.1.10.

Suppose M is a closed subspace of a normed linear space X over a field F. Let $[x] = \{y \in X\colon\ y \sim x\}$, where $x \in X$ and $y \sim x$ if and only if $y - x \in M$. Clearly, "$\sim$" defines an equivalence relation on X. Let X/M

denote the set of all equivalence classes. For $\alpha \in F$ and $[x]$, $[y]$ in X/M, we define

$$[x] + [y] = [x + y] \quad \text{and} \quad \alpha[x] = [\alpha x].$$

These operations are well defined and make X/M a vector space. Let $\| [x] \|_1 = \inf_{y \in M} \| x - y \|$. One can easily check that $(X/M, \| \cdot \|_1)$ is a normed linear space, usually called the *quotient of X by M*. If Φ is the natural map from X onto X/M defined by $\Phi(x) = [x]$, then Φ is a continuous, linear, and open mapping. The linearity of Φ is trivial. The continuity of Φ follows from the fact that $\| [x_n - x] \|_1 \leq \| x_n - x \|$. Also, if $V = \{y \in X: \| y - x \| < r\}$, then $\Phi(V) = \{[y] \in X/M: \| [y] - [x] \|_1 < r\}$. This implies that Φ is open.

In many cases, several properties of the quotient space X/M are strongly related to similar properties of the space X. Problem 6.1.9 and the next proposition will illustrate two of them.

Proposition 6.2. Let M be a closed linear subspace of a normed linear space X. Then X is complete if and only if M and X/M are complete. ▮

Proof. For the "if" part, let M and X/M be complete. Let $(x_n)_{n=1}^\infty$ be a Cauchy sequence in X. Then $([x_n])_{n=1}^\infty$ is Cauchy in X/M and therefore there exists $[y] \in X/M$ such that $[x_n] \to [y]$ in X/M as $n \to \infty$. This means that $\inf_{z \in M} \| x_n - y - z \| \to 0$ as $n \to \infty$. Hence there exists a subsequence (n_k) of positive integers and a sequence (z_k) in M such that $x_{n_k} - y - z_k \to 0$ as $k \to \infty$. This means that the sequence $(z_k)_{k=1}^\infty$ is Cauchy in M and therefore there exists $z \in M$ such that $z_k \to z$ as $k \to \infty$ so that $x_n \to y + z$ as $n \to \infty$. The "if" part is proved.

For the "only if" part, we will use Theorem 6.3. Let X be complete. Then M, being a closed subspace, is also complete. To prove the completeness of X/M, let $\sum_{n=1}^\infty \| [x_n] \|_1 < \infty$. We are finished if we can show that there exists $[y] \in X/M$ such that $[y] = \lim_{k \to \infty} \sum_{n=1}^k [x_n]$. Let $y_n \in M$ be chosen such that $\| x_n + y_n \| \leq \| [x_n] \|_1 + 1/2^n$, for each positive integer n. Then $\sum_{n=1}^\infty \| x_n + y_n \| < \infty$. Since X is complete, there exists $y \in X$ such that $y = \lim_{k \to \infty} \sum_{n=1}^k (x_n + y_n)$. Since the natural map Φ is continuous and linear,

$$[y] = \lim_{k \to \infty} \sum_{n=1}^k [x_n + y_n] = \lim_{k \to \infty} \sum_{n=1}^k [x_n]. \quad ▮$$

Problems

× **6.1.1.** Let X be a normed linear space over a field F. Let $y \in X$ and $\alpha \in F$, $\alpha \neq 0$. Show that the mappings $x \to x + y$ and $x \to \alpha \cdot x$ are homeomorphisms of X onto itself.

× **6.1.2.** Let X be a linear space (of dimension n) over a field F. Let $\{z_1, z_2, \ldots, z_n\}$ be a basis of X. If $x = \sum_{i=1}^{n} \alpha_i z_i$, $\alpha_i \in F$ and $\| x \| = \sup_{1 \le i \le n} | \alpha_i |$, then show that $(X, \| \cdot \|)$ is a Banach space.

× **6.1.3.** Let X be a normed linear space. If $S_x(r) = \{y \in X: \| y - x \| < r\}$, then show that $\overline{S_x(r)} = \{y \in X: \| y - x \| \le r\}$.

× **6.1.4.** Let A and B be two subsets of a normed linear space X. Let

$$A + B = \{x + y: x \in A, y \in B\}.$$

Show that (a) $A + B$ is open whenever either A or B is open; and (b) $A + B$ is closed whenever A is compact and B is closed. (Note that $A + B$ need not be closed even if A and B are both closed.)

× **6.1.5.** Prove that $C_1[0, 1]$, the linear space of complex-valued continuous functions on $[0, 1]$, is a Banach space under the uniform norm.

× **6.1.6.** Show that a finite-dimensional subspace of an infinite-dimensional normed linear space X is nowhere dense in X.

× **6.1.7.** Use Problem 6.1.6 and the Baire Category Theorem to prove that an infinite-dimensional Banach space cannot have a countable Hamel basis.

× **6.1.8.** Let X be a normed linear space and f be the mapping defined by $f(x) = rx/(1 + \| x \|)$. Show that f is a homeomorphism from X onto $\{x: \| x \| < r\}$. [Hint: $f^{-1}(y) = y/(r - \| y \|)$.]

× **6.1.9.** Let M be a closed subspace of a normed linear space X. Show that X is separable if and only if M and X/M are both separable.

× **6.1.10.** (i) Show that the sum of two closed subspaces of a normed linear space is closed whenever one of the subspaces is finite dimensional. [Hint: Let A be a closed subspace and B be a finite-dimensional subspace of a normed linear space X. If Φ is the natural map from X onto X/A, then $\Phi^{-1}(\Phi(B)) = A + B$.]

(ii) Let M be a closed subspace of a normed linear space X. Show that $M + N$ is closed for every subspace N if $\dim(X/M)$ is finite.

(iii) Let $X = l_2$ and for each k, let $e_k = (0, \ldots, 0, 1, 0, 0, \ldots)$, where the only nonzero entry is 1 and at the kth position. Let M be the closed subspace of X spanned by the elements $(e_{2k})_{k=1}^{\infty}$ and N be the closed subspace spanned by the elements $(e_{2k} + (1/k)e_{2k-1})_{k=1}^{\infty}$. Show that $M + N$ is dense in X, but does not contain the element $(1, 0, \frac{1}{3}, 0, \frac{1}{5}, \ldots)$.

× **6.1.11.** Show that a normed linear space whose closed unit ball is totally bounded is finite dimensional.

6.1.12. Let X be an infinite-dimensional linear space over F. Let (x_n) be an infinite set contained in H, a Hamel basis for X. If $x \in X$ and $x = \sum a_\alpha x_\alpha$, $x_\alpha \in H$ and $a_\alpha \in F$, then define $\| x \|_1 = \sup | a_\alpha |$ and $\| x \|_2 = \sum | a_\alpha |$. Show that these two norms are not equivalent. [Hint: Let $x_n \in H$ for $n \geq 1$ and $y_n = (1/n)(x_1 + x_2 + \cdots + x_n)$. Then $\| y_n \|_1 \to 0$ as $n \to \infty$; $\| y_n \|_2 = 1$ for each n.] See Example 6.12 later for $\| \cdot \|_2$.

× **6.1.13.** Prove that $L_p(X, \mathscr{A}, \mu)$, $p \geq 1$, is a Banach space under the norm $\| f \|_p = (\int | f |^p \, d\mu)^{1/p}$.

× **6.1.14.** Prove that $L_\infty(X, \mathscr{A}, \mu)$ is a Banach space under the norm $\| f \|_\infty = \text{ess sup} \, | f |$.

6.1.15. Prove that a subset $A \subset l_p$, $1 \leq p < \infty$, is relatively compact if and only if A is bounded and $\sum_{i=n}^{\infty} | x_i |^p \to 0$ as $n \to \infty$ uniformly for all $(x_1, x_2, \ldots) \in A$.

6.1.16. A *Schauder basis* for a normed linear space X is a sequence (u_n) in X such that for every x in X, there is a unique sequence of scalars (a_n) such that $x = \sum_{n=1}^{\infty} a_n u_n$, i.e., $\lim_{n\to\infty} \| x - \sum_{i=1}^{n} a_i u_i \| = 0$. Prove[†] the following assertions:

(i) A normed linear space with a Schauder basis is separable.

(ii) Let e_n denote the element (x_i) in l_p, where $x_i = 0$ for $i \neq n$ and $x_n = 1$. Then the sequence (e_n) is a Schauder basis for l_p, $1 \leq p < \infty$.

(iii) The sequence of polynomials $\{x^n : n \geq 0\}$ is not a Schauder basis for $C[0, 1]$ with the "sup" norm. (Hint: See if a continuous function, not differentiable at 0, can be represented by a uniformly convergent power series on $[0, 1]$)

(iv) Consider $C[0, 1]$ with the "sup" norm. Define a sequence (f_n) in $C[0, 1]$ as follows: Let (t_i) be the sequence of dyadic points in $[0, 1]$ given, respectively, by

$$0, 1, \tfrac{1}{2}, \tfrac{1}{4}, \tfrac{3}{4}, \tfrac{1}{8}, \tfrac{3}{8}, \tfrac{5}{8}, \tfrac{7}{8}, \tfrac{1}{16}, \tfrac{3}{16}, \ldots$$

Let $f_1(t) = 1$, $f_2(t) = t$ for each t in $[0, 1]$. For $n > 2$, let $f_n(t_i) = 0$ for $i < n$, $= 1$ for $i = n$, and f_n be linear between any two consecutive points among the first n dyadic points. Let $g \in C[0, 1]$. Then the sequence $\sum_{i=1}^{n} a_i f_i$, where the a_i's are given by

$$a_n = g(t_n) - \sum_{i=1}^{n-1} a_i f_i(t_n), \qquad a_1 = g(0),$$

converges uniformly to g on $[0, 1]$. [Thus, the sequence (f_n) is a Schauder basis for $C[0, 1]$.]

[†] For these and other information on Schauder bases, the serious reader should consult R. C. James, *Amer. Math. Monthly*, Nov. 1982, pp. 625–640.

(v) Let X be a Banach space and (e_n) be a Schauder basis of X. Define $|||\, x \,|||$ by $|||\, x \,||| = \sup\{\| \sum_{i=1}^{n} x_i e_i \|: n \geq 1\}$ if $x = \sum_{i=1}^{\infty} x_i e_i$. Then this defines a norm in X. The space X is complete in this new norm, and the identity mapping from $(X, |||\cdot|||)$ to $(X, \|\cdot\|)$ is continuous.

× **6.1.17.** Let Y be a finite-dimensional subspace of a normed linear space X. Let $x \in X - Y$. Show that there exists $y \in Y$ such that

$$\| x - y \| = \inf\{\| x - z \|: z \in Y\}.$$

(The element y is called a *best approximation* of x in Y.) Also, show that the set of best approximations of x in Y is convex.

× **6.1.18.** A normed linear space X is called *strictly convex* if for any x, y in X such that $\| x \| = \| y \| = 1$ and $x \neq y$, we have $\| \frac{1}{2}(x + y) \| < 1$. Prove that X is strictly convex if and only if for any x, y in X, $\| x + y \| = \| x \| + \| y \|$ implies that $x = ay$ or $y = ax$ with $a > 0$.

× **6.1.19.** Let X and Y be as in Problem 6.1.17. Suppose also that X is strictly convex. Show that any element x in X has a unique best approximation in Y.

6.1.20. Show that the L_p spaces, $1 < p < \infty$, are strictly convex under usual norm, whereas L_1 and L_∞ (with usual norms) are not. (Use the remark on the validity of equality in the Hölder's Inequality for equality in the Minkowski's Inequality.)

{Remark: Notice that $C[a, b]$, the real-valued continuous functions with the "sup" norm is not strictly convex. However, we may remark that for each positive integer n, every element in $C[a, b]$ has a unique best approximation in the finite-dimensional subspace of polynomials of degree not exceeding n. In fact, this assertion remains true for any finite-dimensional subspace Y satisfying the following condition (known as the Haar condition): each nonzero y in Y has at most $m - 1$ zeros in $[a, b]$, where $m = \dim Y$. Also, the Haar condition, it turns out, is necessary and sufficient for the uniqueness of the best approximation in Y.}

6.1.21. Let X be a strictly convex normed linear space and Y be a finite-dimensional subspace of X. For each x in X, let $T(x)$ be the best approximation of x in Y. Prove that T is continuous. (Note that T need not be linear.)

6.1.22. *Uniformly Convex Normed Linear Spaces.* A normed linear space X is called *uniformly convex* if for sequences (x_n) and (y_n) in X such that $\| x_n \| \leq 1$, $\| y_n \| \leq 1$, and $\| \frac{1}{2}(x_n + y_n) \| \to 1$ as $n \to \infty$, we have $\lim_{n\to\infty} \| x_n - y_n \| = 0$. [Notice that if the norm satisfies the parallelogram law, i.e., $\| x + y \|^2 + \| x - y \|^2 = 2\| x \|^2 + 2\| y \|^2$ for every x, y in X, then X is uniformly convex.] Prove that uniformly convex spaces are strictly convex. Also prove that a normed linear space X is uniformly convex if and only if for every ε, $0 < \varepsilon \leq 2$, there exists a $\beta(\varepsilon) > 0$ such that $\| x \| \leq 1$, $\| y \| \leq 1$, and $\| x - y \| \geq \varepsilon$ imply $\| \frac{1}{2}(x + y) \| \leq 1 - \beta(\varepsilon)$.

6.1.23. *Best approximations in normed linear spaces.* Let X be a uniformly convex normed linear space and A be a nonempty convex subset of X such that A is complete in the norm of X. Show that for any x in X, there exists a unique y in A such that

$$\| x - y \| = \inf\{\| x - z \| : z \in A\}.$$

[Hint: Let $y_n \in A$ be such that $\| x - y_n \| \to a(x) \equiv \inf\{\| x - z \| : z \in A\}$. Let $a(x) > 0$. Write z_n for $x - y_n$. Then, $\lim_{n\to\infty} \| z_n \| = \lim_{n,m\to\infty} \| \frac{1}{2}(z_n + z_m) \| = a(x)$. Let $z_n' = z_n / \| z_n \|$. Use uniform convexity to show that $\lim_{n,m\to\infty} \| z_n' - z_m' \| = 0$.]

6.1.24. Prove that the L_p, $1 < p < \infty$, spaces are uniformly convex. {Hint: For $p \geq 2$, use the inequality

$$| \tfrac{1}{2}(a + b) |^p + | \tfrac{1}{2}(a - b) |^p \leq \tfrac{1}{2} | a |^p + \tfrac{1}{2} | b |^p,$$

where a and b are any two complex numbers, to prove that for f, g in L_p,

$$(\| \tfrac{1}{2}(f + g) \|_p)^p + (\| \tfrac{1}{2}(f - g) \|_p)^p \leq \tfrac{1}{2}(\| f \|_p)^p + \tfrac{1}{2}(\| g \|_p)^p.$$

For $1 < p < 2$, the proof of uniform convexity is less trivial. The reader is referred to the paper of J. A. Clarkson [*Trans. Am. Math. Soc.* **40** (1936)].}

6.1.25. Prove that a normed linear space is not separable if and only if it contains an uncountable set of pairwise disjoint balls of radius 1. [Hint for the "only if" part: For each positive integer n, let $\mathscr{F}_n$ be a maximal (with respect to inclusion) family of pairwise disjoint balls of radius $1/n$; $\mathscr{F}_n$ exists by Zorn's Lemma. If $\mathscr{F}_n$ is countable for each n, then the space is separable.]

6.1.26. Let X and Y be two normed linear spaces and $T: X \to Y$ be a linear operator such that Ker T is closed in X. For $[x]$ in $X/\text{Ker } T$, define $T_0([x]) = T(x)$. Prove the following assertions:

(i) T_0 is an isomorphism of $X/\text{Ker } T$ onto $T(X) \subset Y$.

(ii) T_0 is continuous if and only if T is continuous, and in this case, $\| T_0 \| = \| T \|$.

(iii) T is open if and only if T_0 is open.

6.1.27. Prove that every separable Banach space X is a quotient space of l_1. [Hint: Define $p: l_1 \to X$ by $(\beta_1, \beta_2, \ldots) \to \sum_{n=1}^{\infty} \beta_n x_n$, where x_n is a countable dense subset of the unit ball of X. Then p is onto, continuous, and linear.] (See Proposition 6.13 in Section 6.3.)

6.1.28. Let X be a vector space, and let A and B be any two Hamel bases for X. Prove that

(i) For each $x \in A$, there exists $y \in B$ such that the set $(A - \{x\}) \cup \{y\}$ is linearly independent;

(ii) Card $A =$ Card B.

[Hint for (ii): It is enough to show that there are $1 - 1$ functions from A into B and from B into A. Consider the nonempty family $\mathscr{F} \equiv \{f: f$ is $1 - 1$, $D_f \subset A$, $R_f \subset B$, and $R_f \cup (A - D_f)$ is linearly independent$\}$, where $D_f =$ the domain of f, and $R_f =$ the range of f. Partially order $\mathscr{F}$ by $\leq$, where $f \leq g$ if $D_f \subset D_g$, and $g(x) = f(x)$ for $x \in D_f$. By Zorn's lemma, $\mathscr{F}$ has a maximal function m. If $D_m = A$, we are done. Suppose that $D_m \neq A$. Then $R_m \neq B$ since $R_m \cup (A - D_m)$ is linearly independent. Let $z \in A - D_m$ and $w \in B - R_m$. The set $R_m \cup \{w\} \cup (A - D_m)$ cannot be linearly independent since otherwise m can be extended by defining $m(z) = w$. Thus, there exist scalars $a_1, a_2, \ldots, a_n$ and vectors $x_1, x_2, \ldots, x_n$ in $R_m \cup (A - D_m)$ such that for some k, $a_k \neq 0$ and $x_k \in A - D_m$ and $w = a_1x_1 + \cdots + a_nx_n$. For a similar reason, the set $R_m \cup \{w\} \cup (A - (D_m \cup \{x_k\}))$ cannot be linearly independent. This contradicts the linear independence of $R_m \cup (A - D_m)$.]

6.2. Bounded Linear Functionals and the Hahn–Banach Theorem

The aim of this section is to prove the Hahn–Banach Theorem. This theorem is one of the most fundamental theorems in functional analysis. It yields the existence of nontrivial continuous linear functionals on a normed linear space, a basic fact necessary for the development of a large portion of functional analysis. Moreover, it is an indispensible tool in the proofs of many important theorems in analysis. (For example, see Proposition 6.6.)

This theorem was first proved by Hahn in 1927 for a normed linear space over the reals, and then by Banach in 1929 for a real linear space (in the absence of any topology). The complex version of this theorem is due to Bohnenblust and Sobczyk in 1938 and, independently, to Soukhomlinoff, also in 1938.

We start this section by introducing the concept of a bounded linear operator.

Definition 6.5. Let X and Y be vector spaces over the same scalar field. Then a mapping T from X into Y is called a *linear operator* if for all

$x_1, x_2 \in X$ and scalars α, β,

$$T(\alpha x_1 + \beta x_2) = \alpha T(x_1) + \beta T(x_2). \quad \blacksquare$$

Examples

6.5. Let X be the real-valued continuous functions defined on $[0, 1]$, under the uniform norm and Y be the reals. Let

$$T(f) = \int_0^1 f(x)\,dx, \quad f \in X.$$

Then T is a *continuous linear operator* from X into Y.

6.6. Let X be the class of real-valued continuously differentiable functions on $[0, 1]$ and Y be the class of real-valued continuous functions on $[0, 1]$, both under the uniform norm. Let

$$T(f) = \frac{df}{dx}, \quad f \in X.$$

Then T is a *linear* operator. But T is *not continuous,* since the sequence $x^n/n \to 0$ in X, but the sequence $T(x^n/n) = x^{n-1}$ does not converge to 0 in Y.

6.7. Let X be a n-dimensional normed linear space over F and let Y be *any* normed linear space over F. Let

$$T\left(\sum_{i=1}^n \alpha_i x_i\right) = \sum_{i=1}^n \alpha_i y_i,$$

where α_i's are scalars, $\{x_1, x_2, \ldots, x_n\}$ is a basis of X, and $y_1, y_2, \ldots, y_n$ are arbitrarily chosen, but fixed elements of Y. Then T is a linear operator.[†] But T is also *continuous,* since for $x = \sum_{i=1}^n \alpha_i x_i$,

$$\begin{aligned} \| T(x) \| &\leq \sum_{i=1}^n \| y_i \| \cdot \sup_{1\leq i\leq n} | \alpha_i | \\ &\leq \sum_{i=1}^n \| y_i \| \cdot K \cdot \| x \|, \end{aligned}$$

where K is a constant such that

$$\sup_{1\leq i\leq n} | \alpha_i | \leq K \left\| \sum_{i=1}^n \alpha_i x_i \right\|,$$

[†] It is also clear that every linear operator from X into Y must be of this form.

for all n-tuples $\{\alpha_1, \alpha_2, \ldots, \alpha_n\}$. Such a K can always be found since in a finite-dimensional space any norm is equivalent to the sup-norm. (See Theorem 6.1.) Note that this example shows that a linear operator from a finite-dimensional normed linear space into any normed linear space is continuous.

6.8. Let T be a mapping from l_1 into F defined by

$$T(x) = \sum_{i=1}^{\infty} x_i, \qquad x = (x_i)_{i=1}^{\infty} \in l_1.$$

Then T is *linear*, but *not continuous* if we define a new norm in l_1 by considering it as a subspace of l_∞. T is not continuous since $T(z_n) = 1$ for all n, where

$$z_n = \Big(\underbrace{\frac{1}{n}, \frac{1}{n}, \ldots, \frac{1}{n}}_{n \text{ terms}}, 0, 0, \ldots\Big),$$

and $\| z_n \|_\infty \to 0$ as $n \to \infty$. Note that $\| x \|_\infty = \sup_{1 \le i < \infty} | x_i |$.

Definition 6.6. A linear operator T from a normed linear space X into a normed linear space Y is called *bounded* if there is a positive constant M such that

$$\| T(x) \| \le M \| x \|, \qquad \text{for all } x \in X. \qquad \blacksquare$$

Proposition 6.3. Let T be a linear operator from a normed linear space X into a normed linear space Y. Then the following are equivalent:

(a) T is continuous at a point.
(b) T is uniformly continuous on X.
(c) T is bounded. ∎

Proof. (a) $\Rightarrow$ (b). Suppose T is continuous at a point x_0. Then given $\varepsilon > 0$, there exists $\delta > 0$ such that $\| x - x_0 \| < \delta \Rightarrow \| Tx - Tx_0 \| < \varepsilon$. Now let y and z be elements in X with $\| y - z \| < \delta$. Then $\| (y - z + x_0) - x_0 \| < \delta$ and therefore $\| T(y - z + x_0) - T(x_0) \| < \varepsilon$, which means that $\| T(y) - T(z) \| < \varepsilon$, by the linearity of T. Hence (b) follows.

(b) $\Rightarrow$ (c). Suppose T is uniformly continuous on X and not bounded. Hence for each positive integer n there exists $x_n \in X$ such that $\| T(x_n) \| > n \cdot \| x_n \|$. This means that $\| T(x_n/n \| x_n \|) \| > 1$. But this contradicts the continuity of T at the origin since $\| x_n/n \| x_n \| \| \to 0$ as $n \to \infty$.

(c) $\Rightarrow$ (a) Boundedness of T trivially implies the continuity of T at the origin. ∎

The bounded linear operators from a normed linear space X into a normed linear space Y, denoted by $L(X, Y)$, form a vector space where addition of vectors and scalar multiplication of vectors are defined by

$$(T_1 + T_2)(x) = T_1(x) + T_2(x), \qquad (\alpha T)(x) = \alpha \cdot T(x).$$

Let us define on this vector space

$$\| T \| = \sup_{x \neq 0} \frac{\| T(x) \|}{\| x \|}.$$

Equivalently,

$$\| T \| = \sup_{\|x\| \leq 1} \| T(x) \| = \sup_{\|x\|=1} \| T(x) \| = \sup_{\|x\|<1} \| T(x) \|.$$

This defines a norm on $L(X, Y)$ and $L(X, Y)$ becomes a normed linear space. The completeness of $L(X, Y)$ in this norm depends upon that of Y. More precisely, we have the following proposition.

Proposition 6.4. *$L(X, Y)$ is a Banach space if and only if Y is complete. (We assume, of course, that $X \neq \{0\}$.)* ∎

Proof. For the "if" part, let (T_n) be a Cauchy sequence in $L(X,Y)$. Then for each $x \in X$, $\| T_n(x) - T_m(x) \| \leq \| T_n - T_m \| \; \| x \|$, so that $\lim_{n\to\infty} T_n(x)$ exists in Y, Y being complete. Let us define $T(x) = \lim_{n\to\infty} T_n(x)$. Then T is a linear operator from X into Y. We wish to show that T is bounded and $\| T_n - T \|$ converges to 0 as $n \to \infty$. To do this, let $\varepsilon > 0$. There exists N such that for $n, m \geq N$, we have $\| T_n - T_m \| < \varepsilon$ or $\| T_n \| \leq \| T_N \| + \varepsilon$. Therefore $\| T(x) \| = \lim_{n\to\infty} \| T_n(x) \| \leq (\| T_N \| + \varepsilon) \cdot \| x \|$. Hence T is bounded. Now, for $x \in X$ and $n \geq N$,

$$\begin{aligned} \| T_n(x) - T(x) \| &= \lim_{m\to\infty} \| T_n(x) - T_m(x) \| \\ &\leq \overline{\lim_{m\to\infty}} \| T_n - T_m \| \cdot \| x \| < \varepsilon \| x \|. \end{aligned}$$

Therefore,

$$\| T_n - T \| = \sup_{\|x\| \leq 1} \| T_n(x) - T(x) \| < \varepsilon,$$

if $n \geq N$. The "if" part of the proof follows. The proof of the "only if" part is an application of the Hahn–Banach theorem and is left to the reader as a problem with hint. See Problem 6.2.4. ∎

Before we present the Hahn–Banach theorem we state a useful proposition providing a criterion for the existence of a bounded inverse of a bounded linear operator.

Proposition 6.5. Suppose T is a linear operator from a normed linear space X into a normed linear space Y. Then the inverse mapping T^{-1} exists and is a bounded linear operator from $T(X)$ into X if and only if there is some $k > 0$ such that $k \| x \| \leq \| T(x) \|$, for all $x \in X$. ▮

The proof of this proposition is routine and left to the reader.

Definition 6.7. When Y is the scalar field F, which is a Banach space over itself under the absolute-value norm, the elements of $L(X, Y)$ are called the *bounded linear functionals* on X. The class $L(X, F)$ is denoted by X^*. (Hence X^* is also a Banach space.) X^* is called the dual of X. ▮

Example 6.9. Suppose X is a n-dimensional normed linear space under the "sup"-norm over the real numbers R. Let T be a bounded linear functional on X. Then $\| T \| = \sup_{\|x\|=1} | T(x) |$. To compute $\| T \|$, let $\{x_1, x_2, \ldots, x_n\}$ be a basis of X and $T(x_i) = r_i$, $1 \leq i \leq n$. Now for $x = \sum_{i=1}^n a_i x_i \in X$,

$$| T(x) | \leq \sup_{1 \leq i \leq n} | a_i | \cdot \sum_{i=1}^{n} | r_i |.$$

Again, if we define $b_i = 1$ if $r_i > 0$, and $b_i = -1$ if $r_i \leq 0$, then, for $x = \sum_{i=1}^n b_i x_i$,

$$\| x \| = \sup_{1 \leq i \leq n} | b_i | = 1$$

and

$$| T(x) | = \sum_{i=1}^{n} b_i r_i = \sum_{i=1}^{n} | r_i |.$$

From this equality and the above inequality, it follows that

$$\| T \| = \sum_{i=1}^{n} | r_i |.$$

We now state and prove the main theorem of this section, which will show, among other things, the nontriviality of X^*, the dual (also called adjoint) of a normed linear space X.

Theorem 6.4. *The Hahn–Banach Theorem* (*Real Version*). Suppose X is a linear space over the *reals* R. Let S be a subspace of X and p be a real-valued function on X with the following properties:

(i) $p(x + y) \leq p(x) + p(y)$;
(ii) $p(\alpha x) = \alpha p(x)$, if $\alpha \geq 0$.

If f is a linear functional on S (i.e., a linear mapping from S into R) such that $f(s) \leq p(s)$ for all $s \in S$, then there exists a linear functional Φ on X such that $\Phi(x) \leq p(x)$ for all $x \in X$ and $\Phi(s) = f(s)$ for all $s \in S$. ∎

Proof. Let $\mathscr{F}$ be the set of all linear functionals g defined on a subspace of X containing S, such that $g(s) = f(s)$ for all $s \in S$ and $g(x) \leq p(x)$, whenever g is defined. Clearly $f \in \mathscr{F}$. We partially order $\mathscr{F}$ by requiring $g \leq h$ if and only if h is a linear extension of g. By the Hausdorff Maximal Principle, there is a maximal chain $\mathscr{F}_0$. We define a functional Φ by setting

$$\text{domain } \Phi = \bigcup_{g \in \mathscr{F}_0} \text{domain } g$$

and

$$\Phi(x) = g(x), \qquad \text{if } x \in \text{domain } g.$$

It is easy to check that domain Φ is a subspace of X and Φ is a linear extension of f. Moreover, Φ is a maximal extension, since if G is a proper extension of Φ, then $\mathscr{F}_0 \cup \{G\}$ will be a chain in $\mathscr{F}$, contradicting the maximality of $\mathscr{F}_0$.

We are finished if we can show that domain $\Phi = X$. This will be shown by showing that any $g \in \mathscr{F}$, with its domain a proper subspace of X, has a proper extension.

Let $y \in X -$ domain g. We wish to extend g to the subspace spanned by y and domain g. Thus we need to define a functional h by

$$h(\alpha y + x) = \alpha h(y) + g(x),$$

$x \in$ domain g and α, a scalar where $h(y)$ is chosen so that $h(\alpha y + x) \leq p(\alpha y + x)$. In particular, we need to choose $h(y)$ so that

$$h(y) \leq p(y + x) - g(x)$$

and

$$-h(y) \leq p(-y - x) + g(x)$$

or

$$-p(-y - z) - g(z) \leq h(y) \leq p(y + x) - g(x),$$

where x and z are elements from domain g. Since, for $x, z \in$ domain g,

$$p(y + x) - g(x) + p(-y - z) + g(z) \geq p(x - z) - g(x - z) \geq 0,$$
$$\sup_z [-p(-y - z) - g(z)] \leq \inf_x [p(y + x) - g(x)].$$

Hence we choose $h(y) = \inf\{p(y + x) - g(x): x \in \text{domain } g\}$. Now the theorem will be proved if we only check that $h(\alpha y + x) \leq p(\alpha y + x)$. To do this, let $\alpha > 0$. Then

$$\begin{aligned} h(\alpha y + x) &= \alpha h(y) + g(x) \\ &\leq \alpha p(y + x/\alpha) - \alpha g(x/\alpha) + g(x) \\ &= p(\alpha y + x). \end{aligned}$$

The cases $\alpha = 0$ and $\alpha < 0$ can similarly be taken care of. ∎

Theorem 6.5. *The Hahn–Banach Theorem (Complex Version).* Let X be a linear space over the complex numbers, S a linear subspace, and p, a real-valued function on X such that $p(x + y) \leq p(x) + p(y)$ and $p(\alpha x) = |\alpha| p(x)$. Let f be a linear functional on S such that $|f(s)| \leq p(s)$ for all $s \in S$. Then there is a linear functional Φ defined on X such that $\Phi(s) = f(s)$ for all $s \in S$ and $|\Phi(x)| \leq p(x)$ for all $x \in X$. ∎

Proof. Let us define the mappings g and h on S, by taking $g(s)$ to be the real part of $f(s)$ and $h(s)$ its imaginary part. Then g and h are linear in the real sense, by considering X (and hence S) as a vector space over the reals.

Now $f = g + ih$. Since, for $s \in S$,

$$g(is) + ih(is) = f(is) = if(s) = ig(s) - h(s),$$

we have $h(s) = -g(is)$. Since $g(s) \leq |f(s)| \leq p(s)$, by Theorem 6.4 we can extend g to a real-valued linear functional G on X considered as a real vector space such that $G(x) \leq p(x)$ for all $x \in X$. Let $\Phi(x) = G(x) - iG(ix)$. Then for $s \in S$, $\Phi(s) = g(s) + ih(s) = f(s)$. It is now easy to check that Φ is linear in the complex sense. Finally for $x \in X$, if α is a complex number with $|\alpha| = 1$ and $\alpha\Phi(x) = |\Phi(x)|$, then $|\Phi(x)| = \Phi(\alpha x) = G(\alpha x) \leq p(\alpha x) = |\alpha| p(x) = p(x)$. ∎

In the rest of the section, we will consider some of the important consequences and a few applications of the Hahn–Banach theorem.

Theorem 6.6. *The Hahn–Banach Theorem (Normed Linear Space Version).* Let X be a normed linear space over a field F and let Y be a subspace of X. If $y^* \in Y^*$, then there exists some $x^* \in X^*$ such that $\| y^* \| = \| x^* \|$ and $y^*(x) = x^*(x)$ for all $x \in Y$. ∎

Proof. Let us define $p(x) = \| y^* \| \| x \|$, $x \in X$. Then $p(x + y) \leq p(x) + p(y)$ and $p(\alpha x) = | \alpha | p(x)$. Also $| y^*(x) | \leq p(x)$, since $\| y^* \| = \sup_{x \neq 0} | y^*(x) | / \| x \|$, for $x \in Y$. Hence by Theorem 6.5, there exists a linear functional x^* on X such that (i) $x^*(x) = y^*(x)$ for all $x \in Y$, and (ii) $| x^*(x) | \leq p(x)$† for all $x \in X$. This means that $| x^*(x) | \leq \| y^* \| \| x \|$, $x \in X$ or $\| x^* \| \leq \| y^* \|$. But then (i) implies that $\| x^* \| = \| y^* \|$. ∎

Corollary 6.3. Let Y be a subspace of a normed linear space X over a field F. Suppose $x \in X$ and $\inf_{y \in Y} \| x - y \| = d > 0$. Then there exists $x^* \in X^*$ such that $\| x^* \| = 1$, $x^*(x) = d$, and $x^*(y) = 0$ for all $y \in Y$. ∎

Proof. Let Z be the subspace spanned by Y and x. We define a linear functional z^* on Z by

$$z^*(\alpha x + y) = \alpha d, \qquad y \in Y \text{ and } \alpha \in F.$$

Then $z^*(y) = 0$ for all $y \in Y$ and $z^*(x) = d$. To show that $z^* \in Z^*$, for $\alpha \neq 0$, $\alpha \in F$ and $y \in Y$,

$$\| \alpha x + y \| = | \alpha | \; \| x + y/\alpha \| \geq | \alpha | d$$

or $| z^*(\alpha x + y) | \leq \| \alpha x + y \|$ for all y in Y. Hence $\| z^* \| \leq 1$. But there exists a sequence $y_k \in Y$ such that $\| x - y_k \| \to d$ as $k \to \infty$. Therefore given $\varepsilon > 0$, $| z^*(\alpha x - \alpha y_k) | = | \alpha | d \geq \| \alpha x - \alpha y_k \| - \varepsilon$ for sufficiently large k. This means that $\| z^* \| \geq 1$. Hence $\| z^* \| = 1$. The corollary now follows by extending z^* by Theorem 6.6. ∎

The next corollary shows that there are sufficiently many bounded linear functionals on a normed linear space to separate points of the space.

Corollary 6.4. Given $x \in X$ ($\neq \{0\}$), a normed linear space over F, there exists $x^* \in X^*$ such that $\| x^* \| = 1$ and $x^*(x) = \| x \|$. In particular, if $x \neq y$, there exists $x^* \in X^*$ such that $x^*(x) - x^*(y) = \| x - y \| \neq 0$. ∎

Proof. The proof follows immediately from Corollary 6.3, by taking $Y = \{0\}$. ∎

† Note that $p(x) = p(-x)$. Thus, for real scalars, $x^*(x) \leq p(x)$ implies that $| x^*(x) | \leq p(x)$.

Corollary 6.5. For any x in a normed linear space X over F,

$$\| x \| = \sup_{\substack{x^* \in X^* \\ \|x^*\|=1}} | x^*(x) |. \qquad \blacksquare$$

We leave the proof to the reader.

Corollary 6.4 above is an important consequence of the Hahn–Banach Theorem. There are many applications of this outstanding theorem. One interesting application is the existence of a finitely additive measure, defined on the class of *all* bounded subsets of R, which is translation invariant and an extension of the Lebesgue measure. We will present this application in this section.

To do this, we first need an extension of the Hahn–Banach Theorem.

Theorem 6.7. *An Extension of the Hahn–Banach Theorem.* Let p be a real-valued function of the linear space X over the reals such that $p(x + y) \leq p(x) + p(y)$ and $p(\alpha x) = \alpha p(x)$, if $\alpha \geq 0$. Suppose f is a linear functional on a subspace S such that $f(s) \leq p(s)$ for all $s \in S$. Suppose also that $\mathscr{F}$ is an Abelian semigroup of linear operators on X (that is $T_1, T_2 \in \mathscr{F}$ implies $T_1T_2 = T_2T_1 \in \mathscr{F}$) such that if $T \in \mathscr{F}$, then $p(T(x)) \leq p(x)$ for all $x \in X$ and $f(T(s)) = f(s)$ for all $s \in S$. Then there is an extension Φ of f to a linear functional on X such that $\Phi(x) \leq p(x)$ and $\Phi(T(x)) = \Phi(x)$ for all $x \in X$. $\blacksquare$

Proof. The proof will be an application of the Hahn–Banach Theorem. First we need to choose a new subadditive function (like p). Let us define

$$q(x) = \inf(1/n)p(T_1(x) + \cdots + T_n(x)), \qquad x \in X,$$

where the infimum is taken over all possible finite subsets $\{T_1, T_2, \ldots, T_n\} \in \mathscr{F}$. To show that this infimum is a real number, we have $q(x) \leq p(x)$. Also since

$$0 = \frac{1}{n}p(0) \leq \frac{1}{n}p(T_1(x) + \cdots + T_n(x)) + \frac{1}{n}p(-T_1(x) - \cdots - T_n(x))$$

and

$$\frac{1}{n}p(- T_1(x) - \cdots - T_n(x)) \leq p(- x),$$

we have

$$\frac{1}{n}p(T_1(x) + \cdots + T_n(x)) \geq - p(- x),$$

and therefore

$$-p(-x) \leq q(x) \leq p(x).$$

Moreover, $q(\alpha x) = \alpha q(x)$, $\alpha \geq 0$. Now we wish to show that $q(x+y) \leq q(x) + q(y)$. So let $x, y \in X$ and $\varepsilon > 0$. Then there exist $\{T_1, T_2, \ldots, T_n\}$ and $\{S_1, S_2, \ldots, S_m\} \in \mathscr{T}$ such that

$$\frac{1}{n} p(T_1(x) + \cdots + T_n(x)) < q(x) + \frac{\varepsilon}{2}$$

and

$$\frac{1}{m} p(S_1(y) + \cdots + S_m(y)) < q(y) + \frac{\varepsilon}{2}.$$

Then

$$\begin{aligned} q(x+y) &\leq \frac{1}{nm} p\Big(\sum_{i=1}^{n} \sum_{j=1}^{m} T_i S_j(x+y)\Big) \\ &\leq \frac{1}{nm} p\Big(\sum_{j=1}^{m} S_j\Big(\sum_{i=1}^{n} T_i(x)\Big)\Big) + \frac{1}{nm} p\Big(\sum_{i=1}^{n} T_i\Big(\sum_{j=1}^{m} S_j(y)\Big)\Big) \\ &\leq \frac{1}{n} p\Big(\sum_{i=1}^{n} T_i(x)\Big) + \frac{1}{m} p\Big(\sum_{j=1}^{m} S_j(y)\Big) \\ &< q(x) + q(y) + \varepsilon. \end{aligned}$$

Since ε is arbitrary, $q(x+y) \leq q(x) + q(y)$. Also, since for $s \in S$

$$f(s) = \frac{1}{n} f(T_1(s) + \cdots + T_n(s)) \leq \frac{1}{n} p(T_1(s) + \cdots + T_n(s)),$$

$f(s) \leq q(s)$. Therefore, by Theorem 6.4, there exists a linear extension Φ of f to all of X such that $\Phi(x) \leq q(x) \leq p(x)$ for all $x \in X$. The proof will be complete if we can show that $\Phi(T(x)) = \Phi(x)$, $x \in X$ and $T \in \mathscr{T}$. To do this, let $x \in X$, $T \in \mathscr{T}$, and n be any positive integer. Then

$$\begin{aligned} q(x - T(x)) &\leq \frac{1}{n} p\Big(\sum_{i=1}^{n} T^i(x - T(x))\Big) \\ &= \frac{1}{n} p(T(x) - T^{n+1}(x)) \\ &\leq \frac{1}{n} [p(x) + p(-x)]. \end{aligned}$$

Letting n approach ∞, $q(x - T(x)) \leq 0$. Since $\Phi(x - T(x)) \leq q(x - T(x))$, we have $\Phi(x) \leq \Phi(T(x))$. Applying this to $-x$, we get $\Phi(x) = \Phi(T(x))$. ∎

[Note that this theorem yields the Hahn–Banach Theorem (real version) when $\mathscr{F}$ consists of the identity operator alone. The complex version of this theorem can also be formulated like that of the Hahn–Banach Theorem and proved.]

We proved in Chapter 3 that it is impossible to define a translation-invariant countably additive measure on the class of all bounded subsets of R, such that the measure of an interval is its length. But interestingly enough, an application of Theorem 6.7 will show that there exists a finitely additive measure having the above properties.

Proposition 6.6. There is a finitely additive measure μ defined for all bounded subsets of R such that

(i) $\mu(A + t) = \mu(A)$, $A \subset R$ and $t \in R$;

(ii) if $A \subset R$ is Lebesgue measurable, then $\mu(A)$ is the Lebesgue measure of A. ∎

Proof. Let X be the linear space over R of all real bounded functions defined on $[0, 1)$, under the natural operations of pointwise addition and scalar multiplication. Let Y be the subspace of all bounded Lebesgue-measurable functions of $[0, 1)$. For $f \in X$, let us define

$$p(f) = \text{l.u.b. } \{f(x)\colon\ x \in [0, 1)\}.$$

Then

(i) $p(f_1 + f_2) \leq p(f_1) + p(f_2), \qquad f_1, f_2 \in X,$

and

(ii) for $\alpha \geq 0$, $p(\alpha f) = \alpha p(f)$, for all $f \in X$.

Let Φ be the linear functional defined on Y by $\Phi(f) = \int_0^1 f(x)\,dx$. Let $\mathscr{F}$ be the Abelian semigroup of linear operators on X defined by

$$\mathscr{F} = \{T_t\colon\ 0 \leq t < 1 \text{ and } T_t[f](x) = f(x \mathbin{\mathring{+}} t),$$

where

$$\begin{aligned} x \mathbin{\mathring{+}} t &= x + t, && \text{if } x + t < 1, \\ &= x + t - 1, && \text{if } x + t \geq 1\}. \end{aligned}$$

Now for $0 \leq t < 1$, an easy computation shows that for any $A \subset [0, 1)$,

$$x \mathbin{\mathring{+}} t \in A \Leftrightarrow x \in A \mathbin{\mathring{+}} (1 - t).$$

Since the Lebesgue measure of a Lebesgue-measurable set A is the same as that of $A \overset{\circ}{+} (1 - t)$, we have

$$\int_0^1 \chi_A(x)\, dx = \int_0^1 \chi_A(x \overset{\circ}{+} t)\, dx.$$

Hence for $f \in Y$ and $0 \leq t < 1$, $\Phi(f) = \Phi(T_t[f])$. Also by the definition of p and T_t, we have

$$p(T_t[f]) \leq p(f), \qquad \text{for all } f \in X,\ 0 \leq t < 1.$$

Hence by Theorem 6.7, there exists a linear extension ψ of Φ defined on X such that $\psi(f) \leq p(f)$ and $\psi(T_t[f]) = \psi(f)$ for all $f \in X$.

For $A \subset [0, 1)$, let us define $\mu(A) = \psi(\chi_A)$. Then since ψ is linear, μ is *finitely additive*. Since $\psi(-\chi_A) \leq p(-\chi_A) \leq 0$, $-\psi(\chi_A) \leq 0$ or $\mu(A) \geq 0$. Also since for $0 \leq t < 1$, $\chi_A(x \overset{\circ}{+} t) = \chi_{A \overset{\circ}{+} (1-t)}(x)$, we have

$$\mu(A) = \psi(\chi_A) = \psi(\chi_A(x \overset{\circ}{+} t)) = \mu(A \overset{\circ}{+} (1 - t)).$$

If $t > \frac{1}{2}$, then $1 - t < \frac{1}{2}$. Hence, for $A \subset [0, \frac{1}{2})$,

$$\mu(A) = \mu(A + s), \qquad \text{for all } s \in [0, \tfrac{1}{2}).$$

Now to extend μ to the class of all bounded subsets on R, consider any $B \subset [n/2, (n + 1)/2)$, where n is any integer. Then $B - (n/2) \subset [0, \frac{1}{2})$. It is the value $\mu(B - n/2)$ we will take as $\mu(B)$. Thus for any bounded set $A \subset [-m, m]$,

$$\mu(A) = \sum_{n=-2m}^{2m-1} \mu\left(A \cap \left[\frac{n}{2}, \frac{n+1}{2}\right)\right).$$

This μ will satisfy the requirements of the theorem. ∎

Before we close this section, we briefly discuss another useful concept —that of separability. A set A in a normed linear space X is called separable if there is a countable subset of A that is dense in A. $C[0, 1]$ is separable under the supremum norm. l_p for $1 \leq p < \infty$ is separable since the set $\{(\alpha_1, \alpha_2, \ldots, \alpha_k, 0, 0, \ldots)\}$ of all sequences where the α_i's are rational is countable and dense in l_p. Also the L_p space of Lebesgue-measurable functions $(1 \leq p < \infty)$ is separable since the polynomials with rational coefficients (being dense in $C[0, 1]$) are dense in L_p. But l_∞, as well as L_∞, is not separable. (See Problem 6.2.7.) Also the L_p $(1 \leq p \leq \infty)$ space over an arbitrary measure space is not, in general, separable. For example, if S is an uncountable set with the counting measure, then for $1 \leq p < \infty$

$$L_p = \left\{f\colon S \to R\colon \sum_{s \in S} |f(s)|^p < \infty\right\};$$

and for $f \in L_p$,

$$\| f \|_p = \left[\sum_{s \in S} | f(s) |^p \right]^{1/p}.$$

If s_1 and s_2 are in S, then

$$\| \chi_{\{s_1\}}(s) - \chi_{\{s_2\}}(s) \|_p = 2^{1/p}.$$

This means that $L_p(S)$ is *not* separable.

It so happens that $l_1^* = l_\infty$ (equality means the existence of a linear isometry onto). This will be discussed in the next section. Therefore l_1^* is not separable, despite the separability of l_1. But the converse situation is different, and once again the Hahn–Banach Theorem helps us clarify the converse.

Proposition 6.7. If the dual X^* of a normed linear space X is separable, then X is also separable. ∎

Proof. Let $(x_k^*)_{k=1}^\infty$ be a countable dense set in X^*. Let $(x_k)_{k=1}^\infty$ be elements of X such that for all k, $\| x_k \| = 1$ and $| x_k^*(x_k) | \geq \frac{1}{2} \| x_k^* \|$. Let Y be the closed subspace spanned by the x_k's. Then Y is separable. We are finished if $Y = X$. If $Y \neq X$, by Corollary 6.3 there exists $x^* \in X^*$ such that $\| x^* \| > 0$ and $x^*(y) = 0$ for all $y \in Y$. Since $(x_k^*)_{k=1}^\infty$ is dense in X^*, there is a subsequence $(x_{k_i}^*)$ such that $\| x_{k_i}^* - x^* \| \to 0$ as $i \to \infty$. But

$$\begin{aligned} \| x_{k_i}^* - x^* \| &\geq | x_{k_i}^*(x_{k_i}) - x^*(x_{k_i}) | \\ &= | x_{k_i}^*(x_{k_i}) | \geq \tfrac{1}{2} \| x_{k_i}^* \|. \end{aligned}$$

Hence $\| x_{k_i}^* \| \to 0$ as $i \to \infty$. This means that $\| x^* \| = 0$, which is a contradiction. ∎

Problems

× **6.2.1.** Let X be a finite-dimensional space with basis $\{x_1, x_2, \ldots, x_n\}$, such that $\| \sum_{i=1}^n \alpha_i x_i \| = \sup_{1 \leq i \leq n} | \alpha_i |$, $\alpha_i \in F$. Define the linear operator A from X into X by

$$A(x_i) = \sum_{j=1}^n \alpha_{ij} x_j, \qquad 1 \leq i \leq n.$$

Then find $\| A \|$.

× **6.2.2.** Let X be as in Problem 6.2.1, with $\| \sum_{i=1}^{n} \alpha_i x_i \| = (\sum_{i=1}^{n} | \alpha_i |^2)^{1/2}$. Then if $f \in X^*$, find $\| f \|$.

× **6.2.3.** Let X be a normed linear space over R and $f: X \rightarrow R$.

(a) Suppose that f is continuous and for all x, y in X, $f(x) + f(y) = f(x + y)$. Prove that $f \in X^*$.

(b) Suppose that f is linear, but discontinuous. Show that $f^{-1}(0)$ is dense in X. (Hint: Note that $Y = f^{-1}(0)$ is not closed since f is discontinuous. Let y be a limit point of Y but not in Y. Then for any x in X, $x = \{x - [f(x)/f(y)]y\} + [f(x)/f(y)]y$.)

(c) Let f be linear. Show that f is continuous if and only if $f^{-1}(0)$ is closed.

(d) Let f be linear and $f^{-1}(0) \neq X$. Then $f^{-1}(0)$ is a maximal linear subspace of X, that is, a linear subspace W such that for any x in $X - W$, $\{x\} \cup W$ spans all of X. (For this result to hold, X needs to be only a linear space.)

[REMARK: Note that if $T: X \rightarrow Y$ is a linear operator between two normed linear spaces X and Y and if $T(X)$ is finite dimensional, then T is continuous if and only if Ker T is closed.]

6.2.4. Show that if $L(X, Y)$ is complete, where X and Y are any two normed linear spaces over F and $X \neq \{0\}$, then Y is complete. {Hint: Let (y_n) be a Cauchy sequence in Y. Then define $T_n \in L(X, Y)$ by $T_n(x) = x^*(x)y_n$, where $x^*(\neq 0) \in X^*$. Show that (T_n) is Cauchy in $L(X, Y)$ and hence $\lim_{n\to\infty} y_n = [1/x^*(x)] \lim_{n\to\infty} T_n(x)$, where $x^*(x) \neq 0$.}

× **6.2.5.** Let $1 \leq p \leq \infty$ and $1/p + 1/q = 1$. For each $g \in L_q$, define $T_g \in L_p{}^*$ by $T_g(f) = \int fg \, d\mu$. Show that for $p > 1$, $\| T_g \| = \| g \|_q$, and that for $p = 1$ this equality holds for all $g \in L_\infty$ if and only if the measure is semifinite. [Hint: If $1 < p < \infty$ and $f = | g |^{q/p} \operatorname{sgn} g$,† then $T_g(f) = \| g \|_q \| f \|_p$.]

× **6.2.6.** For any fixed $f(t) \in C[0, 1]$ (under the uniform norm), let Φ be the linear functional on $C[0, 1]$ defined by

$$\Phi(g(t)) = \int_0^1 f(t)g(t) \, dt.$$

Then show that Φ is bounded and find $\| \Phi \|$.

× **6.2.7.** Prove that l_∞ as well as $L_\infty[0, 1]$ is not separable. [Hint: Let $x_k \in l_\infty$ and $x_k = (x_k{}^1, x_k{}^2, \ldots)$. Define $x = (\alpha_1, \alpha_2, \ldots)$ by $\alpha_k = 0$, if $| x_k{}^k | \geq 1$; $= 1 + | x_k{}^k |$, if $| x_k{}^k | < 1$. Then $x \in l_\infty$ and $\| x - x_k \|_\infty \geq 1$ for all k.]

† By definition, $\operatorname{sgn} \alpha = 0$ if $\alpha = 0$ and $\operatorname{sgn} \alpha = \alpha/| \alpha |$ if $\alpha \neq 0$.

× **6.2.8.** If S is a linear subspace of a Banach space X, the *annihilator* S° of S is defined to be $S^\circ = \{x^* \in X^*: x^*(s) = 0 \text{ for all } s \in S\}$.† If T is a subspace of X^*, then $^\circ T = \{x \in X: x^*(x) = 0 \text{ for all } x^* \in T\}$. Show that

(i) S° is a closed subspace of X^*;

(ii) $^\circ(S^\circ) = \bar{S}$;

(iii) if S is closed, then S^* is linearly isometric to X^*/S° and S° is linearly isometric to $(X/S)^*$.

× **6.2.9.** For $1 < p < \infty$, show that l_p^* is linearly isometric to l_q, $1/p + 1/q = 1$. [Hint: Let

$$e_k = (\underbrace{0, 0, \ldots, 0, 1}_{k}, 0, 0, \ldots).$$

If $x^* \in l_p^*$, let $x^*(e_k) = b_k$. Show that $b = (b_1, b_2, \ldots) \in l_q$ and $\| x^* \| = \| b \|_q$.]

× **6.2.10.** Show that l_1^* is linearly isometric to l_∞.

× **6.2.11.** Let c be the space of all sequences of complex numbers $(x_1, x_2, \ldots)$ such that $\lim_{n\to\infty} x_n$ exists with natural addition and scalar multiplication. For an element $x = (x_1, x_2, \ldots) \in c$, let $\| x \| = \sup_n | x_n |$. Then show that

(i) c is a Banach space;

(ii) c^* is linearly isometric to l_1.

[Hint: Let $e_k = (0, 0, \ldots, 0, 1, 0, \ldots)$ as in Problem 6.2.9 for $1 \leq k < \infty$, $e_0 = (1, 1, 1, \ldots)$, $x^*(e_k) = b_k$, $1 \leq k < \infty$, and $x^*(e_0) = m$, where $x^* \in c^*$. Show that $(d_1, d_2, \ldots) \in l_1$, where $d_1 = m - \sum_{k=1}^\infty b_k$, $d_{i+1} = b_i$, $1 \leq i < \infty$ and $\| x^* \| = \sum_{i=1}^\infty | d_i |$.]

(iii) A subset $A \subset c$ is relatively compact if and only if A is bounded and $\sup_{i,j\geq n} | x_i - x_j | \to 0$ as $n \to \infty$, uniformly for all $x \equiv (x_i) \in A$.

× **6.2.12.** Let c_0 be the subspace of c (see Problem 6.2.11 above) such that $\lim_{n\to\infty} x_n = 0$, if $(x_1, x_2, \ldots) \in c_0$. Show that c_0^* is linearly isometric to l_1.

Further show that a subset $A \subset c_0$ is relatively compact if and only if A is bounded and $\sup_{i\geq n} | x_i | \to 0$ as $n \to \infty$, uniformly for all $x \equiv (x_i) \in A$.

6.2.13. Consider the linear space L_p, $0 < p < 1$, in the Lebesgue measure space on $[0, 1]$, with the pseudometric

$$d(f, g) = \int_0^1 | f(t) - g(t) |^p \, dt.$$

† Although A° is also used to denote the interior of a set A in this text, the meaning of the symbol should be clear from the context.

Then show the following:

(i) (L_p, d), $0 < p < 1$, is a pseudometric linear space [that is, a vector space with metric (pseudo) topology where the vector addition and scalar multiplication are continuous].

(ii) f can be written as $g + h$, $d(g, 0) = d(h, 0) = \frac{1}{2}d(f, 0)$. {Hint: Let $g_x = f \cdot \chi_{[0,x]}$, $h_x = f \cdot \chi_{[x,1]}$. Then $d(g_x, 0) + d(h_x, 0) = d(f, 0)$. But $d(g_x, 0)$ is a continuous function from $[0, 1]$ onto $[0, d(f, 0)]$. Use the Intermediate Value theorem.}

(iii) The only continuous linear functional on L_p is the zero functional. [Hint: Let $\Phi \in L_p^*$ with $\Phi(f) = 1$. Then, by part (ii), there exists $g \in L_p$ with $\Phi(g) \geq \frac{1}{2}$, $d(g, 0) = \frac{1}{2}d(f, 0)$. Let $g_1 = 2g$. Then $\Phi(g_1) \geq 1$, $d(g_1, 0) = 2^{p-1}d(f, 0)$. Continue the process to get $g_1, g_2, \ldots$ with $\Phi(g_n) \geq 1$ and $d(g_n, 0) = 2^{n(p-1)}d(f, 0)$.]

6.2.14. *The Volterra Fixed Point Theorem.* Let X be a Banach space and T be a bounded linear operator on X such that $\sum_{n=1}^{\infty} \| T^n \| < \infty$. Prove that the transformation S defined by

$$S(x) = y + T(x), \qquad y \text{ a fixed element in } X,$$

has a unique fixed point given by

$$x_0 = y + \sum_{n=1}^{\infty} T^n(y).$$

[Hint: For any $z \in X$, consider $x_1 = z$ and $x_{n+1} = T(x_n)$; show that (x_n) is Cauchy in X and $\lim_{n\to\infty} x_n = 0$.]

6.2.15. *An Application of Problem 6.2.14.* Show that the Volterra equation

$$f(t) = g(t) + \int_0^t K(s, t) f(s)\, ds,$$

where $g \in C[0, 1]$ and $K \in C([0, 1] \times [0, 1])$ are given functions, has a unique solution $f \in C[0, 1]$. [Hint: Define T on the Banach space $C[0, 1]$ with "sup" norm by

$$T(f)(t) = \int_0^t K(s, t) f(s)\, ds.$$

Then

$$\| T \| \leq M = \sup_{0 \leq s, t \leq 1} | K(s, t) | \quad \text{and} \quad \| T^n \| \leq M^n/n!.$$

Now use Problem 6.2.14.]

6.2.16. *Existence of Banach Limits.* (*An Application of the Hahn–Banach Theorem*). Banach limits are linear functionals Φ on the space l_∞

bounded sequences of real numbers $(x_n)_{n=0}^{\infty}$ satisfying these conditions:

(i) $\Phi[(x_n)] \geq 0$ if $x_n \geq 0$ for $0 \leq n < \infty$;

(ii) $\Phi[(x_{n+1})] = \Phi[(x_n)]$;

(iii) $\Phi[(1)] = 1$, $(1) = (1, 1, 1, \ldots)$.

Prove the following assertions:

(1)
$$p[(x_n)] = \lim_{n\to\infty} \left(\sup_j \frac{1}{n} \sum_{i=0}^{n-1} x_{i+j} \right)$$

exists and is a sublinear functional on l_∞. [Hint: If

$$a_n = \sup_j \frac{1}{n} \sum_{i=0}^{n-1} x_{i+j},$$

then $a_{km} \leq a_m$ and $(r + km)a_{r+km} \leq ra_r + kma_m$ or $\lim_{k\to\infty} \sup a_{r+km} \leq a_m$ for $r = 1, 2, \ldots, m$. Hence $\lim \sup a_n \leq a_m$ for each m or $\lim a_n$ exists.]

(2) For $(x_n) \in c(\subset l_\infty)$, define

$$f[(x_n)] = \lim_{n\to\infty} x_n \leq p[(x_n)].$$

(3) By Theorem 6.4, there is an extension Φ of f to l_∞ and Φ is a Banach limit. {Hint: $|p[(x_{n+1}) - (x_n)]| = |\lim_n[\sup_j n^{-1}(x_{j+n} - x_j)]| \leq \lim_n 2n^{-1} \sup_j |x_j| = 0$; also $|p[(x_n) - (x_{n+1})]| = 0$.}

(4) The maximal value of Banach limits on (x_n) is $p(x_n)$. (Hint: First, for any Banach limit L, $L[(x_n)] \leq p[(x_n)]$. To see this, by (i) $L[(\sup_j z_j - z_n)] \geq 0$, and therefore $\sup_j z_j \geq L[(z_n)]$; then if $z_n = (1/m)\sum_{i=0}^{m-1} x_{i+n}$, $L[(x_n)] = L[(z_n)] \leq \sup_j z_j$ for each m and therefore letting $m \to \infty$, $\leq p[(x_n)]$. Conversely, given (x_n), there exists a Banach limit L such that $L[(x_n)] = p[(x_n)]$. Indeed, if $(x_n) \notin c$, define L on c as f [in (2)], and as in the proof of Theorem 6.4, $L[(x_n)] = \inf_{(y_n)\in c}\{p[(x_n + y_n)] - \lim y_n\}$; then as in Theorem 6.4, L can be extended to l_∞ and $L[(x_n)] = p[(x_n)]$, since $p[(x_n + y_n)] = p[(x_n)] + \lim y_n$.)

(5) The minimal value of Banach limits on (x_n) is

$$\lim_{n\to\infty} \left(\inf_j \frac{1}{n} \sum_{i=0}^{n-1} x_{i+j} \right).$$

(6) A necessary and sufficient condition in order that all Banach limits on (x_n) agree and be equal is that $\lim_{n\to\infty}(1/n)\sum_{i=0}^{n-1} x_{i+j} = s$ uniformly in j. Hence on convergent sequences, Banach limits agree with limits of the sequences.

The formulation of the above results in the form they appear is due to L. Sucheston [40].

6.2.17. Let $(X, \mathscr{A}, \mu)$ be a finite measure space and let S be the linear space of all real-valued measurable functions (where two functions agreeing a.e. are equal) with the topology of convergence in measure {equivalently, with the metric $d(f, g) = \int [|f - g|/(1 + |f - g|)]\, d\mu$}. Prove the following assertions:

(i) If A is an atom in $\mathscr{A}$, then $\Phi_A(f) = \int_A f\, d\mu$ defines a continuous linear functional on S.

(ii) If Φ is a continuous linear functional on S, then $\Phi(f\chi_A) = 0$ for all $f \in S$ if A contains no atoms.

(iii) Let $\mathscr{F}$ be an (at most countable) collection of pairwise disjoint atoms of $\mathscr{A}$ and $T = \{\Phi_A : A \in \mathscr{F}\}$. Then the class S^* of all continuous linear functionals on S is precisely the linear span of T. [This result is due to T. K. Mukherjee and W. H. Summers.]

(iv) S^* contains a nonzero element if and only if $\mathscr{A}$ contains an atom.

6.2.18. *Another Application of the Hahn–Banach Theorem.* The classical moment problem can be stated as follows: Given a sequence of real numbers (a_n), when does there exist a real-valued function g of bounded variation on $[0, 1]$ such that $\int_0^1 x^n\, dg(x) = a_n$, $n = 0, 1, 2, \ldots$? Show that this problem can be answered in an abstract setup as follows:

Let X be a normed linear space, $(x_\lambda)_{\lambda \in \Lambda}$ be elements in X, and $(a_\lambda)_{\lambda \in \Lambda}$ be scalars. Then the following are equivalent:

(i) There exists $x^* \in X^*$ such that $x^*(x_\lambda) = a_\lambda$ for each $\lambda \in \Lambda$.

(ii) There exists a positive number M such that

$$|\sum a_\lambda b_\lambda| \leq M \cdot \|\sum b_\lambda x_\lambda\|,$$

for every subset $(b_\lambda)_{\lambda \in \Lambda}$ of scalars with all but finitely many of the b_λ's zero.

[Hint: (ii) $\Rightarrow$ (i). Let Y be the subspace spanned by $(x_\lambda)_{\lambda \in \Lambda}$. Define $x^* \in Y^*$ by $x^*(\sum b_\lambda x_\lambda) = \sum b_\lambda a_\lambda$; show that x^* is well defined because of (ii). Then use the Hahn–Banach Theorem.]

6.2.19. Let X and Y ($\neq \{0\}$) be normed linear spaces, and $\dim X = \infty$. Show that there is at least one and therefore infinitely many unbounded linear operators $T: X \to Y$. (Hint: Use the fact that X has a Hamel basis, i.e., a linearly independent subset that spans X, and a linear operator on X is uniquely determined by its values on a Hamel basis.)

6.2.20. Let X be a linear space over R and $Y \subset X$. Show that the following are equivalent:

(a) Y is a hyperplane (that is, $Y = x_0 + Z$, where $x_0 \in X$ and Z is a maximal[†] linear subspace of X).

(b) There exists a linear functional $f: X \to R$, $f \neq 0$, and a real number r such that $Y = \{x \in X: f(x) = r\}$.

[Hint: For (a) $\Rightarrow$ (b), define $f(ax + z) = a$, where $x \in X - Z$ and $z \in Z$. Note that $\{x\} \cup Z$ spans X. Then, $f(x_0) = r$. For (b) $\Rightarrow$ (a), let y be such that $f(y) = 1$; then, take $x_0 = ry$ and $Z = f^{-1}(0)$. Use 6.2.3(d).]

× **6.2.21.** Let X be a linear space over F and let $K \subset X$ be a convex absorbing set. ("Absorbing" means that for each x in X, there exists $a > 0$ such that $x \in aK$.) For $x \in X$, write

$$p_K(x) = \inf\{a: a > 0 \text{ and } x \in aK\}.$$

Show that for x, y in X

(i) $p_K(x) \geq 0$;
(ii) $p_K(0) = 0$;
(iii) $p_K(x + y) \leq p_K(x) + p_K(y)$;
(iv) $p_K(ax) = ap_K(x)$ for $a \geq 0$.

The functional p_K is called the Minkowski functional of K.

6.2.22. Let X be a linear space over R, $K \subset X$ a convex absorbing set, and L a hyperplane such that $K \cap L = \emptyset$. Show that there exists a linear functional x^* on X such that

(i) $L = \{x: x^*(x) = 1\}$;
(ii) $K \subset \{x: x^*(x) < 1\}$;
(iii) for $x \in X$, $x^*(x) \leq p_K(x)$, where p_K is as in Problem 6.2.21.

[Hint: Use Problem 6.2.20 to establish (i). For (ii), let $x, y \in K$ such that $x^*(x) < 1 < x^*(y)$. Then for s in $[0, 1]$, $sx + (1 - s)y \in K$ and $sx^*(x) + (1 - s)x^*(y)$, as a function of s, attains the value 1 for some s_0. Note that $s_0x + (1 - s_0)y \in K \cap L$.]

6.2.23. *A Geometric Form of the Hahn–Banach Theorem.* Let X be a linear space over R and $K \subset X$ be a convex absorbing set. If $L_0 \subset X$ is such that $L_0 = x_0 + W$ and $K \cap L_0 = \emptyset$, where W is a linear subspace, then there exists a hyperplane L and a linear functional x^* on X such that

(i) $L = \{x: x^*(x) = 1\}$;
(ii) $L_0 \subset L$;
(iii) $x^*(x) \leq p_K(x)$, $x \in X$;
(iv) $K \subset \{x: x^*(x) \leq 1\}$.

[†] That is, a proper subspace Z such that for any $z \in X - Z$, the subspace spanned by $Z \cup \{z\}$ is X.

(Hint: Note that $x_0 \notin W$, since $0 \in K$. Let $\{x_0\} \cup W$ span the subspace V. Then L_0 is a hyperplane in V and $K \cap V \cap L_0 = \emptyset$. Use Problem 6.2.22 and Theorem 6.4.)

6.2.24. Let X be a normed linear space over R and $L \subset X$ be a hyperplane such that $L = \{x: x^*(x) = 1\}$ for some linear functional x^* on X. Let G be an open subset of X such that $G \subset \{x: x^*(x) \leq 1\}$. Show that (i) L is closed, (ii) $\bar{G} \subset \{x: x^*(x) \leq 1\}$ and x^* is continuous, and (iii) $G \subset \{x: x^*(x) < 1\}$. (We remark that the results outlined in this and the next problem hold in the more general context of topological vector spaces. See Problem 6.7.29.) [Hint: If L is not closed, then L is dense and therefore $\{x: x^*(x) = 2\}$ is also dense and must intersect G. For (ii), use Problem 6.2.3(c).]

6.2.25. Let X be a normed linear space over R and $K \subset X$ be an open convex set containing 0. Suppose also that $K \cap L_0 = \emptyset$, where $L_0 = x_0 + W$ and W is a linear subspace. Show that there exists a closed hyperplane L and a continuous linear functional x^* on X such that

(i) $L = \{x: x^*(x) = 1\}$;

(ii) $L_0 \subset L$;

(iii) $K \subset \{x: x^*(x) < 1\}$.

(Hint: Use Problems 6.2.23 and 6.2.24.)

6.3. The Open Mapping Theorem, the Closed Graph Theorem, and the Principle of Uniform Boundedness

In this section we will consider three basic principles of functional analysis, which rank in importance with the Hahn–Banach Theorem of the previous section. They provide the foundation for many far-reaching modern results in diverse disciplines of analysis such as ergodic theory, the theory of differential equations, integration theory, and so on. The first, called the Open Mapping Theorem, asserts that certain continuous linear mappings between Banach spaces map open sets into open sets. The second is called the Closed Graph Theorem and asserts that a linear map between two Banach spaces that has a closed graph is continuous. The third is the Principle of Uniform Boundedness, which asserts that a pointwise bounded family of continuous linear mappings from one Banach space to another is uniformly bounded. One form of the Open Mapping and the Closed Graph Theorems was first proved by Banach in 1929; a more general form is due to Schauder in 1930. The principle of uniform boundedness was proved for bounded linear functionals on a Banach space by Hahn in 1922, for continuous linear mappings between Banach spaces by Hildebrandt in 1923, and also by Banach and Steinhaus (in a more general case) in 1927. The Principle of Uniform boundedness is sometimes called the Banach–Steinhaus Theorem.

This section is based on the concept of a closed linear operator. We start with its definition.

Let X and Y be normed linear spaces over the same scalars. Then $X \times Y$ is the normed linear space of all ordered pairs (x, y), $x \in X$ and $y \in Y$, with the usual definitions of addition and scalar multiplication and norm defined by $\| (x, y) \| = \max\{ \| x \|, \| y \| \}$.

Definition 6.8. Let $T: D \to Y$ be a linear operator, where D is a subspace of X. Then T is called *closed* if its *graph* $G_T = \{(x, T(x)): x \in D\}$ is a closed subspace of $X \times Y$. Equivalently, T is closed if and only if the following condition holds: Whenever $x_n \in D$, $x_n \to x \in X$, and $T(x_n) \to y \in Y$, then $x \in D$ and $y = T(x)$. ∎

Examples

6.10. *A Linear Operator That Is Closed But Not Continuous.* Let $X = Y = C[0, 1]$ and $D = \{x \in X: d[x(t)]/dt \in C[0, 1]\}$. Define $T: D \to Y$ by $T(x) = dx/dt$. Then if $x_n(t) = t^n$, $\| x_n(t) \| = 1$; but

$$\| T(x_n) \| = \| n \cdot t^{n-1} \| = n \to \infty.$$

Hence T is not bounded. The reader can easily check that T is closed. Note that D is not complete.

6.11. *A Linear Operator That Is Continuous But Not Closed.* Let D be a nonclosed subspace of any normed linear space X. Let $Y = X$ and $i: D \to Y$ be the identity map. Then i is clearly continuous but not closed.

Clearly we have the following proposition.

Proposition 6.8. Let $T: D \to Y$, $D \subset X$, be a continuous linear operator. Then if D is a closed subspace of X, T is closed. Conversely, if Y is complete and T is closed, then D is a closed subspace of X. ∎

We leave the proof to the reader.

Example 6.12. *A Closed Linear Operator from a Banach Space into a Normed Linear Space That Is Not Continuous.* Let X be any infinite-dimensional Banach space and let H be a Hamel basis for X. We may and do assume that the elements in H have norm 1. Let Y be the vector space X with a new norm $\| \cdot \|_1$ given by

$$\left\| \sum_{i=1}^{n} \alpha_i x_i \right\|_1 = \sum_{i=1}^{n} | \alpha_i |, \qquad x_i \in H.$$

For each $x \in X$, $\| x \|_1 \geq \| x \|$. Let (y_n) be an infinite sequence in H. If

$$z_n = \sum_{k=1}^{n} \left(\frac{1}{k^2} \right) \cdot y_k,$$

then for $n < m$,

$$\| z_n - z_m \|_1 = \sum_{k=n+1}^{m} \frac{1}{k^2} \to 0, \qquad \text{as } n, m \to \infty.$$

Hence (z_n) is Cauchy in Y as well as in X. For any z in Y, $z = \sum_{i=1}^{p} \alpha_i h_i$ for some positive integer p with $h_i \in H$, for $i = 1, 2, \ldots, p$. Hence for $n \geq k$, $\| z_n - z \|_1 \geq 1/k^2$ if k is large enough for $y_k \neq h_i$, $i = 1, 2, \ldots, p$. This means that (z_n) does not converge to any z in Y, so that Y is not complete. Now if we consider the identity map $i: X \to Y$, then i is clearly closed, one-to-one, onto, and has a continuous inverse. However, i is *not* continuous, since in that case Y has to be complete.

Example 6.13. *A Closed Linear Operator from a Normed Linear Space onto a Banach Space That Is Not an Open Map.* Consider Example 6.12 and the identity map $i: Y \to X$. Then i is a closed linear map that is not open.

Now we state and prove the Open Mapping Theorem.

Theorem 6.8. *The Open Mapping Theorem.* Let X be a Banach space and Y be a normed linear space of the second category. If T is a closed linear operator from a linear subspace D in X onto Y, then T is an open map. ∎

Before we prove this theorem, we prove the following lemma.

Lemma 6.1. Let T be a closed linear operator from a linear subspace D of a Banach space X into a normed linear space Y. Suppose that $\overline{T(G)}$ has a nonempty interior whenever G is a relatively open subset of D containing 0. Then T is open and onto. ∎

Proof. Let G be open in D and $0 \in G$. Then there is an open ball $V_r = \{x \in D: \| x \| < r\}$ such that $V_r - V_r \subset G$. Therefore

$$\overline{T(V_r)} - \overline{T(V_r)} \subset \overline{T(V_r) - T(V_r)} \subset \overline{T(V_r - V_r)} \subset \overline{T(G)},$$

so 0 is an interior point of $T(G)$, by the assumptions in the lemma. Notice that if $x \in G$, then for some $s > 0$, $x + V_{2s} \subset G$, so that $T(x) + T(V_{2s}) \subset T(G)$. Thus, it is enough to prove that $\overline{T(V_r)} \subset T(V_{2r})$ for $r > 0$.

Let $r > 0$ and $k_n \to 0$ such that for each $n \geq 1$

$$\{y \in Y: \| y \| < k_n\} \subset \overline{T(V_{(r/2^n)})}.$$

Let $y \in \overline{T(V_r)}$. There exists $y_1 \in T(V_r)$ such that $\| y - y_1 \| < k_1$. Then $y - y_1 \in \overline{T(V_{r/2})}$, and there exists $y_2 \in T(V_{r/2})$ such that $\| y - y_1 - y_2 \| < k_2$. It is clear that there exist $(y_n)_{n=1}^{\infty}$ such that $y_n \in T(V_{(r/2^{n-1})})$ and $\| y - \sum_{i=1}^{n} y_i \| < k_n$. Let $y_n = T(x_n)$ and $\| x_n \| < r/2^{n-1}$ so that $\sum_{i=1}^{\infty} \| x_i \| < 2r$. By Theorem 6.3, there exists x in X such that $\sum_{i=1}^{n} x_i \to x$. Since T is closed, $y = Tx$ and $\| x \| < 2r$. ∎

Proof of Theorem 6.8. Let $r > 0$. Then $D = \bigcup_{n=1}^{\infty} nV_r$ and $Y = \bigcup_{n=1}^{\infty} nT(V_r)$. ($V_r$ is the open r-ball in D.) Since Y is of the second category, there is a positive integer n such that $nT(V_r)$ is not nowhere dense. Thus $\overline{T(V_r)}$ has a nonempty interior, and T is open by Lemma 6.1. ∎

For bounded linear maps, the property of being open is equivalent to the continuity or boundedness of the inverse map when the inverse exists. In general, we have the following result.

Proposition 6.9. Let T be a bounded linear map from a normed linear space X onto a normed linear space Y. Then T is open if and only if there exists $k > 0$ such that for each $y \in Y$, there exists $x \in X$ with $T(x) = y$ and $\| x \| \leq k \| y \|$. ∎

Proof. Recall from Section 6.1 that the map Φ: $X \to X/K$, where $K = T^{-1}(0)$, is continuous and open, so all open sets in the quotient space X/K are of the form $\Phi(U)$, U being an open subset of X. Then the map T_0: $Y \to X/K$ defined by $T_0(y) = x + K$, $T(x) = y$, is well-defined and linear. Since $T_0^{-1}(\Phi(U)) = T(U)$, T_0 is continuous if and only if T is open. Recalling the norm in X/K, it follows that T_0 is continuous if and only if there exists $a > 0$ such that

$$\inf_{z \in K} \| x - z \| \leq a \| y \|, \qquad y = T(x).$$

The rest of the proof is left to the reader. ∎

Proposition 6.9 leads to a partial converse of Theorem 6.8.

Proposition 6.10. If T is a bounded linear operator from a Banach space X *onto* a normed linear space Y, then the "openness" of T implies the completeness of Y. ∎

Proof. Let (y_n) be a Cauchy sequence in Y. Then we can find a sequence (n_k) of positive integers such that $n_k < n_{k+1}$ and for each k, $\| y_{n_{k+1}} - y_{n_k} \| < 1/2^k$. If T is open, by Proposition 6.9 there exists $N > 0$ such that we can find x_k with $T(x_k) = y_{n_{k+1}} - y_{n_k}$ and $\| x_k \| \leq N \| y_{n_{k+1}} - y_{n_k} \|$. Then $\sum_{k=1}^{\infty} \| x_k \| < \infty$. Since X is complete, there exists $x \in X$ such that $x = \lim_{n\to\infty} \sum_{k=1}^{n} x_k$. Since T is continuous and linear, $\sum_{k=1}^{n} T(x_k) \to T(x)$, which means that $y_{n_{k+1}} \to y_{n_1} + T(x)$. Since (y_n) is Cauchy, $y_n \to y_{n_1} + T(x)$. ∎

Now as applications of the Open Mapping Theorem, we present the next three propositions and also the Closed Graph Theorem. Proposition 6.12 is also needed for Theorem 7.10.

Proposition 6.11. Let X be a vector space that is complete in each of the norms $\| \cdot \|$ and $\| \cdot \|_1$. Suppose there exists $k > 0$ such that

$$\| x \| \leq k \| x \|_1$$

for each $x \in X$. Then the norms are equivalent. ∎

Proof. If i is the identity map from $(X, \| \cdot \|_1)$ *onto* $(X, \| \cdot \|)$, then i is clearly closed and linear. By the Open Mapping Theorem, i is also open and so i^{-1} is continuous. ∎

Proposition 6.12. Let X and Y be Banach spaces. Then the set of all surjective maps in $L(X, Y)$ is open in $L(X, Y)$. ∎

Proof. Let T be a surjective map in $L(X, Y)$. Then T is open by Theorem 6.8. Let $S \in L(X, Y)$ and $\| T - S \| < 1/2k$, where k is a real number for the map T with the property as stated in Proposition 6.9. We claim that S is surjective. To prove this, let $y \in Y$ and $\| y \| \leq 1$. Then by Proposition 6.9, there exists $x \in X$, $T(x) = y$ and $\| x \| \leq k$. Let $y_1 = T(x) - S(x)$. Then, $\| y_1 \| \leq \frac{1}{2}$ and there exists $x_1 \in X$, $T(x_1) = y_1$ and $\| x_1 \| \leq k/2$. Let $y_2 = T(x_1) - S(x_1)$. Then, $\| y_2 \| \leq 1/2^2$. Continuing inductively, we find x_n such that $T(x_n) = y_n$ and

$$y_{n+1} = T(x_n) - S(x_n), \qquad \| y_n \| \leq 1/2^n, \qquad \| x_n \| \leq k/2^n.$$

Then, $y = S(x) + S(x_1) + \cdots + S(x_n) + y_{n+1}$. Write $z = x + \sum_{n=1}^{\infty} x_n$. Then $S(z) = y$ and our claim is proven. ∎

Proposition 6.13. Let X be a separable Banach space. Then there exists a closed linear subspace L of the Banach space l_1 such that X is topologically isomorphic to the quotient space l_1/L. ∎

Proof. Let (x_n) be a dense sequence in the unit ball of X. We define the mapping T: $l_1 \to X$ by $T(y) = \sum a_n x_n$, $y = (a_n)_{n=1}^{\infty}$. Clearly, $\| T(y) \| \leq \| y \|_1$. Let $L = T^{-1}\{0\}$. Now the mapping τ defined by $\tau(y + L) = T(y)$ is a well-defined continuous linear one-to-one mapping from l_1/L into X. If T is surjective, then τ is surjective, in which case an application of the Open Mapping Theorem will finish the proof. So it suffices to show that T is surjective.

Suppose $x \in X$ and $\| x \| \leq 1$. By an induction argument, we can find a subsequence (x_{n_i}) such that $x = \sum_{i=0}^{\infty} x_{n_i}/2^i$. [Choose x_{n_0} such that $\| x - x_{n_0} \| < \frac{1}{2}$; then, choose $n_1 > n_0$ such that $\| 2(x - x_{n_0}) - x_{n_1} \| < \frac{1}{2}$ and so on.] Let $y = (a_k)$ be defined so that $a_k = 1/(2^i)$ if $k = n_i$, $= 0$ if $k \neq n_i$ for all i. Then $y \in l_1$ and $T(y) = x$. It follows that T is surjective. ∎

Theorem 6.9. *The Closed Graph Theorem.* A closed linear operator T mapping a Banach space X into a Banach space Y is continuous. ∎

Proof. Since T is closed, $G_T = \{(x, T(x)): x \in X\}$ is a closed subspace of the Banach space $X \times Y$. We define $P_1[(x, T(x))] = x$ and $P_2[(x, T(x))] = T(x)$. Then P_1 and P_2 are both continuous maps from G_T *onto* X and into Y, respectively. By the Open Mapping Theorem, P_1 is open. Since P_1 is one-to-one, P_1^{-1} is continuous and therefore, $P_2 P_1^{-1} = T$ is continuous. ∎

Actually, we could state the Closed Graph Theorem in a more general form.

Theorem 6.10. A closed linear operator T mapping a normed linear space X of the second category into a Banach space Y is continuous. ∎

Proof. We will briefly outline the proof. Let $M = \{x \in X \mid T(x) = 0\}$. Then M is a closed subspace of X, since T is closed. If $X = M$, the theorem is trivial. Suppose $X \neq M$. Then the quotient space X/M is of the second category (see Problem 6.3.3). Define T_0: $X/M \to Y$ by $T_0(\Phi(x)) = T(x)$, where Φ is the natural map from X onto X/M. Then T_0 is well defined, one-to-one, closed, and linear. (See Problem 6.3.4.) Therefore, the mapping T_0^{-1}: $T(X)(\subset Y) \to X/M$ is again closed. By the Open Mapping Theorem, T_0^{-1} is open and therefore T_0 is continuous. This means that T is continuous. ∎

By means of quotient space arguments as used in the proof of Theorem 6.10, it is possible to give a proof of the Open Mapping Theorem for the case when D is closed, using Theorem 6.10 as a starting point. This means that the Open Mapping and the Closed Graph Theorems are two different forms of the same theorem.

Next, we present an incomplete normed linear space of the second category.

Example 6.14. Let X be the Lebesgue-integrable functions on $[0, 1]$ with the L_1 norm. Let (r_n) be the rationals in $(0, 1]$, and x_n be the characteristic function of $[0, r_n]$. Then A, the set containing the x_n's, is linearly independent and hence can be extended to a Hamel basis H of X. Since X is complete, H is uncountable. (See Problem 6.1.7.) Let (y_n) be a sequence of elements in $H - A$. Let Y_n be the subspace spanned by $H - \{y_n, y_{n+1}, \ldots\}$. Then since $X = \bigcup_{n=1}^{\infty} Y_n$, for some $n = p$, Y_p is *of the second category*. Also Y_p is not complete, since Y_p contains A and the span of A is dense in X (the step functions being dense in L_1).

Finally, we come to the Principle of Uniform Boundedness. We need first a lemma.

Lemma 6.2. Let S be a nonempty set and Y be a normed linear space. Let $B(S, Y) = \{f\colon S \to Y \mid \sup_{s \in S} \| f(s) \| < \infty\}$. If Y is complete and $\| f \| = \sup_{s \in S} \| f(s) \|$, then $B(S, Y)$ is a Banach space with this norm. ∎

The proof of this lemma is left as Problem 6.3.5.

Theorem 6.11. *The Principle of Uniform Boundedness.* Let S be a family of bounded linear operators from a Banach space X into a Banach[†] space Y. Suppose that for $x \in X$ there is a constant $M(x)$ such that $\sup_{T \in S} \| T(x) \| \leq M(x) < \infty$. Then there exists $M > 0$ such that

$$\sup\{\| T \| : T \in S\} < M.$$ ∎

Proof. We define the linear operator A from X into $B(S, Y)$ by $A(x)[T] = T(x)$. Then A is well defined since $\sup_{T \in S} \| T(x) \| < \infty$. A is also closed. (Why?) By the Closed Graph Theorem, A is continuous. Therefore,

[†] Completeness of Y is not needed in this theorem. One can consider $B(S, \hat{Y})$, where $\hat{Y}$ is the completion of Y. See Proposition 6.15. Also it suffices to let X be a normed linear space of the second category.

$\sup_{\|x\| \leq 1} \| A(x) \| < M$, for some $M > 0$. But since $\| A(x) \| = \sup_{T \in S} \| T(x) \|$ we have $\sup_{\|x\| \leq 1} \sup_{T \in S} \| T(x) \| < M$ or $\sup_{T \in S} \| T \| < M$. ∎

Remarks on Theorem 6.11 and Some Applications

6.1. Theorem 6.11 need *not* be true if X is not complete. Let $X = \{(a_i) \in l_2 \colon a_i = 0$ for all but finitely many $i\}$. Then $X \subset l_2$. We define $T_n \colon X \to l_2$ linearly by

$$T_n(e_i) = \begin{cases} 0, & i \neq n, \\ ne_n, & i = n, \end{cases}$$

where

$$e_i = (\underbrace{0, 0, \ldots, 0, 1}_{i}, 0, 0, \ldots).$$

Then for $x = \sum_{i=1}^{k} a_i e_i$, $T_n(x) = 0$ for $n > k$ and therefore $\sup_{1 \leq n < \infty} \| T_n(x) \|$ is finite. But $\| T_n \| = n$ and $\sup_{1 \leq n < \infty} \| T_n \| = \infty$.

6.2. From Theorem 6.11, it follows that for $S \subset X^*$, X a Banach space, $\sup_{x^* \in S} \| x^* \| < \infty$ whenever for each $x \in X$, $\sup_{x^* \in S} | x^*(x) |$ is finite.

6.3. If $S \subset X$ a normed linear space such that $\sup_{x \in S} | x^*(x) | < \infty$ for each $x^* \in X^*$, then $\sup_{x \in S} \| x \| < \infty$. This immediately follows by applying Theorem 6.11 to the family of operators $[J(x)]_{x \in S}$ from X^* into F (the scalar field), J being the natural map from X into X^{**}, that is $[J(x)](x^*) = x^*(x)$, $x^* \in X^*$.

6.4. *Analyticity of a Banach Space Valued Function of a Complex Variable.* Let X be a complex Banach space, G an open set in the complex plane, and f a function from G into X. Then f is called analytic on G if f is differentiable at each point $\lambda_0 \in G$, that is,

$$\left\| \frac{f(\lambda) - f(\lambda_0)}{\lambda - \lambda_0} - f'(\lambda_0) \right\| \to 0,$$

as $\lambda \to \lambda_0$ for some function $f' \colon G \to X$. We will apply the principle of uniform boundedness (actually Remark 6.3 above) to show that a necessary and sufficient condition for f to be differentiable on G is that $x^*(f(\lambda))$ be differentiable on G for each $x^* \in X^*$.

Since the "necessary" part is trivial, we show only the "sufficient" part. Let $x^*(f(\lambda))$ be differentiable on G and $\lambda_0 \in G$. There exists $r > 0$ such that

$$| \lambda - \lambda_0 | < r \Rightarrow \lambda \in G.$$

By Cauchy's integral formula (in elementary complex analysis), we have

$$x^*(f(\lambda)) = \frac{1}{2\pi i} \int_C \frac{x^*(f(z))}{z - \lambda} \, dz, \tag{6.1}$$

where $|\lambda - \lambda_0| < r$ and C is the positively oriented circle $|z - \lambda_0| = r$. It follows from equation (6.1) after a simple calculation that for $\lambda \neq \mu$, $|\lambda - \lambda_0| \leq \frac{1}{2}r$ and $|\mu - \lambda_0| \leq \frac{1}{2}r$ (and therefore, $|z - \lambda| \geq \frac{1}{2}r$ and $|z - \mu| \geq \frac{1}{2}r$ for $z \in C$),

$$x^*\left(\left[\frac{f(\lambda) - f(\lambda_0)}{\lambda - \lambda_0} - \frac{f(\mu) - f(\lambda_0)}{\mu - \lambda_0}\right] \Big/ \lambda - \mu\right)$$

$$= \frac{1}{2\pi i} \int_C \frac{x^*(f(z))}{(z - \lambda)(z - \mu)(z - \lambda_0)}\, dz. \tag{6.2}$$

Since $M = \sup\{\| x^*(f(z)) \| : z \in C\} < \infty$, the absolute value of the left-hand side of equation (6.2) does not exceed $4M/r^2$. Applying Remark 6.3 above, we have

$$\left\| \frac{f(\lambda) - f(\lambda_0)}{\lambda - \lambda_0} - \frac{f(\mu) - f(\lambda_0)}{\mu - \lambda_0} \right\| \leq K \cdot |\lambda - \mu|,$$

for some constant K, whenever $0 < |\lambda - \lambda_0| \leq r/2$ and $0 < |\mu - \lambda_0| \leq r/2$. Since X is complete, it follows that f is differentiable at λ_0.

• **6.5.** *Divergence of the Fourier Series.* By the Fourier series of a function f in $L_1[-\pi, \pi]$, we mean the series

$$\sum_{k=-\infty}^{\infty} \hat{f}(k) e^{ikt},$$

where the Fourier transform $\hat{f}$ of f is defined by

$$\hat{f}(k) = \frac{1}{2\pi} \int_{-\pi}^{\pi} f(s) e^{-iks}\, ds, \qquad k \in Z.$$

Such series were invented originally to serve as tools to solve problems in heat conduction, the theory of oscillation, and various other fields. The many problems that arose to determine whether the Fourier series of f converges to f or whether f is determined by its Fourier series gave rise to an important branch of analysis, known as harmonic analysis. We will not consider the convergence of the Fourier series here. In the next chapter (Remark 7.4), we shall show that the Fourier series of f in $L_2[-\pi, \pi]$ converges to f in L_2-norm. Actually a much more nontrivial result holds, namely, that the Fourier series of such a function converges almost everywhere. This result was first conjectured by N. N. Lusin in 1951 and then proven in 1966 by L. Carleson [8]. Later on, in 1968, the result was extended to $L_p[-\pi, \pi]$, $1 < p \leq \infty$, by R. A. Hunt [17]. We will not go into the details of these results here. Here we will consider the divergence of the Fourier series to show another application of Theorem 6.11.

We prove the following: There exists a continuous function f on $[-\pi, \pi]$ such that the Fourier series of f diverges at 0. The set of all such functions is of the second category in $C_1[-\pi, \pi]$.†

Proof. We write $S_n f(x) = \sum_{k=-n}^{n} \hat{f}(k)e^{ikx}$. Then we have, for any f in $C_1[-\pi, \pi]$,

$$S_n f(x) = \frac{1}{2\pi}\int_{-\pi}^{\pi} f(t)D_n(x-t)\,dt,$$

where

$$D_n(t) = \sum_{k=-n}^{n} e^{ikt} = \frac{\sin(n+\frac{1}{2})t}{\sin(\frac{1}{2}t)}, \qquad \text{if } e^{it} \neq 1,$$

$$= 2n+1, \qquad \text{if } e^{it} = 1.$$

For each nonnegative integer n, we define

$$x_n^*(f) = \frac{1}{2\pi}\int_{-\pi}^{\pi} f(t)D_n(t)\,dt, \qquad f \in C_1[-\pi, \pi].$$

Then x_n^* is a bounded linear functional on $C_1[-\pi, \pi]$ with norm $(1/2\pi)$ $\| D_n \|_1$. (The reader should verify this.) Now, suppose that the Fourier series of every $f \in C_1[-\pi, \pi]$ converges at $x = 0$. Then for each $f \in C_1[-\pi, \pi]$, $\sup_n | x_n^*(f) | < \infty$. By Theorem 6.11, it follows that

$$\sup_n \| x_n^* \| = \frac{1}{2\pi} \sup_n \| D_n \|_1 < \infty.$$

But we have

$$\| D_n \|_1 = \int_{-\pi}^{\pi} \left| \frac{\sin(n+\frac{1}{2})t}{\sin(\frac{1}{2}t)} \right| dt = 4\int_0^{\pi/2} \left| \frac{\sin(2n+1)t}{\sin t} \right| dt$$

$$\geq 4\int_0^{\pi/2} \left| \frac{\sin(2n+1)t}{t} \right| dt \geq \frac{8}{\pi} \sum_{k=0}^{2n} \frac{1}{k+1}.$$

Thus, $\sup_n \| D_n \|_1 = \infty$, a contradiction.

To complete the proof, we assume that the set $X \subset C_1[-\pi, \pi]$ of functions f such that $\sup_n | \sum_{k=-n}^{n} \hat{f}(k) | < \infty$ is of the second category. Noting that Theorem 6.11 holds even when X is a normed linear space of the second category, it follows as before that $\sup_n \| x_n^* \| < \infty$, which is a contradiction. The proof is complete. ∎

† The complex-valued continuous functions on $[-\pi, \pi]$ with the sup-norm.

Before we close this section, we present a theorem on projection, an application of the Closed Graph Theorem. The reader will find projection operators extremely useful in Hilbert space theory.

Definition 6.9. A bounded linear operator P from a normed linear space X onto a subspace M of X is called a *projection* if $P^2 = P$.† ∎

Remarks

6.6. If P is a projection and $P \neq 0$, then $\| P \| \geq 1$ since

$$\| P \| = \| P^2 \| \leq \| P \| \cdot \| P \|.$$

In fact, $\| P \|$ can be greater than 1. Let $X = R^2$ with $\| (x, y) \| = | x | + | y |$ and $M = \{(x, x) \mid x \in R\}$, a closed subspace of X. We define $P((x, y)) = (y, y)$. Then P is a projection onto M. But $P((0, 1)) = (1, 1)$ and therefore, $\| P \| > 1$.

6.7. There is always a projection from a normed linear space X onto any of its finite-dimensional subspaces M. To see this, let $x_1, x_2, \ldots, x_n$ be the basis of $M \subset X$. Define $x_i^* \in X^*$ by

$$x_i^*(x_j) = \begin{cases} 1, & i = j, \\ 0, & i \neq j. \end{cases}$$

Let $P(x) = \sum_{i=1}^{n} x_i^*(x)x_i$. Then P is a projection.

Proposition 6.14. Let M be a closed subspace of a Banach space X. Then there exists a projection P from X onto M if and only if there is a closed subspace N of X such that $X = M + N$, $M \cap N = \{0\}$. ∎

Proof. For the "only if" part, let $N = P^{-1}(\{0\})$, where P is a projection of X onto M. Since $x = Px + (x - Px)$ and $x - Px \in N$, $X = M+N$. Also $z \in M \cap N$ implies $P(z) = 0$ and $z = P(x)$ for some $x \in X$, which means $z = 0$.

For the "if" part, let $X = M + N$, $M \cap N = \{0\}$. Then for each $x \in X$, there exist unique $m \in M$ and $n \in N$ such that $x = m + n$. We define $P(x) = m$. Then P is linear and $P^2 = P$. To show that P is bounded, it is sufficient to show that it is closed (because of the Closed Graph Theorem). Let $x_k \to x$ and $P(x_k) \to y$. Let $x_k = m_k + n_k$, $m_k \in M$ and

† In a Hilbert space, projection operators, by definition, need to satisfy an additional condition. See Section 7.4.

$n_k \in N$. Also let $x = m + n$, $m \in M$ and $n \in N$. Then $P(x_k) = m_k \to y \in M$. Hence $n_k = (m_k + n_k) - m_k \to m + n - y$. This means that $m + n - y \in N$ or $m - y \in M \cap N = \{0\}$. Therefore $y = m = P(x)$ and P is closed. ∎

Problems

6.3.1. The notations are as in Lemma 6.2. Show that $S_y(r) \subset TS_X(1)$, if $S_Y(r) \subset TS_X(1/(1-\varepsilon))$ for all $\varepsilon \in (0, 1)$.

× **6.3.2.** Prove Proposition 6.8 and Proposition 6.9.

6.3.3. Let X be a normed linear space of the second category and M be a closed proper subspace of X. Show that X/M is of the second category.

× **6.3.4.** Let T be a closed linear operator from a normed linear space X into a normed linear space Y. Let $M = T^{-1}\{0\}$. Show that there is a unique one-to-one, closed, and linear operator T_0: $X/M \to Y$ such that $T_0 \circ \Phi = T$, where Φ is the natural map from X onto X/M. Also show that T_0^{-1}: $T(X)(\subset Y) \to X/M$ is a closed linear operator.

× **6.3.5.** Prove Lemma 6.2.

× **6.3.6.** Let $T_n \in L(X, Y)$, where X is a Banach space and Y is any normed linear space. If for each $x \in X$, $T(x) = \lim_{n\to\infty} T_n(x)$, then show that $T \in L(X, Y)$.

× **6.3.7.** Let A be a linear operator from X into Y such that $y^* \circ A$ is continuous for each $y^* \in Y^*$. Show that $A \in L(X, Y)$. (X and Y are normed linear spaces which are not necessarily complete.)

× **6.3.8.** If A is a bounded linear operator from D (a dense subspace of a normed linear space X) into a Banach space Y, then show that there exists a unique bounded linear operator B from X into Y with $\| A \| = \| B \|$ and $B(x) = A(x)$, $x \in D$. [Note that completeness is essential here; for consider $D = Y =$ the polynomials on [0, 1], $X = C[0, 1]$ and i: $D \to D$, the identity map.]

× **6.3.9.** Let $A \in L(X, Y)$. The adjoint of A is an operator A^*: $Y^* \to X^*$ defined by $A^*y^*(x) = y^*(A(x))$, $x \in X$. Show the following:

(i) $\| A^* \| = \| A \|$.
(ii) $(\alpha A + \beta B)^* = \alpha A^* + \beta B^*$ for all $\alpha, \beta \in F$ and $A, B \in L(X, Y)$.
(iii) If $A \in L(X, Y)$, $B \in L(Y, Z)$, then $(BA)^* = A^*B^*$.
(iv) If $A \in L(X, Y)$, A is onto and A^{-1} exists and belongs to $L(Y, X)$, then $(A^{-1})^* = (A^*)^{-1}$.

Note that if $X = Y = l_p$, $1 \le p < \infty$, then a bounded linear operator A on l_p can be represented by an infinite matrix (a_{ij}). If e_i is that element in l_p where the ith entry is 1 and every other entry is 0, and if $e_i^* \in l_p^*$ such that

$e_i^*(e_j) = 1$ if $i = j$, $= 0$ if $i \neq j$, then $a_{ij} = e_i^*(A(e_j))$. If $A(x) = y$ and $x = \sum x_i e_i$, then $e_i^*(y) = \sum_j a_{ij} x_j$. Since l_p^* can be identified with l_q, where $pq = p + q$, the adjoint A^* can be easily verified to be represented by the transpose of the infinite matrix (a_{ij}) in the same sense as above.

In the next few problems and later on, the range of A is denoted by $R(A)$ or R_A.

× **6.3.10.** Let $A \in L(X, Y)$, A be one-to-one and X, Y both Banach spaces. Then $R(A)$ is closed if and only if there exists $C > 0$ such that $\| x \| \leq C \| A(x) \|$ for each $x \in X$.

6.3.11. If X and Y are Banach spaces and $A \in L(X, Y)$, then $R(A)$ is closed in Y if and only if there exists $C > 0$ such that $\inf\{\| x - y \|: A(y) = 0\} \leq C \| A(x) \|$ for each $x \in X$.

× **6.3.12.** Let $S \subset X$. Then S° (the *annihilator* of S) $= \{x^* \in X^*: x^*(y) = 0$ for each $y \in S\}$. $S^\circ = X^*$ when S is empty. Similarly, for $E \subset X^*$, ${}^\circ E = \{x \in X: x^*(x) = 0$ for each $x^* \in E\}$. ${}^\circ E = X$, when E is empty. Show the following:

(i) S° and ${}^\circ E$ are both closed subspaces.
(ii) For $S \subset X$, ${}^\circ(S^\circ)$ is the closed subspace spanned by S.
(iii) $[R(A)]^\circ = N(A^*)$, the null space of A^*
(iv) $\overline{R(A)} = {}^\circ[N(A^*)]$.
(v) ${}^\circ[R(A^*)] = N(A)$.
(vi) $R(A^*) \subset [N(A)]^\circ$.

× **6.3.13.** Let $A \in L(X, Y)$, X and Y both Banach spaces. If $R(A)$ is closed in Y, then show that $R(A^*) = [N(A)]^\circ$ and hence is closed in X^*. [Hint: Use Problem 6.3.12 (vi) and the following. For $x^* \in [N(A)]^\circ$, define $f: R(A) \to F$ by $f(A(x)) = x^*(x)$. Show, by Problem 6.3.11, $| f(A(x)) | \leq C \| x^* \| \cdot \| A(x) \|$, $C > 0$. Extend f to $y^* \in Y^*$ so that $A^*(y^*) = x^*$.]

6.3.14. Let $A \in L(X, Y)$. Then show that $R(A^*) = X^*$ if and only if A^{-1} exists and is continuous.

6.3.15. Let $A \in L(X, Y)$, Y complete and $R(A) = Y$. Then show that A^* has a continuous inverse.

6.3.16. Prove that $L_2[0, 1]$ is of the first category in $L_1[0, 1]$. (Hint: The identity map from L_2 into L_1 is continuous, but *not* onto. Use the Open Mapping Theorem.)

6.3.17. *Joint Continuity of a Separately Continuous Bilinear Function.* Suppose X and Y are Banach spaces and $T: X \times Y \to R$ is a mapping such that for each $x \in X$ and each $y \in Y$, the functions $y \to T(x, y)$ and $x \to T(x, y)$ are bounded linear functionals. Show that T is continuous on $X \times Y$. (Hint: Use the Principle of Uniform Boundedness.)

6.3.18. *Convex Functions and the Principle of Uniform Boundedness.* Let $\mathscr{F}$ be the smallest σ-algebra containing the open sets of a Banach space X. Then every real-valued convex function defined on X and measurable with respect to $\mathscr{F}$ is continuous. Using this result, the Principle of Uniform Boundedness can be proven as follows. The function $p(x) = \sup\{\| T_\lambda(x) \|: \lambda \in \Lambda\}$, where $\{T_\lambda: \lambda \in \Lambda\}$ is a family of bounded linear operators from X into a normed linear space Y such that $p(x)$ is a real-valued function on X, is a lower semicontinuous convex function and therefore continuous. Then there exists $\delta > 0$ such that $\| x \| \leq \delta \Rightarrow | p(x) - p(0) | \leq 1$. This means that for each $\lambda \in \Lambda$ $\| T_\lambda \| \leq 1/\delta$.

6.3.19. *Another Application of the Principle of Uniform Boundedness.* Let X be the Banach space (with "sup" norm) of periodic continuous functions f on R with period 2π. For $f \in X$, let $T_n f(t) = n(f(t + 1/n) - f(t))$. Then by Corollary 1.2 (Chapter 1), $\lim_{n\to\infty} T_n f(t) = f'(t)$ exists for all f in a dense subset of X. However, the set of nondifferentiable functions is of the second category in X. [Notice that $\| T_n \| = 2n$. If $D \subset X$ is the set of functions differentiable at 0 in R and D is of the second category, then the Principle of Uniform Boundedness applies to the sequence $x^* \circ T_n$, where $x^*(f) = f(0)$.]

6.3.20. *Equivalent Norms in* $C[0, 1]$. Any complete norm $\| \cdot \|$ in $C[0, 1]$, where $\lim_{n\to\infty} \| f_n - f \| = 0 \Rightarrow \lim_{n\to\infty} f_n(t) = f(t)$ for all $t \in [0, 1]$, is equivalent to the usual "sup" norm. {Here the Closed Graph Theorem can be used to show that the identity map from $C[0, 1]$ with "sup" norm into $(C[0, 1], \| \cdot \|)$ is continuous; then Proposition 6.11 applies.}

★ **6.3.21.** *Necessary and Sufficient Condition for a Sequence to be a Schauder*† *Basis in a Banach Space* (*An Application of the Open Mapping Theorem*). Let X be a Banach space and (e_n) a sequence in X such that the linear span of (e_n) is dense in X and no e_n is 0. Prove that (e_n) is a Schauder basis for X if and only if there is a positive k_0 such that for all positive integers n and p and scalars (a_i),

$$k_0 \left\| \sum_{i=1}^{n+p} a_i e_i \right\| \geq \left\| \sum_{i=1}^{n} a_i e_i \right\|. \qquad (**)$$

[Hint: Use Problem 6.1.16(v) and the norm $||| \cdot |||$ defined there. By the

† See Problem 6.1.16 for definition.

open mapping theorem, T^{-1} is continuous, where T is the identity map from $(X, |||\cdot|||)$ to $(X, \|\cdot\|)$. This proves the "only if" part. For the "if" part, assume (**) and let Y be the linear span of (e_n). Define e_i^* on Y by

$$e_i^*\left(\sum_{j=1}^{n} a_j e_j\right) = a_i \quad \text{if } n \geq i,$$
$$= 0 \quad \text{otherwise.}$$

By (**), e_i^* is continuous, and can be extended by continuity to all of X. Given any subsequence (m_k), there exists $(n_k) \subset (m_k)$ such that $y_{n_k} = \sum_{i=1}^{n_k} a_{i,k} e_i$ and $\| x - y_{n_k} \| \leq 1/(k_0 2^{k+2})$. For $i \leq n_k$, $e_i^*(y_{n_k}) = a_{i,k}$. Use (**) to show that

$$\left\| \sum_{i=1}^{n_k} a_{i,k+p} e_i - \sum_{i=1}^{n_k} a_{i,k} e_i \right\| \leq 1/2^{k+1},$$

so that for each $p \geq 1$,

$$\left\| \sum_{i=1}^{n_k} e_i^*(y_{n_{k+p}}) e_i - x \right\| \leq 1/2^k.$$

Now let p go to infinity, and then, let k go to infinity.]

★ **6.3.22.** *A Schauder Basis for* $L_p[0, 1]$, $1 \leq p < \infty$. Let k and r be positive integers such that $k \leq 2^{r-1}$. Partition $[0, 1]$, into 2^r equal subintervals $I_{j,r}$, $1 \leq j \leq 2^r$. (Thus, for $r = 2$, $I_{1,2} = [0, \frac{1}{4})$, $I_{2,2} = [\frac{1}{4}, \frac{1}{2})$, $I_{3,2} = [\frac{1}{2}, \frac{3}{4})$, and $I_{4,2} = [\frac{3}{4}, 1]$.) Define the sequence of functions (g_i) on $[0, 1]$ by

$$g_{2^{r-1}+k} = 1, \quad \text{on } I_{2k-1,r},$$
$$= -1, \quad \text{on } I_{2k,r},$$
$$= 0, \quad \text{otherwise,}$$

where $1 \leq k \leq 2^{r-1}$. Notice that all characteristic functions of the intervals $I_{j,r}$ are contained in the linear span of the g_i's, which is, therefore, dense in $L_p[0, 1]$. Use condition (**) in Problem 6.3.21 to show that (g_i) is a Schauder basis for $L_p[0, 1]$, $1 \leq p < \infty$.

6.3.23. *Continuation of Problem* 6.3.20. Prove the following assertions in $C[0, 1]$:

(i) The norm $\| f \| = \sum_{n=1}^{\infty} (1/2^n) \mid f(t_n) \mid$, where (t_n) is a dense sequence in $(0, 1)$, is incomplete. (Does convergence in this norm imply pointwise convergence?)

(ii) The norm $\| f \| = \sup_{t\in[0,1]} \mid tf(t) \mid + \mid f(0) \mid$ is incomplete.

(iii) There are at least two complete norms in $C[0, 1]$ that are not equivalent. What if $C[0, 1]$ is replaced by an infinite-dimensional vector space? [Hint for (iii): $C[0, 1]$, as well as $L_2[0, 1]$, has a Hamel basis with cardinality c. In fact, a Hamel basis of any separable Banach space has the cardinality c (under the assumption of the continuum hypothesis). Let (f_n) be a uniformly bounded linearly independent sequence in $C[0, 1]$ such that $\| f_n \|_2 \to 0$, but $f_n \not\to 0$ in $C[0, 1]$. Extend (f_n) to a Hamel basis H_1 of $C[0, 1]$ and also to a Hamel basis H_2 of $L_2[0, 1]$. It is now easy to transport the L_2-norm to $C[0, 1]$ via a linear bijection.]

6.3.24. Let S be a closed subspace of $L_1[0, 1]$. If $S \subset L_p[0, 1]$ for some $p > 1$, then what can be said about S?

6.4. Reflexive Banach Spaces and the Weak Topology

In the theory of Banach spaces an often useful concept is that of reflexivity, which is based upon a characterization of a class of bounded linear functionals. The notion and properties of reflexive Banach spaces, along with some representation theorems for bounded linear functionals on certain well-known normed linear spaces, is the subject of this section. One reason for the study of such spaces is that a large number of useful results that are not true in general Banach spaces are valid in such spaces; also, one comes across a wide class of such spaces in theory as well as in practice. For instance, the L_p $(1 < p < \infty)$ spaces[†] will be shown to belong to this class.

Among other things, we also show in this section the interplay between reflexivity and a basic convergence concept, the concept of weak convergence. The topics of weak convergence and weak topology are essential in functional analysis. They find an immense number of applications in various contexts in the theory of differential equations and in the calculus of variations.

Definition 6.10. The *natural map* J of a normed linear space X into its second conjugate space $X^{**}[= L(X^*, F)]$ is defined by

$$[J(x)](x^*) = x^*(x), \qquad x^* \in X^*.$$

If the range of J is all of X^{**}, then X is called *reflexive*. ∎

[†] In this section, for convenience the L_p spaces are taken over the reals.

Remark 6.8. A reflexive normed linear space is complete. The reason is that

$$\| J(x) \| = \sup_{\|x^*\|=1} | x^*(x) | = \| x \|,$$

by Corollary 6.5; and therefore J is a linear isometry from X onto X^{**}, which is complete. However, the existence of a linear isometry from X *onto* X^{**} does *not* guarantee the reflexivity of X. For an example demonstrating this fact, the serious reader is referred to R. C. James.†

Remark 6.9. A finite-dimensional normed linear space X is reflexive. The reason is that $\dim X = \dim X^* = \dim X^{**}$, and a one-to-one linear operator between finite-dimensional spaces of the same dimension is also *onto*.

Proposition 6.15. Every normed linear space X is a dense subspace of a Banach space. ▮

Proof. The idea is simple. Simply identify X with $J(X)$ and note that $\overline{J(X)}$ is a Banach space. The details are slightly messy. Let $\hat{X} = X \cup (\overline{J(X)} - J(X))$. Denoting vector addition in X^{**} by $\mathring{+}$, we define addition in $\hat{X}$ as follows:

(i) $x + y$ is the same as it is in X, if $x \in X$ and $y \in X$;

(ii) $x + y = J(x) \mathring{+} y$, if $x \in X$ and $y \in \hat{X} - X$;

(iii) $x + y = x \mathring{+} y$, if $x \in \hat{X} - X$, $y \in \hat{X} - X$ and $x \mathring{+} y \notin J(X)$;

(iv) $x + y = J^{-1}(x \mathring{+} y)$, if $x \in \hat{X} - X$, $y \in \hat{X} - X$ and $x \mathring{+} y \in J(X)$.

We define scalar multiplication in X similarly so that $\hat{X}$ is a vector space with X as a subspace. For $x \in X$, let its norm in X be its norm in $\hat{X}$. For $x \in \hat{X} - X$, we define its norm to be its norm in X^{**}. It follows easily that $\hat{X}$ is a normed linear space with X as a dense subspace. The map $\Phi: \hat{X} \to \overline{J(X)}$ defined by

$$\begin{aligned} \Phi(x) &= J(x), \text{ if } x \in X; \\ &= x, \text{ if } x \in \hat{X} - X, \end{aligned}$$

is a surjective linear isometry. This means that $\hat{X}$ is complete. ▮

Before we get involved with the properties of reflexive spaces, we should consider some interesting examples of such spaces. To this end, we study

† R. C. James, A nonreflexive Banach space isometric with its second conjugate, *Proc. Nat. Acad. Sci. U.S.A.* **37**, 174–177 (1951).

first the conjugate (or dual) spaces of some important Banach spaces. Let us recall the following four results, which were proven in Chapter 4 of Part A and appeared there as Theorem 4.8, Corollaries 4.4, 4.5, and 4.6. These results involve L_p, $1 \leq p < \infty$, spaces. For simplicity, we consider only real-valued L_p functions.

Theorem 6.12. *Riesz Representation Theorem.* Let $(X, \mathscr{A}, \mu)$ be a σ-finite measure space and Φ be a bounded linear functional on L_p, $1 \leq p < \infty$. If $1/p + 1/q = 1$, then there is a *unique* element $g \in L_q$ such that

$$\Phi(f) = \int fg \, d\mu, \qquad f \in L_p,$$

and

$$\| \Phi \| = \| g \|_q. \qquad \blacksquare$$

Corollary 6.6. Let $(X, \mathscr{A}, \mu)$ be any measure space and Φ be a bounded linear functional on L_p, $1 < p < \infty$. Then there is a unique element $g \in L_q$, $1/p + 1/q = 1$ such that

$$\Phi(f) = \int fg \, d\mu, \qquad \text{for all } f \in L_p,$$

and $\| \Phi \| = \| g \|_q$. ∎

Corollary 6.7. For $1 < p < \infty$, $L_p{}^*$ is linearly isometric to L_q, $1/p + 1/q = 1$. ∎

Corollary 6.8. For $1 \leq p < \infty$, $l_p{}^*$ is linearly isometric to l_q, $1/p + 1/q = 1$. ∎

Corollary 6.9. For $1 < p < \infty$, L_p is reflexive. ∎

Proof. Let J be the natural map from L_p into $L_p{}^{**}$. To show that J is surjective, let $x^{**} \in L_p{}^{**}$. Let χ be the map from L_q into $L_p{}^*$, $1/p + 1/q = 1$, defined by

$$[\chi(g)](f) = \int fg \, d\mu, \qquad f \in L_p, \; g \in L_q.$$

Then by Corollary 6.6 χ is a surjective linear isometry. Let us define the map

$$x^* = x^{**} \circ \chi.$$

Then $x^* \in L_q^*$. Hence by Corollary 6.6, there exists $h \in L_p$ such that $x^*(g) = \int gh\, d\mu$, $g \in L_q$. Let $y^* \in L_p^*$. Then there exists $g_0 \in L_q$ such that $\chi(g_0) = y^*$. Now clearly

$$x^{**}(y^*) = x^*(g_0) = \int g_0 h\, d\mu = [\chi(g_0)](h) = y^*(h) = [J(h)](y^*).$$

Hence J is surjective. ∎

Corollary 6.10. For $1 < p < \infty$, l_p is reflexive. However, l_1 is *not* reflexive. ∎

Proof. The first part follows from Corollary 6.9. For the second part suppose l_1 is reflexive. Then l_1^{**}, being homeomorphic to l_1, is separable and therefore, by Proposition 6.7, l_1^* (and hence l_∞) is separable, which is a contradiction. ∎

We have seen above that the dual of L_p, $1 \leq p < \infty$, is (linearly isometric to) L_q in a σ-finite measure space. Unfortunately, such a representation does not hold for the bounded linear functionals on L_∞. Consider the Lebesgue measure space on $[0, 1]$. Let Φ be the linear functional on $C[0, 1] \subset L_\infty$ defined by $\Phi(f) = f(0)$. Let Φ_0 be the continuous extension of Φ (possible by the Hahn–Banach Theorem) to L_∞. Then $\| \Phi_0 \| \geq \| \Phi \| = 1$. Suppose there is a $g \in L_1$ such that $\Phi_0(f) = \int_0^1 f(t)g(t)\, dt$ for all $f \in L_\infty$. Let us define $f_n \in C[0, 1]$ by

$$f_n(t) = \begin{cases} 0, & 1/n \leq t \leq 1, \\ 1 - nt, & 0 \leq t \leq 1/n. \end{cases}$$

Then $\int_0^1 f_n(t)g(t)\, dt \to 0$ as $n \to \infty$, but $\Phi_0(f_n) = f_n(0) = 1$, which is a contradiction.

From the above results it is clear that there are plenty of infinite-dimensional reflexive spaces and therefore development of a theory for such spaces is in order and naturally will be very useful.

So far we have explored to some extent the class of all bounded linear functionals on a normed linear space. The concept of reflexivity is based upon characterizing a class of *bounded* linear functionals. Why is it that we do not consider with equal interest the class of all *linear* (not necessarily bounded) functionals on a normed linear space? The next result sheds some light on this question. First we need a definition.

Definition 6.11. Let X be a vector space over F, and let X' be the vector space (under natural operations) of all linear functionals on X—called the *algebraic dual* of X. Let J be the natural map from X into X'' (the algebraic dual of X') defined by

$$[J(x)](x') = x'(x), \qquad x' \in X'.$$

Then if J is onto, X is called *algebraically reflexive*. ∎

Proposition 6.16. A vector space X over F is algebraically reflexive if and only if X is finite-dimensional. ∎

Proof. We will only prove the "only if" part. Let H be the Hamel basis of X. Suppose X is infinite dimensional. Then H is infinite and let $H = \{x_i : i \in I\}$, I being an infinite indexed set.

Let us define $x_i' \in X'$ by

$$x_i'(x_j) = \begin{cases} 1, & i = j, \\ 0, & i \neq j, \end{cases}$$

and

$$x_i'(\sum \alpha_j x_j) = \sum \alpha_j x_i'(x_j).$$

Then $A' = \{x_i' : i \in I\}$ is linearly independent in X'. (Why?). Let H' be a Hamel basis of X' containing A'. Let us define $x'' \in X''$ such that

$$x''(x_i') = \beta_i, \qquad i \in I,$$
$$x''(x') = 0, \qquad x' \in H' - A',$$

where infinitely many of the β_i's are nonzero. If J is onto, then there exists $x \in X$ such that $[J(x)]x' = x'(x) = x''(x')$ for all $x' \in X'$. However $x_i'(x) = x''(x_i')$ for all $i \in I$. If $i \in I$ is such that the term x_i is missing in the unique representation of x as a linear combination of elements of H, and $x''(x_i') = \beta_i \neq 0$, then $x_i'(x) = 0 \neq \beta_i = x''(x_i')$, which is a contradiction. ∎

We see therefore from Proposition 6.16 that the concept of algebraic reflexivity is not very useful since the study of algebraically reflexive spaces is nothing more than the study of finite-dimensional spaces.

Now let us reconsider the space c (see Problem 6.2.11), a subspace of l_∞. By Problem 6.2.11, c^* is linearly isometric to l_1 and therefore c^{**} is linearly isometric to l_1^*. (Why?) If c is reflexive, then c^{**} is separable since c is separable. But this contradicts the nonseparability of l_1^* (or l_∞). Hence c as well as c^* (or l_1) is *not* reflexive. Here arises a natural question: Does the reflexivity of a Banach space X imply that of X^*? The following proposition answers this.

Proposition 6.17. A Banach space X is reflexive if and only if X^* is reflexive. ▮

Proof. Suppose X is reflexive. To prove that the natural map J: $X^* \to X^{***}$ is onto, let $x^{***} \in X^{***}$. We define

$$x^* = x^{***} \circ J_X,$$

where J_X is the natural map from X onto X^{**}. Then $x^* \in X^*$ and $J(x^*)[J_X(x)] = J_X(x)[x^*] = x^*(x) = x^{***}[J_X(x)]$. Since $J_X(X) = X^{**}$, $x^{***} = J(x^*)$ and J is onto.

To prove the converse, let X^* be reflexive. If X is not reflexive, then $J_X(X)$ is a closed proper subspace of X^{**}. By Corollary 6.3 there exists $x^{***} \in X^{***}$ such that $x^{***}(x^{**}) \neq 0$ for some $x^{**} \in X^{**} - J_X(X)$ and $x^{***}[J_X(x)] = 0$ for each $x \in X$. Since $J(X^*) = X^{***}$, there exists $x^* \in X^*$ such that $J(x^*) = x^{***}$. Therefore, $0 = J(x^*)[J_X(x)] = J_X(x)[x^*] = x^*(x)$ for each $x \in X$, which means that $x^* = 0$; and therefore $x^{***} = 0$, which is a contradiction. ▮

We will not try to determine the space l_∞^* in this book, since this determination is nontrivial; and we will not need it in our discussion. However, knowing that c is *not* reflexive, we can assert that l_∞ is also *not* reflexive. The following proposition makes it possible.

Proposition 6.18. Every closed subspace of a reflexive Banach space is reflexive. ▮

Proof. Let Y be a closed subspace of a reflexive Banach space X. Let J_Y: $Y \to Y^{**}$ and J_X: $X \to X^{**}$ be the natural maps. Given $y^{**} \in Y^{**}$, we define $x^{**} \in X^{**}$ by $x^{**}(x^*) = y^{**}(x^*|_Y)$. Since J_X is onto, there exists $x \in X$ such that $J_X(x) = x^{**}$. We claim that $x \in Y$ and $J_Y(x) = y^{**}$. If $x \notin Y$, then by Corollary 6.3, there exists $x^* \in X^*$ such that $x^*(x) \neq 0$ and $x^*(y) = 0$ for each $y \in Y$. Then $x^{**}(x^*) = y^{**}(x^*|_Y) = 0$ or $J_X(x)[x^*] = x^*(x) = 0$, which is a contradiction. Hence $x \in Y$. The rest is left to the reader. ▮

Next we consider a very important nonreflexive Banach space—the space $C[a, b]$ of (real valued) continuous functions under the uniform (sup) norm. It is clear that for any function g of bounded variation on $[a, b]$

$$\varphi(f) = \int_a^b f(t)\, dg(t)$$

defines a bounded linear functional φ on $C[a, b]$. Actually we will see in what follows that *every* bounded linear functional on $C[a, b]$ has the above form. This fact was first discovered by F. Riesz in 1909. Later, in 1937, it was extended by Banach to the case of a compact metric space (instead of $[a, b]$) and then by Kakutani in 1941 to the case of a compact Hausdorff space (instead of $[a, b]$). Kakutani considered signed measures instead of functions of bounded variation. (See Chapter 5.)

Let us denote by $\mathrm{BV}[a, b]$ the Banach space of real functions of bounded variation on $[a, b]$ with the norm

$$\| g \| = V(g) + | g(a) |,$$

where $V(g)$ denotes the total variation of g on $[a, b]$. (The reader should convince him- or herself that $\mathrm{BV}[a, b]$ is a Banach space—a fact that is neither difficult nor trivial to verify.) We will denote by $B[a, b]$ the Banach space of all real bounded functions on $[a, b]$ with the usual supremum norm.

Theorem 6.13. (*Riesz*).† For every bounded linear functional Φ on $C[a, b]$ there is a function g of bounded variation such that, for each $f \in C[a, b]$,

$$\Phi(f) = \int_a^b f(t)\, dg(t)$$

and $\| \Phi \| = V(g)$. ∎

Proof. Since $C[a, b] \subset B[a, b]$, by the Hahn–Banach Theorem there exists a continuous linear functional Φ_0 on $B[a, b]$ extending Φ such that $\| \Phi_0 \| = \| \Phi \|$. We define for $s \in (a, b]$

$$z_s(t) = \begin{cases} 1, & a \leq t \leq s, \\ 0, & s < t \leq b, \end{cases}$$

$$z_a(t) \equiv 0.$$

Define $g(s) = \Phi_0(z_s)$. We claim that $V(g) \leq \| \Phi_0 \| < \infty$. Clearly if

$$a = t_0 < t_1 < t_2 < \cdots < t_{n-1} < t_n = b$$

† The reader may note that a similar result holds in $C_1[a, b]$ when the scalars are complex numbers.

and

$$k_i = \operatorname{sgn}[g(t_i) - g(t_{i-1})], \quad 1 \leq i \leq n,$$

then we have

$$\begin{aligned}\sum_{i=1}^{n} |g(t_i) - g(t_{i-1})| &= \sum_{i=1}^{n} k_i[\Phi_0(z_{t_i}) - \Phi_0(z_{t_{i-1}})] \\ &= \Phi_0\Big(\sum_{i=1}^{n} k_i[z_{t_i} - z_{t_{i-1}}]\Big) \\ &\leq \| \Phi_0 \|,\end{aligned}$$

since the function inside the parenthesis has norm 1. Hence $V(g) \leq \| \Phi_0 \|$. Now to complete the proof of the theorem, let $f \in C[a, b]$. For $a = t_0 < t_1 < \cdots < t_{n-1} < t_n = b$, let $h(t) = \sum_{i=1}^{n} f(t_i)[z_{t_i}(t) - z_{t_{i-1}}(t)]$. Then we have

$$\begin{aligned}| h(t) - f(t) | &= | f(t_i) - f(t) |, \quad & t_{i-1} < t \leq t_i, \\ &= | f(t_1) - f(t) |, \quad & t = a,\end{aligned}$$

and

$$\Phi_0(h) = \sum_{i=1}^{n} f(t_i)[g(t_i) - g(t_{i-1})].$$

It is clear now that in the limit when $n \to \infty$ and $\max_{1 \leq i \leq n} | t_i - t_{i-1} | \to 0$,

$$\Phi(f) = \Phi_0(f) = \int_a^b f(t)\, dg(t).$$

Since $| \int_a^b f(t)\, dg(t) | \leq \| f \|\, V(g)$, $\| \Phi \| \leq V(g)$. The proof is complete. ∎

The above theorem does not provide a one-to-one correspondence between bounded linear functionals Φ on $C[a, b]$ and functions of bounded variation on $[a, b]$, as the following lemma shows.

Lemma 6.3. Let $g \in \mathrm{BV}[a, b]$. Let h be defined by

$$h(t) = \begin{cases} g(t + 0) - g(a), & a < t < b, \\ g(b) - g(a), & t = b, \\ 0, & t = a. \end{cases}$$

Then $h \in \mathrm{BV}[a, b]$ and for each $f \in C[a, b]$

$$\int_a^b f(t)\, dg(t) = \int_a^b f(t)\, dh(t)$$

and $V(h) \leq V(g)$. ∎

The proof of this lemma is left to the reader.

To provide a one-to-one correspondence between $C^*[a, b]$ and a suitable subspace of $\mathrm{BV}[a, b]$, we need the following definition.

Definition 6.12. A function $g \in \mathrm{BV}[a, b]$ is called *normalized* if $g(a) = 0$ and $g(t+0) = g(t)$, $a < t < b$. ∎

The collection of normalized functions of bounded variation on $[a, b]$ is denoted by $\mathrm{NBV}[a, b]$. The next two results will show that $\mathrm{NBV}[a, b]$ will provide us with the one-to-one correspondence desired above. We need another lemma.

Lemma 6.4. Let $g \in \mathrm{BV}[a, b]$ such that $\int_a^b f(t)\, dg(t) = 0$ whenever $f \in C[a, b]$. Then $g(a) = g(b)$ and for $a < t < b$, $g(t-0) = g(t+0) = g(a)$. ∎

Proof. Clearly if $f \equiv 1$, then $\int_a^b dg(t) = 0$ and therefore $g(a) = g(b)$. For $a \leq c < b$ and $0 < h < b - c$, let us define

$$f(t) = \begin{cases} 1, & a \leq t \leq c, \\ 1 - (t-c)/h, & c \leq t \leq c+h, \\ 0, & c+h \leq t \leq b. \end{cases}$$

Then $f \in C[a, b]$ and

$$0 = \int_a^b f(t)\, dg(t) = g(c) - g(a) + \int_c^{c+h} f(t)\, dg(t).$$

Integrating by parts and simplifying,

$$0 = -g(a) + \frac{1}{h} \int_c^{c+h} g(t)\, dt, \qquad 0 < h < b - c,$$

and therefore by letting $h \to 0$, $g(c+0) = g(a)$. Similarly, $g(c-0) = g(a)$ for $a < c < b$. ∎

Theorem 6.14. The dual space of $C[a, b]$ is linearly isometric to $\mathrm{NBV}(a, b]$. ∎

Proof. For $\Phi \in C^*[a, b]$, by Theorem 6.13 we can find $g \in \mathrm{BV}[a, b]$ such that for each $f \in C[a, b]$ we have

$$\Phi(f) = \int_a^b f(t)\, dg(t),$$

and $\| \Phi \| = V(g)$. Let h be as defined in Lemma 6.3. Then $h \in \mathrm{NBV}[a, b]$, $V(h) \leq V(g)$ and for $f \in C[a, b]$

$$\Phi(f) = \int_a^b f(t)\, dh(t).$$

Then since $\| \Phi \| \leq V(h)$, $V(h) = \| \Phi \|$. If we define $T(\Phi) = h$, then T is a linear isometry from $C^*[a, b]$ *onto* $\mathrm{NBV}[a, b]$. That T is well defined (that is, there is a unique $h \in \mathrm{NBV}[a, b]$ with the above properties) is guaranteed by Lemma 6.4. ∎

Finally in this section, we wish to consider a new notion of convergence of a sequence of vectors in a normed linear space—called weak convergence. The reason for considering this topic is that a very important interplay exists between reflexivity and weak convergence, as will be clear in what follows.

Definition 6.13. A sequence (x_n) in a normed linear space X is said to converge *weakly* to $x \in X$ if for every $x^* \in X^*$, $x^*(x_n) \to x^*(x)$ as $n \to \infty$. ∎

Remarks

6.10. If $x_n \to x$ in X, then $x_n \xrightarrow{w} x$, i.e., x_n converges weakly to x.

6.11. The converse to Remark 6.10 need not be true. For instance, if $X = l_p$ $(1 < p < \infty)$ and $x_n = (0, 0, \ldots, 0, \underset{n}{1}, 0, \ldots)$, then $\| x_n \|_p = 1$ for each n while $x_n \xrightarrow{w} 0$. [Recall that for each $x^* \in X^*$ there is $b = (b_1, b_2, \ldots) \in l_q$, $1/p + 1/q = 1$ such that $x^*(x_n) = b_n$.]

6.12. If $x_n \xrightarrow{w} x$, then $\sup_{1 \leq n < \infty} \| x_n \| < \infty$. The reason is that if $J(x_n)[x^*] = x^*(x_n)$, then $[J(x_n)]_{n=1}^{\infty}$ becomes a pointwise bounded family of bounded linear operators from X^* into F; and therefore by the Principle of Uniform Boundedness $\sup_{1 \leq n < \infty} \| J(x_n) \| = \sup_{1 \leq n < \infty} \| x_n \| < \infty$.

6.13. If X is finite dimensional, then $x_n \xrightarrow{w} x$ if and only if $x_n \to x$. See Problem 6.4.3.

6.14. If a sequence (f_n) in $C[a, b]$ converges weakly to $f \in C[a, b]$, then the sequence is uniformly bounded and for $a \leq t \leq b$, $\lim_{n\to\infty} f_n(t) = f(t)$. This is because of Remark 6.12 above and because $y_t^*(f_n) \to y_t^*(f)$, where $y_t^* \in C^*[a, b]$ is defined by $y_t^*(h) = h(t)$. The converse is also true by Theorem 6.13.

6.15. If $x_n = (x_{n1}, x_{n2}, \ldots) \in l_p$, $z = (z_1, z_2, \ldots) \in l_p$ with $1 < p < \infty$, then $x_n \xrightarrow{w} z$ if and only if $\lim_{n\to\infty} x_{ni} = z_i$, $1 \leq i < \infty$, and $\sup_{1 \leq n < \infty} \| x_n \|_p < \infty$. See Problem 6.4.4.

6.16. A normed linear space X is called *weakly sequentially complete* if every weak Cauchy sequence $(x_n) \in X$ [that is, $(x^*(x_n))$ is Cauchy for each $x^* \in X^*$] converges weakly to some element $x \in X$. A reflexive Banach space X is weakly sequentially complete. To prove this, let (x_n) be a weak Cauchy sequence in X. Let $T(x^*) = \lim_{n\to\infty} x^*(x_n)$. Then T is linear; and by Remark 6.12 above, $\| T \| \leq \sup_{1\leq n<\infty} \| x_n \| < \infty$ and therefore $T \in X^{**}$. If X is reflexive, there exists $x \in X$ such that $J(x) = T$, J being the usual natural map from X onto X^{**}. Hence $\lim_{n\to\infty} x^*(x_n) = x^*(x)$.

6.17. $C[0, 1]$ is *not* weakly sequentially complete and hence *not* reflexive. Consider $x_n(t) = (1 - t)^n$. Then $x_n(t)$ cannot converge pointwise to a continuous function and hence by Remark 6.14 cannot converge weakly. But for $y^* \in C^*[0, 1]$, there exists $g \in \text{NBV}[0, 1]$ such that for $z_{mn} = x_m - x_n$

$$y^*(z_{mn}) = \int_0^1 z_{mn}(t)\, dg(t) = \int_0^{t_1} z_{mn}(t)\, dg^*(t) + \int_{t_1}^1 z_{mn}(t)\, dg(t),^\dagger$$

where $g^*(t) = g(t + 0), 0 \leq t < 1$. The first integral can be made arbitrarily small by taking t_1 (>0) sufficiently close to 0, $g^*(t)$ being continuous from the right; and the second integral can be made arbitrarily small {since $x_n(t) \to 0$ uniformly in $[t_1, 1]$} by taking n and m sufficiently large. This means that (x_n) is weakly Cauchy in $C[0, 1]$.

6.18. In a reflexive Banach space X, each bounded (that is, norm bounded) sequence has a weakly convergent subsequence.

To prove this, let (x_n) be a sequence in X such that $\sup_{1\leq n<\infty} \| x_n \| < \infty$. Consider the closed linear subspace Y spanned by the x_n's. Then Y is separable. Y is also reflexive by Proposition 6.18. Therefore Y^{**} is separable, and by Proposition 6.7 Y^* is also separable. Let $(y_m^*)_{m\in N}$ be dense in Y^*. Since the sequence $y_1^*(x_n)$ is bounded, there exists a subsequence $(x_{1,n})_{n\in N}$ such that $y_1^*(x_{1,n})$ converges as $n \to \infty$. Similarly, there exists a subsequence $(x_{2,n}) \subset (x_{1,n})$ such that $y_2^*(x_{2,n})$ converges as $n \to \infty$. Using induction we can find for each positive integer k a subsequence $(x_{k+1,n}) \subset (x_{k,n})$ such that for each i, $1 \leq i \leq k + 1$, $y_i^*(x_{k+1,n})$ converges as $n \to \infty$. If $(x_{nn})_{n=1}^\infty$ is the diagonal sequence, then it is easy to show that for each i, $y_i^*(x_{nn})$ converges; and therefore for each $y^* \in Y^*$, $y^*(x_{nn})$ converges. We define

$$y^{**}(y^*) = \lim_{n\to\infty} y^*(x_{nn}).$$

† Note that for $f \in C[0, 1]$ with $f(0) = 0$, $\int_0^t f\, dg = \int_0^t f\, dg^*$ for all t in $(0, 1]$.

Then $y^{**} \in Y^{**}$, since $\sup_{1 \le n < \infty} \| x_n \| < \infty$. Since Y is reflexive, there exists $y \in Y$ such that $y^*(y) = y^{**}(y^*)$ whenever $y^* \in Y^*$. Hence $x_{nn} \xrightarrow{w} y$.

6.19. It is clear from the previous remark that in a *reflexive* Banach space if the weak convergence of a *sequence* implies its convergence, then its closed unit ball must be compact, and consequently the space must be finite dimensional. This is not true in a nonreflexive space, as the following remark shows.

6.20. In l_1, a nonreflexive space, the weak convergence of a sequence implies its (strong) convergence. If this were not the case, there would be an $\varepsilon > 0$ and a sequence of elements $x_n = (x_1{}^n, x_2{}^n, x_3{}^n, \ldots)$ in l_1 such that for each $(w_1, w_2, \ldots) \in l_\infty$, we have

$$\sum_{j=1}^{\infty} w_j x_j{}^n \to 0, \qquad \text{as } n \to \infty,$$

and

$$\sum_{j=1}^{\infty} | x_j{}^n | > \varepsilon, \qquad 1 \le n < \infty. \tag{6.3}$$

Choosing the w_k's properly, it follows easily that for each k, $\lim_{n\to\infty} x_k{}^n = 0$. Let $m_0 = n_0 = 1$. Then the sequence (m_k, n_k) is defined inductively as follows: n_k is the smallest integer $n > n_{k-1}$ such that

$$\sum_{j=1}^{m_{k-1}} | x_j{}^{n_k} | < \frac{\varepsilon}{5}, \tag{6.4}$$

and m_k is the smallest integer $m > m_{k-1}$ such that

$$\sum_{j=m_k}^{\infty} | x_j{}^{n_k} | < \frac{\varepsilon}{5}. \tag{6.5}$$

We define $w = (w_1, w_2, \ldots)$ by

$$w_j = \begin{cases} \operatorname{sgn} x_j{}^{n_k}, & m_{k-1} \le j < m_k, \\ \operatorname{sgn} x_j{}^{n_{k+1}}, & m_k \le j < m_{k+1}, \text{ etc.} \end{cases}$$

Then by inequalities (6.4) and (6.5) we have

$$\left| \sum_{j=1}^{\infty} w_j x_j{}^{n_k} - \sum_{j=1}^{\infty} | x_j{}^{n_k} | \right| \le 2 \sum_{j=1}^{m_{k-1}} | x_j{}^{n_k} | + 2 \sum_{j=m_k}^{\infty} | x_j{}^{n_k} | < \frac{4\varepsilon}{5}.$$

Therefore it follows from (6.3) that for each k,

$$\left| \sum_{j=1}^{\infty} w_j x_j{}^{n_k} \right| \ge \frac{\varepsilon}{5},$$

which contradicts the first statement in (6.3).

From the above remarks the reader has some ideas about weak convergence in a normed linear space and its connection with reflexivity of the space. This connection will be more distinct if we study what is called the weak topology of a normed linear space X.

The *weak topology* of X is the weakest topology on X such that every element in X^* is continuous. Clearly a basis for the weak topology on X consists of sets of the form

$$\{x: |f_i(x) - f_i(x_0)| < \varepsilon, \ i = 1, 2, \ldots, n\},$$

where $x_0 \in X$, $\varepsilon > 0$, and $f_i \in X^*$. Then it is easily proven that the weak topology is contained in the metric topology of X so that every weakly closed set is strongly closed. (The reader should verify this.)

That the converse is not true is clear from Remark 6.11. However, the following proposition holds.

Proposition 6.19. A linear subspace Y of a normed linear space X is weakly closed if and only if it is strongly closed. ∎

Proof. Suppose Y is strongly closed and $y \notin Y$. Then $\inf_{z \in Y} \| y - z \| > \delta > 0$. Hence by Corollary 6.3 there is $x^* \in X^*$ such that $x^*(y) \neq 0$ and $x^*(z) = 0$ for each $z \in Y$. This means that $A = \{w \in X: | x^*(w) | > 0\}$ is an open set in the weak topology, but $A \cap Y$ is empty. Hence y is not a weak-closure point of Y. The rest is clear. ∎

In Remark 6.20 we have seen that in l_1 weak convergence of a sequence is equivalent to its strong convergence. However, the topology of a topological space is not determined by the concept of convergence of a sequence unless the space is first countable. We will see that in l_1, as well as in any infinite-dimensional normed linear space, the weak topology is *properly* contained in the strong topology. The following theorem demonstrates this.

Theorem 6.15. The weak topology of a normed linear space coincides with its strong topology if and only if the space is finite dimensional. ∎

Proof. We prove only the "only if" part. We will prove that if the open unit ball in X is weakly open, then X^* (and therefore X) is finite dimensional.

Suppose $S = \{x \in X: \| x \| < 1\}$ is weakly open. Then there exist $x_1^*, \ldots, x_n^*$ in X^* and positive real numbers $r_1, \ldots, r_n$ such that

$$\{x \in X: | x_i^*(x) | < r_i, 1 \leq i \leq n\} \subset S.$$

If $\bigcap_{i=1}^{n}\{x \in X: x_i^*(x) = 0\}$ $(= A$, say) contains x_0, then for every real number r, $rx_0 \in A \subset S$. This implies that $x_0 = 0$, so $A = \{0\}$. We claim that $\{x_1^*, \ldots, x_n^*\}$ spans X^*. To prove this let $x^* \in X^*$. We define for $x \in X$, $T(x) = (x_1^*(x), \ldots, x_n^*(x)) \subset F^n$, where F is the scalar field. We also define h from $T(X)$ into F by $h(T(x)) = x^*(x)$. Then h is well defined since $T(x) = T(y)$ implies $x - y \in A = \{0\}$ or $x = y$. Since h is a linear functional on $T(X) \subset F^n$ and $T(X)$ [and therefore $T(X)^*$] is finite dimensional, we may assume that there exist $h_1, h_2, \ldots, h_m$ $(1 \leq m \leq n)$ in $T(X)^*$ such that $h_i\,(t_1, t_2, \ldots, t_n) = t_i$, $1 \leq i \leq m$ and $h = \sum_{i=1}^{m} a_i h_i$, $a_i \in F$. Then $x^* = \sum_{i=1}^{m} a_i x_i^*$. Hence X^* is finite dimensional. ∎

Before we close this section, we will present characterizations of a reflexive Banach space in terms of the weak compactness and also weak sequential compactness of its closed unit ball. We will do this by introducing another useful concept called the weak* topology, a topology of X^*.

We know that the weak topology in X^* is the weakest topology in X^* such that each element in X^{**} is continuous. However, this topology turns out to be less useful than the topology in X^* generated by the elements in $J(X)$, J being the natural map from X into X^{**}. This latter topology is called the weak* topology for X^* and is clearly weaker than its weak topology. A base for the weak* topology is given by the sets of the form

$$\{f \in X^*: |f(x_i) - f_0(x_i)| < \varepsilon,\ i = 1, \ldots, n\},$$

where $x_1, x_2, \ldots, x_n \in X$, $\varepsilon > 0$, and $f_0 \in X^*$.

If X is reflexive, then $J(X) = X^{**}$ and therefore the weak topology for X^* and its weak* topology coincide. The usefulness of the weak* topology[†] stems mainly from the following basic theorem. A sequential form of this important theorem for separable Banach spaces was proved by Banach in 1932. Alaoglu proved the theorem in the following general form in 1940.

Theorem 6.16. *The Banach–Alaoglu Theorem.* The closed unit ball in X^* is compact in its weak* topology. ∎

Proof. Let $S^* = \{f \in X^*: \|f\| \leq 1\}$. If $f \in S^*$, then $f(x) \in \{c \in F: |c| \leq \|x\|\} = I_x$, say. Then we can think of S^* as a subset of $P = \prod_{x \in X} I_x$, which is the set of all functions f on X with $f(x) \in I_x$, given the usual product topology. The topology which S^* inherits as a subset of P is the weak*

[†] The reader should verify by using the Hahn–Banach Theorem that the weak and the weak*-topology are both Hausdorff.

topology of S^*. Since P is compact by Tychonoff's theorem (Theorem 1.5), S^* will be compact if it is closed as a subset of P.

Let f be a point of closure of S^* in P. Then $f: X \to F$ and $|f(x)| \leq \|x\|$. Now for $x, y \in X$ and $\alpha, \beta \in F$, the set

$$V = \{g \in P: |g(x) - f(x)| < \varepsilon, \ |g(y) - f(y)| < \varepsilon, \\ |g(\alpha x + \beta y) - f(\alpha x + \beta y)| < \varepsilon\}$$

is an open subset of P containing f and hence $V \cap S^* \neq \varnothing$. Since for $g \in V \cap S^*$, g is linear, $|f(\alpha x + \beta y) - \alpha f(x) - \beta f(y)| < \varepsilon(1 + |\alpha| + |\beta|)$. Since this inequality holds for every $\varepsilon > 0$, f is linear and therefore $f \in S^*$. ∎

If X is reflexive, then X and X^{**} can be identified, and therefore the weak topology on X can be regarded as the weak* topology of X^{**}. Hence by Theorem 6.16 the closed unit ball in X is weakly compact. The converse is also true. To prove this we need the following lemma.

Lemma 6.5. Let $S \subset X$ and $S^{**} \subset X^{**}$ be defined as follows:

$$S = \{x: \|x\| \leq 1\} \quad \text{and} \quad S^{**} = \{x^{**}: \|x^{**}\| \leq 1\}.$$

Then $J(S)$ is dense in S^{**} with the weak* topology of X^{**}, where J is the natural map from X into X^{**}. ∎

Proof. Let $x_0^{**} \in S^{**}$, $\varepsilon > 0$, and $x_1^*, \ldots, x_n^* \in X^*$. The lemma will be proved if we can find $x_0 \in S$ such that

$$|x_i^*(x_0) - x_0^{**}(x_i^*)| < \varepsilon, \qquad i = 1, 2, \ldots, n. \tag{6.6}$$

Let $Y = \cap \{x \in X: x_i^*(x) = 0, i = 1, 2, \ldots, n\}$. Then Y is a closed subspace of X. Let $Y^\circ = \{x^* \in X^*: x^*(y) = 0 \text{ for each } y \in Y\}$. Then by an argument similar to that used in the proof of Theorem 6.15 (see also Problem 6.4.7), Y° is the closed subspace of X^* spanned by $x_1^*, \ldots, x_n^*$. Consider the mapping $T: (X/Y)^* \to Y^\circ$ by $T(V^*) = x^*$, where $x^*(x) = V^*([x])$. T is well defined and an onto linear isometry. (The reader can easily verify this.) Hence $(X/Y)^*$ (and therefore X/Y) is finite dimensional. Define $V^{**} \in (X/Y)^{**}$ by $V^{**}(V^*) = x_0^{**} \circ T(V^*)$. Then $\|V^{**}\| \leq 1$ since T is an isometry. Since X/Y is reflexive (being finite dimensional), there exists $V \in X/Y$ such that $V^{**}(V^*) = V^*(V)$ for each $V^* \in (X/Y)^*$ and $\|V\| = \|V^{**}\| \leq 1$. Let us choose k such that $\sup_{1 \leq i \leq n} \|x_i^*\| < k$. Then there exists $x \in V$ such that

$$\| x \| \leq \| V \| + \varepsilon/k \leq 1 + \varepsilon/k,$$

and $V^*(V) = x^*(x)$, where $T(V^*) = x^*$. This means that for each $x^* \in Y^\circ$, $x^*(x) = x_0^{**}(x^*)$. If $x_0 = [k/(k+\varepsilon)]x$, then $\| x_0 \| \leq 1$; and for $x^* \in Y^\circ$,

$$\begin{aligned} | x^*(x_0) - x_0^{**}(x^*) | &= | x^*(x_0) - x^*(x) | \\ &\leq \| x^* \| \, \| x_0 - x \| < (\varepsilon/k) \| x^* \|. \end{aligned}$$

It follows that x_0 satisfies the inequalities (6.6). ∎

Theorem 6.17. A normed linear space X is reflexive if and only if the closed unit ball in X is weakly compact. ∎

Proof. The "only if" part follows easily from the Banach–Alaoglu Theorem. To prove the "if" part we see that the natural map J from X (with weak topology) into X^{**} (with its weak* topology) is a linear homeomorphism onto $J(X) \subset X^{**}$. Hence if the closed unit ball S of X is weakly compact, $J(S)$ is compact (and hence closed) in the weak* topology of X^{**}. By Lemma 6.5, $J(S) = S^{**} = \{x^{**} \in X^{**}: \| x^{**} \| \leq 1\}$. Since J is linear, $J(X) = X^{**}$. ∎

Our next theorem in this section is the well-known *Eberlein–Šmulian Theorem*, one of the most remarkable results in Banach spaces. In 1940, V. L. Šmulian proved that the weak countable compactness of the weak closure of a subset A of a Banach space X implies its weak sequential compactness. W. F. Eberlein, in 1947, proved that the subset A is weakly compact if and only if it is weakly closed and weakly sequentially compact. Note that these results are nontrivial since the weak topology need not be even first countable. It is relevant here only to mention (without proof) at this point that the weak topology of the closed unit ball of a Banach space X is a metric topology if and only if X^* is separable.

Definition 6.14. A set $A \subset X^*$, where X is a normed linear space, is called *total* if whenever $x^*(x) = 0$ for every $x^* \in A$, then $x = 0$. ∎

If X is a separable Banach space, then X^* contains a *countable total* set; for if (x_n) is a dense sequence in $\{x: \| x \| = 1\}$ and $x_n^* \in X^*$ with $x_n^*(x_n) = \| x_n^* \| = 1$, then (x_n^*) is a *total set* and for all x in X, $\| x \| = \sup_n | x_n^*(x) |$.

The proof of the Eberlein–Šmulian Theorem that we present here is due to R. Whitley. First we need a lemma.

Lemma 6.6. Let X be a normed linear space such that X^* contains a countable total set. Then the weak topology on a weakly compact subset of X is *metrizable.* ▮

Proof. Let (x_n^*) be total and $\| x_n^* \| = 1$. Let

$$d(x, y) = \sum_{n=1}^{\infty} \frac{1}{2^n} \mid x_n^*(x - y) \mid$$

and let A be a weakly compact subset of X. By Remark 6.3, A is bounded since $x^*(A)$ is compact for $x^* \in X^*$. Let i be the identity map from A (with weak topology) onto A (with the metric topology induced by d). i is clearly continuous and therefore a homeomorphism; for if B is a weakly closed subset of A, then B is weakly compact and $i(B)$ is compact in the metric d and therefore closed, in this metric. ▮

Corollary 6.11. Let $A \subset X$, a normed linear space. Let B, the weak closure of A, be compact in the weak topology. Then B is also sequentially compact in the weak topology. ▮

Proof. Let (a_n) be a sequence from B, and let $\overline{\text{sp}}(a_n)$ be the (norm) closure of the linear subspace spanned† by (a_n). By Proposition 6.19 $\overline{\text{sp}}(a_n)$ is weakly closed and therefore $B \cap \overline{\text{sp}}(a_n)$ is a weakly compact subset of the separable normed linear space $\overline{\text{sp}}(a_n)$. By Lemma 6.6, the weak topology on $B \cap \overline{\text{sp}}(a_n)$ is metrizable and therefore sequentially compact by Proposition 1.22 of Chapter 1. The rest is clear. ▮

Theorem 6.18. *The Eberlein–Šmulian Theorem.* The following are equivalent for any subset A of a Banach space X:

(a) The weak closure of A is weakly compact.
(b) Any sequence in A has a weakly convergent (in X) subsequence.
(c) Every countable infinite subset of A has a weak-limit point in X. ▮

Proof. (a) ⇒ (b) by Corollary 6.11, and (b) ⇒ (c) trivially. We establish only (c) ⇒ (a). So we assume (c). Since $x^*(A)$ is a bounded set of

† Sometimes the subspace spanned by a set E is also denoted by $[E]$.

scalars for each $x^* \in X^*$, by Remark 6.3 A is bounded. If J is the natural map of X into X^{**}, $J(A)$ is bounded; and therefore by Theorem 6.16, $w^*(J(A))$, the weak* closure of $J(A)$, is compact in the weak* topology of X^{**} (i.e., the topology induced by X^*). Since J is a homeomorphism from X (with the weak topology) onto $J(X)$ (with the weak* topology), it is sufficient to show that $w^*(J(A)) \subset J(X)$.

To show this, let $x^{**} \in w^*(J(A))$. We will use induction. Let $x_1^* \in X^*$ with $\| x_1^* \| = 1$. Now there is $a_1 \in A$ with $| (x^{**} - J(a_1))(x_1^*) | < 1$. Let E_2 be the finite-dimensional subspace spanned by x^{**} and $x^{**} - J(a_1)$. Since the surface of the closed unit ball in E_2 is compact, there are $(y_i^{**})_{i=2}^{n_2} \in E_2$ with $\| y_i^{**} \| = 1$ such that for any $y^{**} \in E_2$ and $\| y^{**} \| = 1$, $\| y^{**} - y_i^{**} \| < \frac{1}{4}$ for some i. Let $x_i^* \in X^*$, $\| x_i^* \| = 1$ be such that $y_i^{**}(x_i^*) > \frac{3}{4}$, $2 \le i \le n_2$. Then for every $y^{**} \in E_2$, we have

$$\max\{| y^{**}(x_i^*) |: 2 \le i \le n_2\} \ge \tfrac{1}{2} \| y^{**} \|.$$

Again there exists $a_2 \in A$ so that

$$\max\{|[x^{**} - J(a_2)](x_i^*) |: 1 \le i \le n_2\} < \tfrac{1}{2}.$$

Then we consider the space E_3 spanned by $x^{**}, x^{**} - J(a_1)$, and $x^{**} - J(a_2)$; we find $(x_i^*)_{i=n_2+1}^{n_3}$ and then choose $a_3 \in A$ as before so that for every $y^{**} \in E_3$, we have

$$\max\{| y^{**}(x_i^*) |: n_2 < i \le n_3\} \ge \tfrac{1}{2} \| y^{**} \|$$

and

$$\max\{| [x^{**} - J(a_3)](x_i^*) |: 1 \le i \le n_3\} < \tfrac{1}{3}.$$

In this way, we continue to construct the sequence a_n.

By (c), the sequence (a_n) obtained above has a weak-limit point x. Clearly $x \in \overline{\text{sp}}(a_n)$ and so $x^{**} - J(x)$ is in the space $\overline{\text{sp}}(x^{**}, x^{**} - J(a_n)$ for $1 \le n < \infty)$. Therefore by the construction of (a_n) above, we have

$$\sup_{1 \le i < \infty} \{|(x^{**} - J(x))(x_i^*)|\} \ge \tfrac{1}{2} \| x^{**} - J(x) \|. \tag{6.7}$$

Also

$$\begin{aligned} | [x^{**} - J(x)](x_i^*) | &\le | [x^{**} - J(a_p)](x_i^*) | + | x_i^*(a_p - x) | \\ &\le 1/p + | x_i^*(a_p - x) |, \qquad \text{for } i \le n_p. \end{aligned}$$

Since x is a weak-limit point of (a_n), it follows that $[x^{**} - J(x)](x_i^*) = 0$ for all i. By inequality (6.7), $x^{**} = J(x)$. The proof is complete. ∎

Remark 6.21. By Theorems 6.17 and 6.18, a Banach space X is reflexive if and only if its closed unit ball is weakly sequentially compact. This is true even for any normed linear space (Problem 6.4.4).

An application of the Eberlein–Šmulian Theorem characterizes weakly compact subsets of $C[0, 1]$. (See Problem 6.4.13.) The Banach–Alaoglu Theorem has already been utilized in obtaining Theorems 6.17 and 6.18. Another application of this theorem is outlined in Problem 6.4.14. In what follows, we present still another application of this theorem in the context of weak*-sequential compactness and then use this result to solve a problem in harmonic analysis.

First, we prove the following useful result.

• **Theorem 6.19.** Let X be a normed linear space and $S^* = \{f \in X^*: \|f\| \leq 1\}$. Then X is separable if and only if the weak* topology on X^* restricted to S^* is a metric topology. ∎

Proof. Suppose X is separable. Let $A \subset X$ be a countable dense subset of X. Then the weak* topology restricted to S^* is the topology on S^* that is induced by A; i.e., as in the proof of Theorem 6.16, the topology induced on S^* is a subset of the set $P_A = \prod_{x \in A} I_x$, where I_x is the set of all scalars c with $|c| \leq \|x\|$, with product topology. Since P_A, as a countable product of metric spaces, is metrizable, the "only if" part of the theorem follows.

To prove the "if" part, we assume that the weak* topology on S^* is metrizable and therefore has a countable local base at 0. Then there exist real numbers r_n and finite subsets $A_n \subset X$ such that

$$\{0\} = \bigcap_{n=1}^{\infty} W_n, \qquad W_n = \{x^* \in S^*: |x^*(x)| < r_n,\ x \in A_n\}.$$

Let $A = \bigcup_{n=1}^{\infty} A_n$. It is clear that $x^* = 0$ whenever $x^* \in S^*$ and $x^*(x) = 0$ for every $x \in A$. Hence, by an application of the Hahn–Banach Theorem, it follows that the closed linear subspace spanned by A is X. Since A is countable, X is separable. ∎

In a metric space, compactness and sequential compactness are equivalent and therefore an application of the Banach–Alaoglu Theorem leads to the following

• **Corollary 6.12.** In a separable normed linear space X, the closed unit ball in X^* is weak*-sequentially compact. ∎

We will now close this section with an application to a problem of harmonic analysis. The problem is to characterize all those operators in $L(X, Y)$, $X = L_1(R)$ and $Y = L_p(R)$ where $1 < p \leq \infty$, which commute with convolution. Note that we have already introduced the notion of convolution in Problem 4.3.21. We recall that for $f \in L_1$ and $g \in L_p$ $(1 \leq p \leq \infty)$, $f * g$ is the convolution of f and g, and

$$f * g(x) = \int f(x - y)g(y)\, dy.$$

• **Theorem 6.20.** Let $T \in L(X, Y)$, where $X = L_1(R)$, $Y = L_p(R)$ and $1 < p \leq \infty$. Then the following are equivalent:

(i) For f and g in L_1, $T(f * g) = T(f) * g$.
(ii) There exists h in L_p such that for f in L_1, $T(f) = h * f$. ∎

Proof. Since (ii) ⇒ (i) by Problem 4.3.21, we prove only that (i) ⇒ (ii). We assume (i) and define the sequence

$$u_n(x) = (n/2)\chi_{[-1/n, 1/n]}(x), \qquad n = 1, 2, 3, \ldots.$$

Then $\| u_n \|_1 = 1$ and for any $f \in C_c(R)$ (and so for $f \in L_1$),

$$\lim_{n\to\infty} \| u_n * f - f \|_1 = 0.$$

Hence

$$\lim_{n\to\infty} \| T(f) - T(u_n) * f \|_p = \lim_{n\to\infty} \| T(f) - T(u_n * f) \|_p$$
$$\leq \| T \| \cdot \lim_{n\to\infty} \| f - u_n * f \|_1 = 0.$$

Since the sequence $T(u_n)$ is bounded in L_p-norm, it follows (after considering L_p as L_q^*, $pq = p + q$) by Corollary 6.12 that some subsequence $T(u_{n_i})$ converges to some h in L_p in the weak* topology of L_p. This means that for $f \in L_1$ and $g \in L_q$, we have

$$\begin{aligned}
\int T(f)(x)g(-x)\, dx &= \lim_{i\to\infty} \int T(u_{n_i}) * f(x)g(-x)\, dx \\
&= \lim_{i\to\infty} \{[T(u_{n_i}) * f] * g\}(0) \\
&= \lim_{i\to\infty} [T(u_{n_i}) * (f * g)](0) \\
&= \lim_{i\to\infty} \int T(u_{n_i})(x) f * g(-x)\, dx
\end{aligned}$$

$$= \int h(x) f * g(-x)\, dx$$
$$= [(h * f) * g](0)$$
$$= \int (h * f)(x) g(-x)\, dx.$$

It follows easily that $T(f) = h * f$ whenever $f \in L_1$. We leave the details to the reader. ∎

Problems

× **6.4.1.** Let $X = \{x_1, x_2\}$ and μ be a measure on 2^X such that $\mu(\{x_1\}) = 1$ and $\mu(\{x_2\}) = \infty$. Show that dim $L_1(\mu) =$ dim $L_1^*(\mu) = 1$, whereas dim $L_\infty(\mu) =$ dim $L_\infty^*(\mu) = 2$.

× **6.4.2.** Prove that BV$[a, b]$ is a Banach space under the norm $\| g \| = V(g) + | g(a) |$. Prove Lemma 6.3.

× **6.4.3.** Show that in a finite-dimensional normed linear space a sequence is convergent if and only if it is weakly convergent.

× **6.4.4.** Prove Remarks 6.15 and 6.21.

× **6.4.5.** Let f be a real-valued measurable function in a σ-finite measure space such that for all g in L_p $(1 \leq p < \infty)$, $fg \in L_1$. Show that $f \in L_q$ where $1/p + 1/q = 1$. What happens when the measure is semifinite? (Hint: Write $X = \cup X_n$, $X_n \subset X_{n+1}$, and $\mu(X_n) < \infty$. Let $f_n(x) = \chi_{X_n}(x) \inf\{| f(x) |, n\}$. Define $T_n(g) = \int f_n g \, d\mu$. Use the uniform boundedness principle.)

6.4.6. Let S be a linear subspace of $C[0, 1]$ which is closed as a subspace of $L_2[0, 1]$. Show that S is finite-dimensional. (Hint: Show that S is closed as a subspace of $C[0, 1]$ and that therefore there exists $k > 0$ such that $\| f \|_\infty \leq k \| f \|_2$ for all f in S. Use this to show that the closed unit ball of S in L_2 is compact.)

× **6.4.7.** Let $g, f_1, f_2, \ldots, f_n$ be linear functionals on a vector space X such that $\bigcap_{i=1}^n \{x: f_i(x) = 0\} \subset \{x: g(x) = 0\}$. Show that g is a linear combination of the f_i. [Hint: Consider the mapping $(g(x), f_1(x), \ldots, f_n(x)) \to (f_1(x), \ldots, f_n(x))$, which is injective.]

× **6.4.8.** Show that if X is reflexive and separable, then so is X^*.

6.4.9. Show that if

$$f_n(t) = nt, \qquad 0 \leq t \leq 1/n,$$
$$= 2 - nt, \qquad 1/n \leq t \leq 2/n,$$

$$= 0, \qquad 2/n \leq t \leq 1,$$

then $f_n \xrightarrow{w} 0$ in $C[0, 1]$; but $f_n \not\to 0$ in $C[0, 1]$.

6.4.10. Let Y be a closed subspace of X. Show that $(X/Y)^*$ is linearly isometric onto $Y^\circ = \{x^* \in X^*: x^*(y) = 0$ for all y in $Y\}$. {Hint: Consider the mapping $T(V^*) = x^*$ where $x^*(x) = V^*([x])$.}

6.4.11. Let Y be a closed subspace of X. Show that X is reflexive if and only if Y and X/Y are reflexive. (Use Problem 6.4.10 for the "only if" part.)

× **6.4.12.** (i) If f is a linear functional on X, then show that f is continuous relative to the weak topology if and only if $f \in X^*$.

(ii) If g is a linear functional on X^*, then show that g is continuous relative to the weak*-topology if and only if there is $x \in X$ such that $g(x^*) = x^*(x)$ for every $x^* \in X^*$.

6.4.13. *Weak Compactness in* $C[0, 1]$. Prove that a subset E of $C[0, 1]$ is weakly compact if and only if E is weakly closed, norm bounded, and every sequence (f_n) in E has a subsequence (f_{n_k}) such that $\lim_{k\to\infty} f_{n_k}(x) = f(x)$ for some f in E and all $x \in [0, 1]$. Also, is a pointwise convergent and uniformly bounded sequence in $C[0, 1]$ weakly convergent?

6.4.14. *Banach Spaces as Spaces of Continuous Functions*. Show that given a real Banach space X there exists a compact Hausdorff space S such that X is linearly isometric onto a closed linear subspace of $C(S)$ with uniform norm. (Hint: Take $S = \{x^* \in X^*: \| x^* \| \leq 1\}$, which is compact in the relative weak*-topology. Let $f_x(x^*) = x^*(x)$ for $x \in X$ and $x^* \in S$. Consider $\varrho(x) = f_x$.)

6.4.15. Let X be an infinite-dimensional Banach space. Prove that the weak closure of $\{x \in X: \| x \| = 1\}$ is the unit ball $\{x \in X: \| x \| \leq 1\}$. [Hint: Suppose that $\| y \| < 1$ and there is no x with norm 1 such that $| x_i^*(x - y) | \geq r > 0$ for $i = 1, 2, \ldots, n$. Since for any nonzero x, there is a real t such that $\| y + tx \| = 1$, the mapping $x \to (x_1^*(x), \ldots, x_n^*(x))$ is injective.]

× **6.4.16.** *Weak Convergence in* L_p, $1 < p < \infty$. Show that a norm bounded sequence (f_n) in L_p converges weakly to f in L_p if $f_n \to f$ in measure.

6.4.17. Prove the following result due to E. Hewitt: Let f be a real-valued measurable function in a semifinite measure space such that $f \notin L_p(\mu)$ for some $p > 1$. Then the set $\{g \in L_q: fg \in L_1$ and $\int fg \, d\mu = 0\}$ is dense in L_q, where $1/p + 1/q = 1$. (Hint: The set $E = \{g \in L_q: fg \in L_1\}$ is dense in L_q since it contains all functions χ_A, where

$$\mu(A) < \infty \quad \text{and} \quad A \subset \{x: n \leq | f(x) | < n + 1\}.$$

Then T, where $T(g) = \int fg\,d\mu$, is not continuous, but linear on E. Now use Problem 6.2.3(b).)

6.4.18. Let X be a compact Hausdorff space. Let $\varrho: X \to C(X)^*$ be defined by $\varrho(x)(f) = f(x)$. Show that ϱ is a homeomorphism from X onto $\varrho(X) \subset C(X)^*$ (with weak*-topology). (Since X need not be sequentially compact, this shows that the Eberlein–Šmulian theorem is false for weak*-compact subsets.)

6.4.19. Show that a Banach space X is reflexive if and only if every total subspace of X^* is dense in X^*. (Hint: For the "if" part, suppose $x^{**} \in X^{**} - J(X)$. Then the subspace $\{x^* \in X^*: x^{**}(x^*) = 0\}$ is not dense in X^*. Show that it is total.)

6.4.20. *Banach–Saks Theorem.* Every weakly convergent sequence (f_n) in $L_2[0, 1]$ has a subsequence $(f_{n_k} = g_k)$ such that the sequence (h_n), $h_n = (1/n)\sum_{k=1}^{n} g_k$, converges in L_2 norm. (Hint: Consider a subsequence (f_{n_k}) such that for $j \geq n_{i+1}$ and $1 \leq k \leq i$, $| \int f_j(x) f_{n_k}(x)\,dx | < 1/2^{i+1}$.) This result remains true in L_p, $p > 1$. The reader can later observe that the proof in L_2 extends easily to any Hilbert space.

6.4.21. Prove that $L_1(\mu)$ is weakly sequentially complete if μ is σ-finite. (Hint: If (f_n) is a weak Cauchy sequence in L_1, then for each measurable set E, $\lim_{n\to\infty} \int_E f_n\,d\mu$ exists. Use Problem 4.3.16 to show that $\nu(E) = \lim_{n\to\infty} \int_E f_n\,d\mu$ defines a bounded signed measure absolutely continuous with respect to μ. Now apply the Radon–Nikodym theorem.)

★ **6.4.22.** *Weak Convergence in L_1.* Prove that in a σ-finite measure space, a sequence (f_n) in L_1 converges weakly to f in L_1 if and only if $\sup_n \| f_n \|_1 < \infty$ and for each measurable set E, the sequence $\int_E f_n\,d\mu$ converges.

★ **6.4.23.** Prove that X^* is weak*-sequentially complete if X is a Banach space. [It is relevant to mention here that though L_1 as well as any reflexive Banach space is weakly sequentially complete, a normed linear space X is weakly complete (that is, every weak Cauchy net converges weakly) if and only if X is finite-dimensional. A similar statement holds for the weak* topology of X^*.]

6.5. Compact Operators and Spectral Notions

To prove some of Fredholm's results on integral equations, F. Riesz devised vector space techniques which easily extend and can be applied to a special class of linear operators called compact operators. These operators are very useful and often find applications in classical integral equations as well as in nonsingular problems of mathematical physics.

In this section we will derive basic properties of compact operators and then consider the Riesz–Schauder theory of such operators. The connection between the classical approximation problem for compact operators (by finite-dimensional operators) and the Schauder-basis problem in Banach spaces will be briefly discussed. We will finally introduce the spectral notions for a bounded linear operator on a Banach space and then consider briefly the spectral theory of compact operators.

Let X and Y be normed linear spaces over the same scalars.

Definition 6.15. A linear operator A from X into Y is called *compact* (or *completely continuous*) if A maps bounded sets of X into relatively compact (that is, having compact closure) sets of Y. ∎

Remarks

6.22. If A is compact then A is continuous.

6.23. If $A \in L(X, Y)$ and $A(X)$ is finite-dimensional, then A is compact.

6.24. An operator $A \in L(X, Y)$ need not be compact. For example, the identity operator on an infinite-dimensional normed linear space is not compact. (See Theorem 6.2.)

Example 6.15. Let $A\colon C[0, 1] \to C[0, 1]$ be defined by

$$Af(x) = \int_0^1 k(x, y) f(y)\, dy,$$

where $k(x, y)$ is a continuous function of (x, y) on $[0, 1] \times [0, 1]$. The reader can easily check that if $S = \{f \in C[0, 1]\colon \|f\|_\infty \leq 1\}$, then $A(S)$ is uniformly bounded and equicontinuous. By the Arzela–Ascoli Theorem $A(S)$ is relatively compact and therefore A is compact.

The next few results give some basic properties of compact operators.

Proposition 6.20. Any finite linear combination of compact operators is compact. ∎

The proof is left to the reader.

Proposition 6.21. Let A and B be in $L(X, X)$ with A compact. Then AB and BA are both compact. ∎

Proof. If S is a bounded set in X, then $AB(S) = A(B(S))$ is relatively compact since $B(S)$ is bounded and A is compact. Also $BA(S) \subset B(\overline{A(S)})$, which is compact since B is continuous and $\overline{A(S)}$ is compact. ∎

Proposition 6.22. Let $A \in L(X, Y)$ and $(A_n)_N$ be a sequence of compact operators from X into Y, where Y is complete and $\lim_{n\to\infty} \| A_n - A \| = 0$. Then A is compact. ∎

Proof. We will show that $A(S)$, for S bounded in X, is totally bounded and hence relatively compact in Y, a complete metric space. Let $\varepsilon > 0$. Then there is a positive integer n such that for each $x \in S$, $\| A_n(x) - A(x) \| < \varepsilon$. Since A_n is compact, $\overline{A_n(S)}$ is compact and hence totally bounded. Therefore there exist $x_1, x_2, \ldots, x_m$ in S such that, for each $x \in S$,

$$\inf_{1 \leq i \leq m} \| A_n(x) - A_n(x_i) \| < \varepsilon.$$

This means that for each $x \in S$

$$\inf_{1 \leq i \leq m} \| A(x) - A(x_i) \| < 3\varepsilon.$$

Hence $A(S)$ is totally bounded. ∎

The following example shows that completeness of Y is essential in Proposition 6.22.

Example 6.16. Let B be the operator from $c_0 (\subset l_\infty)$ into l_2 defined by

$$B((\alpha_k)) = (\alpha_k / k).$$

Let $X = c_0$, $Y =$ the range of B, and $A \in L(X, Y)$ be defined by $A(x) = B(x)$. However, A is not compact. To see this, let

$$x_n = (\underbrace{1, 1, 1, \ldots, 1}_{n}, 0, 0, \ldots) \in c_0;$$

then the sequence $A(x_n) = (1, 1/2, \ldots, 1/n, 0, 0, \ldots)$ converges to $(1/k)_{k \in N}$ in l_2. However, $(1/k)_{k \in N} \notin Y$, since if $B((\alpha_k)) = (1/k)$ then $(1, 1, 1, \ldots) \in c_0$, which is a contradiction. Hence the sequence $A(x_n)$ cannot have a convergent subsequence in Y or A is *not* compact. Nevertheless if we define $A_n \in L(X, Y)$ by

$$A_n((\alpha_k)) = (\beta_k),$$

where

$$\beta_k = \begin{cases} \alpha_k / k, & 1 \leq k \leq n, \\ 0, & k > n, \end{cases}$$

then the A_n's are compact and $\lim_{n\to\infty} \| A_n - A \| = 0$.

Proposition 6.22 above shows that the compact operators form a closed linear subspace of $L(X, Y)$, when Y is complete. Also every operator in $L(X, Y)$ with finite-dimensional range is compact. It is therefore natural to ask the following:

(A) Is every compact operator T in $L(X, Y)$ a limit in the norm of operators with finite-dimensional range?

This is the famous *approximation problem* in Banach spaces and formerly was one of the most widely known unsolved problems in the theory of Banach spaces. The approximation problem was studied in detail by Grothendieck.[†] He conjectured that (A) is not true in general. Only recently has it been solved in the negative by Per Enflo.[‡]

This problem is really a problem of the structure of Banach spaces. To clarify this let us say that a Banach space Y is said to have the *approximation property* if for every compact set $K \subset Y$ and every $\varepsilon > 0$, there is $P \in L(Y, Y)$, depending on K and ε, with finite-dimensional range such that $\| P(x) - x \| \leq \varepsilon$ for every $x \in K$. Now if Y has the approximation property, then the answer to (A) is affirmative. Indeed, if T is compact and K is the compact set $\overline{T(S_X)}$, where S_X is the closed unit ball of X, then $\dim PT(X) < \infty$ and $\| PT - T \| \leq \varepsilon$.

Also it is clear that a Banach space has the approximation property if every separable subspace of it has this property. To see this, one has to consider the closed subspace (separable) Y_K spanned by the compact set K in the definition above and then use the Hahn–Banach Theorem to extend P from an operator in $L(Y_K, Y_K)$ to one in $L(Y, Y)$. It can be easily verified that this is possible since the range of P is finite dimensional. This shows that (A) is related to the classical basis problem:

(B) Does every separable Banach space Y have a *Schauder basis* $(x_i)_{i=1}^{\infty}$ (that is, can every $y \in Y$ be written uniquely in the form $\sum_{i=1}^{\infty} \lambda_i x_i$)?

Actually, if Y has a Schauder basis, then Y has the approximation property. To see this, let $(x_i)_{i=1}^{\infty}$ be the Schauder basis of Y and $P_n \in L(Y, Y)$ be defined by $P_n(\sum_{i=1}^{\infty} \lambda_i x_i) = \sum_{i=1}^{n} \lambda_i x_i$. Then $\| P_n(x) - x \| \to 0$ uniformly as $n \to \infty$ on every compact subset of Y (see Problem 6.5.2), and, consequently, Y has the approximation property. Hence a positive answer to (B) must give a positive answer to (A). Per Enflo (in the paper mentioned above) solved (A) in the negative [and therefore also (B) in the negative]

[†] A. Grothendieck, *Can. J. Math.* **7**, 552–561 (1955).
[‡] Per Enflo, *Acta Math.* **130**, 3–4, 309–317 (1973).

by giving an example of a separable reflexive Banach space that does not have the approximation property.

In the next chapter, we will see that every separable subspace of a Hilbert space has a Schauder basis, and as such Problem (A) has an affirmative solution in a Hilbert space. A different proof of this fact will also be considered there.

We now leave Problem (A) and consider other basic properties of a compact operator.

Proposition 6.23. If A is a compact operator, then the range of A is separable. ∎

Proof. The range of A is contained in $\bigcup_{n=1}^{\infty} \overline{A(S_n)}$, where $S_n = \{x: \| x \| \leq n\}$. Since this is a countable union of compact sets, and a compact metric space is separable, the result follows. ∎

Proposition 6.24. Let $A \in L(X, Y)$, where X is infinite dimensional and A is compact. Then A^{-1}, if it exists, is not bounded. ∎

The proof is left to the reader.

Let us now recall the definition of an adjoint operator (Problems 6.3.9, 6.3.12). If $A \in L(X, Y)$, then $A^* \in L(Y^*, X^*)$ is defined by

$$A^*(y^*) = y^* \circ A.$$

Then $\| A^* \| = \| A \|$. The concept of the adjoint of a bounded linear operator A is useful in obtaining information about the range and inverse of A, as was outlined in Problems 6.3.12–6.3.15. There is a further duality between A and A^* as the following theorem shows.

Theorem 6.21. Let $A \in L(X, Y)$. If A is compact, then A^* is compact. Conversely, if A^* is compact and Y complete, then A is compact. ∎

Proof. Let A be compact. It suffices to show that $A^*(W)$, for W bounded in Y^*, is totally bounded in X^*. Let $\varepsilon > 0$ and $S = \{x \in X: \| x \| \leq 1\}$. Since A is compact, there exist $x_1, x_2, \ldots, x_n \in X$ such that, for each $x \in S$,

$$\inf_{1 \leq i \leq n} \| A(x) - A(x_i) \| < \varepsilon. \tag{6.8}$$

Let B be defined from Y^* into F^n by

$$B(y^*) = (y^* \circ A(x_1), \ldots, y^* \circ A(x_n)),$$

then $B \in L(Y^*, F^n)$. Hence $B(W)$ is totally bounded. Thus there exist $y_1^*, \ldots, y_m^* \in W$ such that for each $y^* \in W$ and each i, $1 \leq i \leq n$,

$$\inf_{1 \leq j \leq m} | y^* \circ A(x_i) - y_j^* \circ A(x_i) | < \varepsilon. \tag{6.9}$$

Now let $y^* \in W$ and let y_j^* for $1 \leq j \leq m$ be such that for each i, $1 \leq i \leq n$,

$$| y^* \circ A(x_i) - y_j^* \circ A(x_i) | < \varepsilon. \tag{6.10}$$

Then

$$\| A^*(y^*) - A^*(y_j^*) \| = \sup_{\|x\| \leq 1} | y^* \circ A(x) - y_j^* \circ A(x) |,$$

which is, by inequalities (6.8) and (6.9), less than or equal to $(2M + 1)\varepsilon$, where M is an upper bound for $\| y^* \|$, $y^* \in W$. This proves that $A^*(W)$ is totally bounded and hence A^* is compact.

Conversely, let A^* be compact. By what we have just proved, A^{**} is compact. Let (x_n) be a bounded sequence in X. Then the sequence $(J(x_n))$ is also bounded in X^{**}, J being the natural mapping from X into X^{**}. Since $A^{**} \in L(X^{**}, Y^{**})$ and A^{**} is compact, there exists a subsequence $[A^{**}(J(x_{n_k}))]_{k \in N}$ which converges in Y^{**}. Now $J_1(A(x_{n_k})) = A^{**}(J(x_{n_k}))$, where J_1 is the natural mapping from Y into Y^{**}. Since J_1 is an isometry, $(A(x_{n_k}))_{k \in N}$ is a Cauchy sequence in Y and therefore must converge since Y is complete. This means that A is compact. ∎

Corollary 6.13.[†] If the range of a compact operator A is complete, then the range of A is finite dimensional. ∎

Proof. Let $A \in L(X, Y)$ be compact and let Z be the range of A. Then if Z is complete, A as an operator in $L(X, Z)$ is compact, and therefore, by Theorem 6.21, A^* is compact. Also, by Problem 6.3.15, A^* has a bounded inverse. Hence by Proposition 6.24 Z^* (and therefore Z) is finite dimensional. ∎

[†] When X is complete, one can prove this noting that A is open onto $A(X)$ (by Theorem 6.8) and that the unit ball in $A(X)$ is compact (by Proposition 6.9).

Next we present the famous Fredholm Alternative Theorem, which was developed by F. Riesz as a tool for the study of linear integral equations.

The proof of this theorem uses the properties of S° (the annihilator of S) as given in Problems 6.3.12 and 6.3.13.

Theorem 6.22. *The Fredholm Alternative Theorem.* Let X be complete and $K \in L(X, X)$ be compact. Then $R(I - K)$ is closed and $\dim N(I - K) = \dim N(I - K^*) < \infty$, where N denotes the null space, R the range, and I the identity operator. In particular, *either* $R(I - K) = X$ and $N(I - K) = \{0\}$, *or* $R(I - K) \neq X$ and $N(I - K) \neq \{0\}$. ∎

Proof. We will prove the theorem in several steps. We will write A for $I - K$.

Step I. In this step we will show that $R(A)$ is closed [or equivalently, there exists $k > 0$ such that for every $x \in X$, $d(x, N(A)) \leq k \| A(x) \|$, where d denotes the usual distance]. (See Problem 6.3.11.)

Suppose $R(A)$ is not closed. Then there exists a sequence $(x_n)_{n \in N} \in X$ such that $d(x_n, N(A)) = 1$ and $\| A(x_n) \| \to 0$ as $n \to \infty$. Therefore, there also exists a sequence (z_n) such that $d(z_n, N(A)) = 1$, $\| A(z_n) \| \to 0$ as $n \to \infty$, and $1 \leq \| z_n \| \leq 2$. Since K is compact, there exists a subsequence $(z_{n_i})_{i \in N}$ such that $K(z_{n_i})$ converges to some $z \in X$. Then z_{n_i}, which is $A(z_{n_i}) + K(z_{n_i})$, also converges to z; or $A(z_{n_i})$ converges to $A(z)$. Hence $A(z) = 0$ or $z \in N(A)$. This is a contradiction, since for each i, $d(z_{n_i}, N(A)) = 1$ and $z_{n_i} \to z$ as $i \to \infty$.

Step II. In this step we will show that $\dim N(A) < \infty$ and $\dim N(A^*) < \infty$.

Consider a sequence $(x_n)_N$ in $N(A)$ with $\| x_n \| \leq 1$. Then for each n, $A(x_n) = 0$ or $K(x_n) = x_n$. Since K is compact, there exists a subsequence $(x_{n_i})_{i \in N}$ such that $K(x_{n_i}) = x_{n_i}$ is convergent. This means that the unit ball (closed) of $N(A)$ (considered as a normed linear space) is compact, and therefore $\dim N(A) < \infty$. Similarly, $\dim N(A^*) < \infty$.

Step III. In this step we will prove that $\dim N(A) = 0$ if and only if $\dim N(A^*) = 0$.

Suppose that $\dim N(A^*) = 0$. Since $R(A)$ is closed by Step I, $R(A) = {}^\circ N(A^*) = X$. (See Problem 6.3.12.) Suppose $\dim N(A) > 0$. Then there exists $x_1 \neq 0$ such that $A(x_1) = 0$. Since $R(A) = X$, there exists a sequence $(x_n)_N$ such that $A(x_{n+1}) = x_n$. Now $A^{n+1}(x_{n+1}) = A^n(x_n) = \cdots = A(x_1) = 0$, and $A^n(x_{n+1}) = A^{n-1}(x_n) = \cdots = A(x_2) = x_1 \neq 0$. This means that for each positive integer n, $N(A^n)$ is a *proper* closed subspace of $N(A^{n+1})$.

By Riesz's lemma (Proposition 6.1), there exists $z_n \in N(A^n)$ such that $\| z_n \| = 1$ and $d(z_n, N(A^{n-1})) > 1/2$. Since $n > m$ implies

$$\begin{aligned} \| K(z_n) - K(z_m) \| &= \| z_n - A(z_n) - z_m + A(z_m) \| \\ &= \| z_n - [z_m - A(z_m) + A(z_n)] \| \\ &\geq d(z_n, N(A^{n-1})) > 1/2, \end{aligned}$$

we have a contradiction to the fact that K is compact. Hence $\dim N(A) = 0$.

Conversely, suppose that $\dim N(A) = 0$. Then $R(A^*) = N(A)^\circ = X^*$ (see Problem 6.3.13). Since $A^* = I - K^*$ and K^* is also compact, by a similar argument to that above, $\dim N(A^*) = 0$.

Step IV. In this step we will show that $\dim N(A) = \dim N(A^*)$, and this will complete the proof of the theorem.

Suppose $\dim N(A) = n > 0$ and $\dim N(A^*) = m > 0$. Let $\{x_1, x_2, \ldots, x_n\}$ be a basis for $N(A)$ and $\{x_1^*, x_2^*, \ldots, x_m^*\}$ be a basis for $N(A^*)$. Then we claim the following:

(i) There exists $x_0^* \in X^*$ such that $x_0^*(x_i) = 0$ for $1 \leq i < n$, and $x_0^*(x_n) \neq 0$.

(ii) There exists $x_0 \in X$ such that $x_i^*(x_0) = 0$ for $1 \leq i < m$, and $x_m^*(x_0) \neq 0$.

(iii) If $A_1 = A - K_0$ where $K_0(x) = x_0^*(x) \cdot x_0$, then $\dim N(A_1) = n - 1$ and $\dim N(A_1^*) = m - 1$.

Statement (i) follows easily. We prove (ii) by an inductive argument. Let $p(k)$ be the statement "There exist $\{z_j : j = 1, 2, \ldots, k\}$ such that for $1 \leq i, j \leq k$, $x_i^*(z_j)$ is equal to zero if $i \neq j$ and is equal to one if $i = j$." Clearly $p(k)$ holds for $k = 1$. Suppose that $p(k)$ holds for $k = m - 1$. Then for $1 \leq i \leq m - 1$, we have

$$x_i^*\left(x - \sum_{j=1}^{m-1} x_j^*(x) z_j\right) = 0$$

for each $x \in X$. If for each $x \in X$, $x_m^*(x - \sum_{j=1}^{m-1} x_j^*(x) z_j) = 0$, then $x_m^* = \sum_{j=1}^{m-1} x_m^*(z_j) x_j^*$, a contradiction. Hence there exists z_m' such that $x_m^*(z_m') = 1$ and for $1 \leq i \leq m - 1$, $x_i^*(z_m') = 0$. Now letting $z_j' = z_j - x_m^*(z_j) z_m'$ for $1 \leq j \leq m - 1$, we have for $1 \leq i, j \leq m$, $x_i^*(z_j')$ equals zero if $i \neq j$ and equals one if $i = j$. The argument is complete and (ii) is established.

To prove (iii), let $x \in N(A_1)$. Then $A(x) = K_0(x) = x_0^*(x)x_0$. Since $x_0 \notin {}^\circ[N(A^*)] = R(A)$ (see Problem 6.3.12), $x_0^*(x) = 0$ and therefore $x \in N(A)$. So we can write $x = \sum_{j=1}^{n} \alpha_j x_j$ for some scalars α_j, and since

$$0 = x_0^*(x) = \sum_{j=1}^{n} \alpha_j x_0^*(x_j) = \alpha_n \cdot x_0^*(x_n),$$

we have $\alpha_n = 0$ or $x = \sum_{j=1}^{n-1} \alpha_j x_j$. This means that $\dim N(A_1) = n - 1$. To prove that $\dim N(A_1^*) = m - 1$, let $x^* \in N(A_1^*)$ or $A^*x^* = K_0^*x^* = x^*(x_0) \cdot x_0^*$. Since $x_0^* \notin N(A)^\circ = R(A^*)$ (see Problem 6.3.13), $x^*(x_0) = 0$ or $x^* \in N(A^*)$. Therefore, we can write $x^* = \sum_{j=1}^{m} \beta_j x_j^*$ for some scalars β_j. Then since $x^*(x_0) = 0$ and $x^*(x_0) = \beta_m x_m^*(x_0)$, $\beta_m = 0$. This means that $\dim N(A_1^*) = m - 1$.

Now the proof of the theorem will follow easily. If $n < m$, then $\dim N(A_1) \neq \dim N(A_1^*)$. Repeating the above process a finite number of times, we end up with an operator $A_n = I - (K + K_0 + \cdots + K_{n-1}) = I -$ (a compact operator) such that $\dim N(A_n) = 0$ and $\dim N(A_n^*) > 0$. But this contradicts the result in Step III. Hence $n \geq m$. Similarly, $n \leq m$. This proves the theorem. ∎

At the end of this section, we will outline some applications of the above theorem in linear integral equations.

Now we consider briefly what is called the spectral theory of linear operators, which is the systematic study of various connections between $T - \lambda I$ and $(T - \lambda I)^{-1}$, where $T \in L(X, X)$, λ is a scalar, and I is the identity operator. A large part of the theory of bounded linear operators is centered around their spectral theory. The most highly developed spectral theory is that for a class of operators called self-adjoint operators on Hilbert spaces (this will be discussed in depth in the next chapter).

If T is a linear operator on C^n, then the eigenvalues of T are complex numbers λ such that $(T - \lambda I)x = 0$ for some nonzero x. The set of all such λ is called the spectrum of T. If λ is not an eigenvalue, then $T - \lambda I$ has an inverse. The study of the spectrum of T often leads to a better understanding of the operator T. For example, if $A = (a_{ij})$ is the $n \times n$ matrix that represents T with respect to some basis in C^n and if A is Hermitian (that is, $a_{ij} = \overline{a_{ji}}$), then it is well known from linear algebra that T can be represented with respect to some orthonormal basis $\{x_1, x_2, \ldots, x_n\}$ by a diagonal matrix so that $T(x_i) = \lambda_i x_i$, $i = 1, 2, \ldots, n$, for some scalars $\lambda_1, \lambda_2, \ldots, \lambda_n$. This gives a very simple representation for such operators in terms of their spectrum. The spectral theory of linear operators in infinite-dimensional spaces is much more involved and extremely important in the study of these operators. (See, for example, for a generalization of the discussion above to the infinite-dimensional situation, the statement of Theorem 7.8 in the next chapter.) In what follows, we define the spectrum of T and show that the spectrum of a bounded linear operator on a Banach space is nonempty, and for compact such operators, the spectrum is at most countable. At the end of this section, we give an application of these results to integral equations.

Definition 6.16. Let $T: X \to X$ be a linear operator on a nonzero complex normed linear space X. The resolvent set $\varrho(T)$ of T is the set

$$\{\lambda \in C: \overline{R(T - \lambda I)} = X \text{ and } T - \lambda I \text{ has a continuous inverse}\}.$$

The complement in the complex plane of the resolvent set of T is called the *spectrum* of T and is denoted by $\sigma(T)$. The *point spectrum* $P\sigma(T)$ consists of all scalars λ in $\sigma(T)$ such that $T - \lambda I$ is not one-to-one. If $\lambda \in P\sigma(T)$ then λ is also called an *eigenvalue* of T and a nonzero $x \in X$ such that $Tx = \lambda x$ is called a corresponding *eigenvector* to λ. The *continuous spectrum* $C\sigma(T)$ consists of all scalars λ in $\sigma(T)$ such that $T - \lambda I$ is one-to-one and $R(T - \lambda I)$ is dense in X but $(T - \lambda I)^{-1}$ is not continuous as an operator from $R(T - \lambda I)$ onto X. The *residual spectrum* $R\sigma(T)$ of T consists of scalars λ in $\sigma(T)$ such that $T - \lambda I$ has an inverse $(T - \lambda I)^{-1}$ which may or may not be continuous but its domain $R(T - \lambda I)$ is not dense in X. ∎

We will deal with the continuous and residual spectrums when we study operators on Hilbert spaces later in the text.

Observe that if X is complete and T is closed, then when $\lambda \in \varrho(T)$ the operator $(T - \lambda I)^{-1}$ is in $L(X, X)$. This follows from Proposition 6.10 since $(T - \lambda I)^{-1}$ is in this case closed and continuous and has therefore a closed domain equal to X.

We denote $(T - \lambda I)^{-1}$ by $R(\lambda, T)$ whenever $(T - \lambda I)^{-1} \in L(X, X)$.

Example 6.17. Let $(X, \mathscr{A}, \mu)$ be a semifinite measure space. If $G \in L_\infty(\mu)$ and $T_G \in L(L_2, L_2)$ is given by $T_G f = Gf$, then

$$\sigma(T_G) = \{\lambda \mid \text{for all } \varepsilon > 0,\ \mu(\{x: |G(x) - \lambda| < \varepsilon\}) > 0\},$$

the so-called *essential range* of G. (See Problem 6.5.21.)

Remarks

6.25. We note that when X is finite dimensional, $\lambda \in \sigma(T)$ if and only if $T - \lambda I$ is not one-to-one, which is true if and only if λ is an eigenvalue of T. Since the eigenvalues of $T - \lambda I$ are the solutions of the equation

$$\det (T_m - \lambda I) = 0,$$

where I is the identity matrix and T_m is the matrix representing T (with respect to some fixed basis of X), it follows that $\sigma(T)$ is nonempty. (Note that every nonconstant polynomial with complex coefficients has a complex root.)

6.26. $\sigma(T)$ can also be defined in *real* normed linear spaces X, but the difficulty is that $\sigma(T)$ can be empty even when X is finite dimensional. For example, let T be defined from R^2 into R^2 linearly by $T((1, 0)) = (0, -1)$ and $T((0, 1)) = (1, 0)$; then $\sigma(T)$ is empty.

6.27. When X is infinite dimensional, there can be elements in $\sigma(T)$ that are not eigenvalues. For example, let $X = l_2$ and T be defined on l_2 by

$$T((x_1, x_2, \ldots)) = (0, x_1, x_2, \ldots).$$

Since for $x \in l_2$ $\| T(x) \| = \| x \|$, T is one-to-one and bounded so that 0 is not an eigenvalue of T. But since $(1, 0, 0, \ldots) \notin$ the range of T, T^{-1} is not defined on X and therefore $0 \in \sigma(T)$.

6.28. When X is infinite dimensional, $\sigma(T)$ can be *uncountably infinite*. For example, let $X = l_2$ and let T be defined on l_2 by

$$T((x_1, x_2, \ldots)) = (x_2, x_3, \ldots).$$

Then $T \in L(X, X)$. If λ is a complex number with $|\lambda| < 1$, then $x = (1, \lambda, \lambda^2, \ldots) \in l_2$ and $T(x) = (\lambda, \lambda^2, \ldots) = \lambda x$. This means that $\sigma(T) \supset \{\lambda : |\lambda| < 1\}$. (The operator T is called the *shift* on l_2.)

Next we present a basic result concerning $\sigma(T)$: for complex scalars, it is always nonempty. First we need the following result.

Proposition 6.25. Let X be complete and $T \in L(X, X)$. Then the resolvent set $\varrho(T)$ is open and if $\lambda, \mu \in \varrho(T)$, then [writing $R_\mu = R(\mu, T)$]

$$R_\mu - R_\lambda = (\mu - \lambda) R_\lambda R_\mu.$$

Moreover, R_λ as a function from $\varrho(T)$ to $L(X, X)$ has derivatives of all orders. (See Remark 6.4 for definition.) ∎

Proof. Suppose $\lambda \in \varrho(T)$ and $|\mu - \lambda| < 1/\| R_\lambda \|$. Then $T - \mu I = (T - \lambda I)[I - (\mu - \lambda)R_\lambda]$. Now $\sum_{n=0}^{\infty}(\mu - \lambda)^n R_\lambda{}^n$ is convergent in $L(X, X)$, and it follows easily that $[I - (\mu - \lambda)R_\lambda]^{-1} = \sum_{n=0}^{\infty}(\mu - \lambda)^n R_\lambda{}^n$. Hence $(T - \mu I)^{-1} \in L(X, X)$ and $\mu \in \varrho(T)$; consequently, $\varrho(T)$ is open. Also for $\lambda, \mu \in \varrho(T)$ we have

$$\begin{aligned} R_\mu - R_\lambda &= R_\lambda[R_\lambda{}^{-1} - R_\mu{}^{-1}]R_\mu \\ &= R_\lambda[T - \lambda I - (T - \mu I)]R_\mu \\ &= (\mu - \lambda)R_\lambda R_\mu. \end{aligned}$$

Hence $(R_\mu - R_\lambda)/(\mu - \lambda) = R_\lambda R_\mu$ or as $\mu \to \lambda$, $(R_\lambda - R_\mu)/(\lambda - \mu) \to R_\lambda{}^2$, in the $L(X, X)$ norm. [Note that $\| R_\mu \| \leq \| R_\lambda \| (1 - |\lambda - \mu| \; \| R_\lambda \|)^{-1}$, when $|\lambda - \mu|$ is sufficiently small.] By induction, it can be shown that R_λ has derivatives of all higher orders. ∎

Proposition 6.26. Let X be complete and $T \in L(X, X)$. Then $|\lambda| > \| T \|$ implies that $\lambda \in \varrho(T)$ and $R_\lambda = -\sum_{n=1}^{\infty}\lambda^{-n}T^{n-1}$. Hence $\sigma(T)$ is a compact subset of the complex plane. ∎

We leave the proof to the reader.

We use Proposition 6.25 and 6.26 to prove that $\sigma(T)$ is nonempty.

Theorem 6.23. If X is complete and $T \in L(X, X)$, then $\sigma(T)$ is not empty. ∎

Proof. We use Liouville's theorem from complex analysis and the Hahn–Banach Theorem. Let $x \in X$ and $x^* \in X^*$. Then the complex-valued function $x^*(R_\lambda(x))$ is, by Proposition 6.25, an analytic (or differentiable) function on $\varrho(T)$. By Proposition 6.26 for $|\lambda| > \| T \|$, we have

$$\| R_\lambda \| \leq \sum_{n=1}^{\infty} \frac{\| T \|^{n-1}}{|\lambda|^n} = \frac{1}{|\lambda|(1 - \| T \|/|\lambda|)} \to 0,$$

as $|\lambda| \to \infty$. This means that $x^*(R_\lambda(x))$ is a bounded function on the entire complex plane if $\sigma(T)$ is empty, since then $\varrho(T)$ is the entire complex plane. By Liouville's theorem, $x^*(R_\lambda(x))$, being a bounded entire function, must be a constant ($= 0$, in this case). An application of Corollary 6.4 then asserts that for all $x \in X$, $R_\lambda(x) = 0$, which is a contradiction. ∎

Now we show that $\lim_{n\to\infty} \| T^n \|^{1/n}$ exists and equals $\sup\{|\lambda| : \lambda \in \sigma(T)\}$. Let C be any circle with origin as center and radius greater than

$\| T \|$. For $x^* \in X^*$ and $x \in X$, we consider the complex-valued analytic function $x^*(R_\lambda(x))$ on $\varrho(T)$. Using the line integral over C taken in a suitable direction, we have from Proposition 6.26

$$\int_C \lambda^n \cdot x^*(R_\lambda(x))\, d\lambda = - \sum_{k=1}^{\infty} x^*(T^{k-1}(x)) \cdot \int_C \lambda^{n-k}\, d\lambda = 2\pi i \cdot x^*(T^n(x)).$$

It follows that if C_0 is any circle containing $\sigma(T)$ in its interior, then

$$x^*(T^n(x)) = \frac{1}{2\pi i} \int_{C_0} \lambda^n \cdot x^*(R_\lambda(x))\, d\lambda,$$

where the line integral is taken in a suitable direction.

This leads to the following

Proposition 6.27. Let X be complete and $T \in L(X, X)$. Let $r(T) = \sup_{\lambda \in \sigma(T)} | \lambda |$. Then $\lim_{n \to \infty} \| T^n \|^{1/n}$ exists and is equal to $r(T)$. [Here $r(T)$ is called the *spectral radius* of T.] ▮

Proof. Let $\varepsilon > 0$, C be the circle $| \lambda | = r(T) + \varepsilon$. Then for $x^* \in X^*$ with $\| x^* \| = 1$ and $x \in X$ with $\| x \| = 1$, we have

$$| x^*(T^n(x)) | = \left| \frac{1}{2\pi i} \int_C \lambda^n \cdot x^*(R_\lambda(x))\, d\lambda \right|$$
$$\leq \frac{1}{2\pi} \cdot [r(T) + \varepsilon]^n \cdot 2\pi(r(T) + \varepsilon) \cdot M(C),$$

where $M(C) = \sup_{\lambda \in C} \| R_\lambda \| < \infty$, since R_λ is a continuous function of λ. This means that

$$\limsup_{n \to \infty} \| T^n \|^{1/n} \leq r(T) + \varepsilon.$$

Since $\varepsilon > 0$ is arbitrary, we have

$$\limsup_{n \to \infty} \| T^n \|^{1/n} \leq r(T).$$

We will now complete the proof by showing that

$$r(T) \leq \liminf_{n \to \infty} \| T^n \|^{1/n}.$$

To show this, we note that $\lambda \in \sigma(T) \Rightarrow \lambda^n \in \sigma(T^n)$. The reason is that

$$T^n - \lambda^n I = (T - \lambda I)A = A(T - \lambda I),$$

where

$$A = \sum_{k=0}^{n-1} \lambda^k T^{n-k-1},$$

so that $\lambda^n \in \varrho(T^n) \Rightarrow T - \lambda I$ is bijective $\Rightarrow \lambda \in \varrho(T)$, by the Open Mapping Theorem. Hence, if $\lambda \in \sigma(T)$, by Proposition 6.26

$$| \lambda^n | \leq \| T^n \|,$$

for each positive integer n. It follows that

$$r(T) \leq \liminf_{n\to\infty} \| T^n \|^{1/n}.$$

This completes the proof. ∎

Our next result is what is called the Spectral Mapping Theorem. This theorem answers a natural question, namely, when the equation $p(T)(x) = y$ has a unique solution for each y in X, where p is a polynomial, $T \in L(X, X)$, and $p(T) = \sum_{k=0}^n a_k T^k$, when $p(x) = \sum_{k=0}^n a_k x^k$. Here $T^0 \equiv I$. It is clear that the equation is solvable if 0 is not in the spectrum of $p(T)$. The Spectral Mapping Theorem answers the question more precisely: The above equation can be solved uniquely for each y in X if and only if no λ in $\sigma(T)$ is a root of p. Here X is a Banach space.

Theorem 6.24. *The Spectral Mapping Theorem.* For $T \in L(X, X)$ and any polynomial p, $p(\sigma(T)) = \sigma(p(T))$. [Here $p(\sigma(T)) = \{p(\lambda): \lambda \in \sigma(T)\}$.] ∎

Proof. Let $\lambda \in \sigma(T)$. Since λ is a root of the polynomial $p(t) - p(\lambda)$, we can write

$$p(t) - p(\lambda) = q(t) \cdot (t - \lambda)$$

and

$$p(T) - p(\lambda)I = q(T)(T - \lambda I) = (T - \lambda I)q(T),$$

for some polynomial $q(t)$. Now if $p(\lambda) \in \varrho(p(T))$, then $T - \lambda I$ is bijective and therefore, by the Open Mapping Theorem, $\lambda \in \varrho(T)$. This proves that $p(\sigma(T)) \subset \sigma(p(T))$. For the opposite inclusion, let $\lambda \in \sigma(p(T))$. Suppose that $\lambda_1, \lambda_2, \ldots, \lambda_n$ are the complex roots of $p(t) - \lambda$. Then

$$p(T) - \lambda I = c \cdot (T - \lambda_1 I) \cdots (T - \lambda_n I), \qquad c \neq 0.$$

If each λ_i is in $\varrho(T)$, then for each i, $(T - \lambda_i I)^{-1} \in L(X, X)$ and therefore, by the above, $\lambda \in \varrho(p(T))$, which is a contradiction. Thus, one of the λ_i must be in $\sigma(T)$. Since this λ_i is a root of $p(t) - \lambda$, it follows that $\lambda \in p(\sigma(T))$. The proof is complete. ∎

Our last theorem in this section (preceding the applications) concerns the spectrum of a compact operator T on a Banach space X. The spectrum of such operators is at most countable and contains 0 when X is infinite dimensional; this follows from Proposition 6.24 and the next result.

Theorem 6.25. Let T be a compact operator on a Banach space X. If $\lambda \neq 0$, then $\lambda \in \varrho(T)$ or λ is an eigenvalue of T. Moreover, $\sigma(T)$ is at most countable and 0 is its only possible limit point. ∎

Proof. Suppose $\lambda \neq 0$ is in the spectrum of T. If $T - \lambda I$ is not one-to-one, then clearly λ is an eigenvalue of T. If $(T - \lambda I)^{-1}$ is bounded, then by Theorem 6.22 $R(T - \lambda I) = X$ and $\lambda \in \varrho(T)$. On the other hand, if $(T - \lambda I)^{-1}$ exists as a function but is not bounded, then by Proposition 6.5, for each positive integer n, there exists an x_n in X with $\| x_n \| = 1$ and with $\| (T - \lambda I)x_n \| < 1/n$. This means $Tx_n - \lambda x_n \to 0$ in X. Since T is compact, Tx_n has a convergent subsequence $(Tx_{n_k})_{k \in N}$ converging to y in X. Since also $Tx_{n_k} - \lambda x_{n_k} \to 0$, it is clear that $\lambda x_{n_k} \to y$. Since T is continuous,

$$Ty = \lim_{k \to \infty} T(\lambda x_{n_k}) = \lambda \lim_{k \to \infty} T(x_{n_k}) = \lambda y.$$

Since $y \neq 0$ as $\| y \| = \lim_{k \to \infty} \| \lambda x_{n_k} \| = | \lambda | \lim_{k \to \infty} \| x_{n_k} \| = | \lambda |$, λ is an eigenvalue of T.

To prove the rest of the theorem, it is sufficient to prove that for any $\varepsilon > 0$, the set

$$P_\varepsilon = \{\lambda \colon | \lambda | \geq \varepsilon \text{ and } \lambda \in \sigma(T)\}$$

is finite. Suppose this is false. Then there exists $\varepsilon > 0$ such that P_ε is infinite. Let $(\lambda_i)_{i=1}^\infty$ be a sequence of distinct eigenvalues in P_ε, with $(x_i)_{i=1}^\infty$ the corresponding eigenvectors. Since eigenvectors corresponding to distinct eigenvalues are linearly independent,† the subspace X_n spanned by $\{x_1, x_2, \ldots, x_n\}$ is properly contained in X_{n+1} spanned by $\{x_1, x_2, \ldots, x_{n+1}\}$. By Riesz's result (Proposition 6.1), there exist $y_n \in X_n$ with $\| y_n \| = 1$ and $\inf_{x \in X_{n-1}} \| y_n - x \| \geq 1/2$. We write $y_n = \sum_{i=1}^n \alpha_i x_i$; then

$$\begin{aligned} \lambda_n y_n - T(y_n) &= \sum_{i=1}^n \alpha_i \lambda_n x_i - \sum_{i=1}^n \alpha_i \lambda_i x_i \\ &= \sum_{i=1}^{n-1} \alpha_i (\lambda_n - \lambda_i) x_i \in X_{n-1}. \end{aligned}$$

† $\sum_{i=1}^n c_i x_i = 0 \Rightarrow \sum c_i \lambda_i x_i = 0 = T(\sum c_i x_i)$. The linear independence of the x_i's follows by induction.

Therefore, for $n > m$,

$$\| T(y_n) - T(y_m) \| = \| \lambda_n y_n - [\lambda_n y_n - T(y_n) + T(y_m)] \| \geq \tfrac{1}{2} | \lambda_n | \geq \varepsilon/2,$$

which is a contradiction to the compactness of T. The theorem follows. ∎

We conclude this section with some applications of the preceding theory to the study of integral equations.

• **Remark 6.29.** *Applications to Integral Equations: The Dirichlet Problem.* A number of problems in applied mathematics and mathematical physics can be reduced to equations of the type

$$f(s) - \lambda \int_a^b K(s, t) f(t)\, dt = g(s), \tag{6.11}$$

where $K(s, t)$ is a complex-valued Lebesgue-measurable function on $[a, b] \times [a, b]$ such that

$$\| K \|_2^2 = \int_a^b \int_a^b | K(s, t) |^2\, ds\, dt < \infty,$$

$f, g \in L_2[a, b]$. Here λ is a nonzero complex number and f is the unknown function. These equations are usually called Fredholm equations of the second kind and the function K is called the kernel of the equation.

In what follows, we will apply Theorems 6.22 and 6.25 to solve the problem of existence of solutions of equation (6.11); the results on integral equations will then be useful in studying a fundamental problem of mathematical physics—the Dirichlet problem.

First, we define the operator T by

$$Tf(s) = \int_a^b K(s, t) f(t)\, dt, \qquad f \in L_2[a, b]. \tag{6.12}$$

It follows from Fubini's Theorem (Theorem 3.7) and the Hölder Inequality (Proposition 3.13) that, for almost all s,

$$| Tf(s) |^2 \leq \int_a^b | K(s, t) |^2\, dt \cdot \int_a^b | f(t) |^2\, dt$$

and therefore

$$\| Tf \|_2 \leq \| K \|_2 \cdot \| f \|_2. \tag{6.13}$$

Thus, T is a bounded linear operator on L_2. We claim that T is compact.

To prove our claim, we assume with no loss of generality that K is a continuous function of (s, t). This is possible since by Lusin's Theorem (see Problem 3.1.13) we can approximate the kernel by a continuous kernel in L_2-norm and then Proposition 6.22 applies. Now by Theorem 1.26, a continuous kernel $K(s, t)$ can be approximated uniformly by kernels $K_n(s, t)$ of the form $\sum_{i=1}^n u_i(s)v_i(t)$. If we define

$$T_n f(s) = \int_a^b K_n(s, t) f(t)\, dt,$$

then

$$T_n f(s) = \sum_{i=1}^{n} u_i(s) \int_a^b v_i(t) f(t)\, dt.$$

This means that T_n is finite dimensional and therefore, by Remark 6.23 T_n is compact. By the same argument as used in obtaining (6.13), we have

$$\| Tf - T_n f \|_2 \leq \| K - K_n \|_2 \cdot \| f \|_2.$$

It follows that $\lim_{n\to\infty} \| T - T_n \| = 0$. By Proposition 6.22, T is compact.

Now we write equation (6.11) as

$$(I - \lambda T) f = g \tag{6.14}$$

or, equivalently,

$$(T - \lambda^{-1} I) f = - \lambda^{-1} g. \tag{6.15}$$

Taking λT as the operator K in Theorem 6.22, we obtain easily the following.

Theorem A. Either the equation (6.14) has a unique solution

$$f = (I - \lambda T)^{-1} g$$

for each $g \in L_2$ or the homogeneous equation

$$f(s) - \lambda \int_a^b K(s, t) f(t)\, dt = 0 \tag{6.16}$$

has a nonzero solution f in L_2. In the latter case, the number of linearly independent solutions of equation (6.16) is finite. ∎

Now an application of Theorem 6.25 gives us immediately the following

Theorem B. The equation (6.16) can have nonzero solutions for at most countably many values of λ. If there is an infinite sequence (λ_n) of such values, then $|\lambda_n| \to \infty$. ▮

To find some more information on equation (6.11), we need to find the adjoint of T. For $h \in L_2$, let h^* denote the linear functional on L_2 defined by

$$h^*(f) = \int_a^b f(t)\overline{h(t)}\,dt, \qquad f \in L_2.$$

Then it can be verified by a simple computation that $T^*h^* = g^*$, where g^* is the linear functional induced by g as above and g is given by

$$g(s) = \int_a^b \overline{K(t, s)}h(t)\,dt. \tag{6.17}$$

Another application of Theorem 6.22 leads to our next result.

Theorem C. The equation

$$f(s) - \bar{\lambda}\int_a^b \overline{K(t, s)}f(t)\,dt = 0 \tag{6.18}$$

and equation (6.16) have the same number of linearly independent solutions. Moreover, if λ^{-1} is an eigenvalue of T, then equation (6.11) has a solution in L_2 for a given g in L_2 if and only if

$$\int_a^b g(t)\overline{f(t)}\,dt = 0,$$

whenever f is a solution of equation (6.18). ▮

Note that this last result follows since $(I - \lambda T)(L_2) = {}^0[N(I - \bar{\lambda}T^*)]$.

We remark that the preceding Fredholm theory is also valid if the kernel $K(s, t)$ is a continuous function on $G \times G$, where G is a compact set in R^n, and the operator T acts on $C(G)$.

We now consider the Dirichlet Problem, a fundamental problem in mathematical physics and one of the oldest problems in potential theory.

The Dirichlet Problem. This is the first boundary-value problem of potential theory. The problem is to find a harmonic function on an open connected set E in R^n, which is continuous on $\bar{E}$ and coincides with a given continuous function g on the boundary of E. The problem originates in the study of various physical phenomena from electrostatics, fluid dynamics, heat conduction, and other areas of physics. During the last hundred years or so, this problem has been studied by many celebrated mathematicians including Dirichlet, Poincaré, Lebesgue, Hilbert, and Fredholm, and many different methods have been discovered for solving this problem. Though this problem is not solvable for all domains E, the existence of solutions has been proven in many important cases. In what follows, we shall consider only the two-dimensional case and show how the Fredholm theory can be applied in showing the existence of a solution under certain general assumptions. We shall assume (without proving) several facts from potential theory, our intent being to give the reader only an idea of the applicability of the Fredholm theory. For a detailed discussion of the Dirichlet Problem, the reader can consult *Partial Differential Equations.*†

Let E be an open connected set in R^2 bounded by and in the interior of a simple closed curve C with continuous curvature [i.e., points of C have rectangular coordinates $x(s)$, $y(s)$ (in terms of arc length s), possessing continuous second derivatives]. A function $u(x, y)$ on E with continuous second-order derivatives and satisfying the equation

$$\Delta u = \frac{\partial^2 u}{\partial x^2} + \frac{\partial^2 u}{\partial y^2} = 0$$

in E is called a *harmonic function* on E.

Let f be a continuous function on C. Then it is a fact from potential theory that the function

$$v(p) = \int_C f(t) \frac{\partial}{\partial n_t} \log\left(\frac{1}{|p-t|}\right) dt \tag{6.19}$$

is a harmonic function in E as well as in $(\bar{E})^c$. Here $\partial/\partial n_t$ represents the derivative in the direction of the interior normal n_t at t. For $s \in C$, let us write

$$v^-(s) = \lim_{\substack{t\to s\\ t\in E}} v(t)$$

and

$$v^+(s) = \lim_{\substack{t\to s\\ t\notin \bar{E}}} v(t).$$

It is known that these limits exist and the following equalities are valid. [Note that the integral in equation (6.19) defines $V(p)$ even when $p \in C$.]

† P. R. Garabedian, *Partial Differential Equations*, John Wiley, New York (1967).

$$\begin{aligned}&\text{(i)}\quad v^-(s) = v(s) + \pi f(s);\\&\text{(ii)}\quad v^+(s) = v(s) - \pi f(s);\\&\text{(iii)}\quad \text{the normal derivative of } v \text{ is continuous on } C.\end{aligned} \tag{6.20}$$

The reader can find detailed proofs of similar equalities in Garabedian's book.

It is clear from equations (6.19) and (6.20) that the function $u(t)$, given by

$$u(t) = \begin{cases} v(t), & t \in E, \\ v^-(t), & t \in C, \end{cases}$$

will be a solution of the Dirichlet problem if we find a solution f of the equation

$$\frac{1}{\pi} g(s) = f(s) + \int_C K(s, t) f(t)\, dt, \tag{6.21}$$

where

$$K(s, t) = \frac{1}{\pi} \frac{\partial}{\partial n_t} \log\left(\frac{1}{|s - t|}\right)$$

and g is defined as the given continuous function on the boundary. Here one can show by straightforward computations and by using the continuous curvature of C that $K(s, t)$ is a continuous function of (s, t), even when $s = t$. To apply the Fredholm theory, we consider the homogeneous equation

$$f(s) + \int_C K(s, t) f(t)\, dt = 0 \tag{6.22}$$

and show that this equation does not have any nonzero continuous solution. To show this, we note that any continuous solution f of equation (6.22) will define, as in equation (6.19), a function F harmonic in E as well as in $(\bar{E})^c$ such that by equation (6.20), $F^-(s) = F(s) + f(s) = 0$, $s \in C$. Since a harmonic function is known to assume its maximum and minimum values on the boundary, the function $F(t) = 0$ for all $t \in E$. This means that $(\partial F/\partial n)^- = 0$ on C, and by equation (6.20) (iii), $(\partial F/\partial n)^+ = 0$ on C. Now using the harmonic property of F in $(\bar{E})^c$, it can be proven by using the classical divergence theorem (or Green's first identity) that

$$\iint_{(\bar{E})^c} \left[\left(\frac{\partial F}{\partial x}\right)^2 + \left(\frac{\partial F}{\partial y}\right)^2\right] dx\, dy = -\int_C F^+(t) \left(\frac{\partial F}{\partial n}\right)^+ dt = 0.$$

This means that

$$\frac{\partial F}{\partial x} = 0 = \frac{\partial F}{\partial y} \qquad \text{in } (\bar{E})^c,$$

and therefore F is constant in $(\bar{E})^c$. Since F, because of its representation as in equation (6.19), is known to be zero at infinity, $F^+(s) = 0$, $s \in C$ and by equation (6.20)

$$f(s) = \tfrac{1}{2}[v^-(s) - v^+(s)] = 0, \qquad s \in C.$$

This proves that equation (6.22) cannot have any nonzero solutions. Since an analog of Theorem A holds also for continuous kernels and for operators T acting on $C(\bar{E})$, the following result is immediate.

Theorem D. For every continuous function g given on the boundary C, the Dirichlet problem has a solution. ∎

Problems

× **6.5.1.** Prove Proposition 6.20.

× **6.5.2.** Suppose $(x_i)_{i=1}^{\infty}$ is a Schauder basis for a Banach space X. Let

$$P_n\Big(\sum_{i=1}^{\infty} \lambda_i x_i\Big) = \sum_{i=1}^{n} \lambda_i x_i.$$

Show that $\| P_n(x) - x \| \to 0$ uniformly on every compact set K as $n \to \infty$.

× **6.5.3.** Prove Proposition 6.24.

× **6.5.4.** Let T be the operator

$$T(f)(x) = \int_0^1 K(x, y) f(y)\, dy$$

from $L_p[0, 1]$ into $L_q[0, 1]$, where $1/p + 1/q = 1$, $1 < p < \infty$, and $K(x, y) \in L_q([0, 1] \times [0, 1])$. Show that T is compact. (Hint: First prove the result when K is continuous; then approximate K by continuous functions and use Proposition 6.22.)

× **6.5.5.** Show that a compact operator $T \in L(X, Y)$ maps weakly convergent sequences onto convergent sequences.

× **6.5.6.** (i) Let T be a bounded linear operator on a reflexive Banach space. If T maps weakly convergent sequences onto convergent sequences, then show T is compact.

(ii) Show that the result of (i) may not be true in a nonreflexive space. (Hint: Look at l_1.)

6.5.7. Consider the following Fredholm integral equation:

$$f(x) = g(x) + \lambda \int_0^1 K(x, y) f(y)\, dy,$$

where $g \in L_2[0, 1]$ and $K \in L_2([0, 1] \times [0, 1])$. Prove that if $g = 0$ implies $f = 0$, then there exists a unique solution of the equation for any $g \in L_2[0, 1]$.

6.5.8. Let X be a compact metric space and μ be a finite measure on it. Let $K(x, y)$ be continuous on $X \times X$, and suppose the only continuous solution of

$$f(x) = \lambda \int K(x, y) f(y)\, d\mu(y)$$

is $f = 0$. Prove that for every continuous function $g(x)$ on X, there exists a unique continuous solution $f(x)$ of the integral equation in Problem 6.5.7.

6.5.9. Consider the Volterra integral equation

$$f(x) = g(x) + \int_0^x K(x, t) f(t)\, dt, \qquad 0 \leq x \leq 1,$$

where $K(x, t)$ is continuous on $[0\ 1] \times [0, 1]$. Prove that for any continuous function g, there exists a unique continuous solution f of the Volterra equation.

6.5.10. Let $T \in L(X, X)$, X a complex Banach space. Show that $\varrho(T) = \varrho(T^*)$ and $R(\lambda, T^*) = [R(\lambda, T)]^*$.

6.5.11. Let X, Y, and Z be Banach spaces, $K \in L(X, Y)$ and $T \in L(Z, Y)$. If K is compact and $T(Z) \subset K(X)$, then prove that T is compact. (Hint: Let $N = K^{-1}(\{0\})$. Then K_0, defined by $K_0(x + N) = K(x)$, is a compact operator from X/N into Y.)

6.5.12. *Weakly Compact Operators.* A linear operator mapping bounded sequences onto sequences having a weakly convergent subsequence is called weakly compact. Prove the following: (i) Weakly compact operators are continuous. (ii) If $T \in L(X, Y)$ and either X or Y is reflexive, then T is weakly compact. (iii) If T is the operator from $L_1[0, 1]$ into $L_p[0, 1]$, $1 \leq p < \infty$, defined by $T(f)(x) = \int_0^1 K(x, y) f(y)\, dy$, where $K(x, y)$ is a bounded measurable function on $[0, 1] \times [0, 1]$, then T is weakly compact. (Hint: Use Problem 6.4.22 for $p = 1$.)

6.5.13. Let $T \in L(X, Y)$ be compact and $Z = T(X)$. Define T_0: $X \to Z$ by $T_0(x) = T(x)$. Is T_0 compact? What if X is reflexive? What if

Z is closed in Y? What happens if "compact" is replaced by "weakly compact"?

6.5.14. If $T \in L(X, X)$ is compact and $T + \lambda I$ is invertible in $L(X, X)$, prove there exists a scalar λ_0 and a compact operator T_0 such that $(T + \lambda I)^{-1} = T_0 + \lambda_0 I$.

6.5.15. Prove that the set of invertible elements of $L(X, X)$, X a Banach space, is open. [Hint: If T is invertible let $S \in L(X, X)$ have $\| S \| < \| T^{-1} \|^{-1}$. Show that $T + S = T[I + T^{-1}S]$ and that $(I + T^{-1}S)^{-1} = \sum_{n=0}^{\infty}(-1)^n(T^{-1}S)^n$.]

6.5.16. *Banach Algebras.* A *Banach algebra* is a complex Banach space A on which an operation of multiplication is defined satisfying for all a, b, and c in A and $\lambda \in C$ the properties $(ab)c = a(bc)$; $a(b + c) = ab + ac$; $(b + c)a = ba + bc$; $\lambda(ab) = (\lambda a)b = a(\lambda b)$; and $\| ab \| \leq \| a \| \| b \|$. If there is an element e in A such that $ae = ea = a$ for all a in A and $\| e \| = 1$ then A is said to have a *unit* e. If $ab = ba$ for all a, b in A, then A is said to be *commutative*.

(i) Show $L(X, X)$, X a complex Banach space, is a commutative Banach algebra with a unit.

(ii) Show $C_1(X)$, the space of continuous complex valued functions on a compact Hausdorff space, is a commutative Banach algebra with a unit.

(iii) If A is a Banach algebra with a unit, define for each $a \in A$, the mapping M_a in $L(A, A)$ by the rule $M_a(b) = ab$. Show that the mapping $a \mapsto M_a$ is a linear isometry from A into $L(A, A)$ which also preserves multiplication.

(iv) If A is an in (iii), the spectrum set $\sigma(a)$ of a in A is $\{\lambda \mid a - \lambda e$ has no inverse in $A\}$. [$a - \lambda e$ has an inverse if there exists b in A such that $b(a - \lambda e) = (a - \lambda e)b = e$.] Prove $\sigma(a) = \sigma(M_a)$. Conclude that $\sigma(a)$ is nonempty and compact.

(v) *Gelfand–Mazur Theorem.* If A is a Banach algebra with unit and if every nonzero element in A has an inverse, then A is isometrically isomorphic to the algebra of complex numbers. Prove this. (Hint: Show that for each a in A there is exactly one λ_a such that $a - \lambda_a e = 0$.) Conclude that A is commutative.

6.5.17. (i) If X is a complex Banach space and if $T \in L(X, X)$, prove that there is a maximal closed subspace M of $L(X, X)$ that contains T, contains I, is closed with respect to composition, and on which composition is commutative. (M is an example of a commutative Banach algebra with unit.)

(ii) Prove $\sigma(T) = \sigma_M(T)$, which is defined to be $\{\lambda \mid T - \lambda I$ has no inverse in $M\}$.

6.5.18. This is a continuation of Problem 6.5.17.

(i) If W is a proper ideal in M (that is, a proper subspace of M such that $AB \in W$ whenever $A \in M$ and $B \in W$), prove that W is contained in a maximal ideal Z which is closed (topologically) and $Z \neq M$. [Hint: Use Zorn's lemma to get a maximal ideal Z which does not contain I. Since $\bar{Z}$ is also an ideal and the set of invertible elements in $L(X, X)$ is open (6.5.15), $Z = \bar{Z}$.]

(ii) Prove M/Z (see Section 6.1) is an algebra in which every nonzero element has an inverse. [Hint: Let $\Phi: M \to M/Z$ be the natural map $\Phi(A) = [A]$. Let V_0 be the ideal $\{AT_0 + S \mid A \in M, S \in Z\}$ for some $T_0 \in M \setminus Z$. Then $V_0 = Z$ so $\Phi(AT_0 + S) = \Phi(I)$ for some A and S or $[A][T_0] = [I]$.

(iii) Prove M/Z is isometrically isomorphic to the algebra of complex numbers. [Hint: Gelfand–Mazur, 6.5.16(v).]

(iv) Prove $\sigma(T) = \{h(T) \mid h$ is a homomorphism (preserving scalar multiplication, addition, and multiplication) from M onto $C\}$. [Hint: If $\lambda \in \sigma(T) = \sigma_M(T)$, then $\{(T - \lambda I)S \mid S \in M\}$ is a proper ideal W of M contained in a maximal proper ideal Z of M. Let $h = i \circ \Phi: M \to M/Z \to C$. Then $h(T - \lambda I) = 0$. Conversely, if $\lambda \notin \sigma(T)$, then for some $S \in M$, $(T - \lambda I)S = I$. Hence for every homomorphism h from M onto C, $h(T - \lambda I) \neq 0$. Here i is the map from (iii).]

6.5.19. Using (ii) of Problem 6.5.17 and (iv) of Problem 6.5.18, prove the following spectral mapping theorems.

(i) If $T \in L(X, X)$ and p is any polynomial of one complex variable, then $\sigma(p(T)) = p(\sigma(T))$.

(ii) If T and S are commuting operators in $L(X, X)$ and $p(x, y)$ is a complex polynomial in two variables, then $\sigma(p(S, T)) \subset p(\sigma(S), \sigma(T)) \equiv \{p(\alpha, \beta): \alpha \in \sigma(S), \beta \in \sigma(T)\}$.

6.5.20. Let T be a linear operator on a normed linear space X. If λ is an eigenvalue of T, let $M_\lambda = \{x \in X: Tx = \lambda x\}$, the eigenspace of T corresponding to λ. Show that if T is closed, then M_λ is a closed linear subspace of X which is invariant under T, that is, $T(M_\lambda) \subset M_\lambda$.

6.5.21. Verify Example 6.17. [Hint: It is sufficient to prove T_G is not invertible if and only if $\mu\{x \mid |G(x)| < \varepsilon\} > 0$ for all $\varepsilon > 0$.]

6.6. Topological Vector Spaces

A natural extension of normed linear spaces are vector spaces where the topology is induced by a family of norms, instead of one norm. Such generalizations include many important examples such as the weak topology in a normed linear space X and the weak*-topology in X^*, and provide proper framework for the consideration of various applications. Vector spaces with such a topology are special cases of a more general class of spaces, called topological vector spaces. The main purpose of this section is to give a brief introduction of topological vector spaces, leading to the basic properties of locally convex spaces and a discussion of the Krein–Milman theorem and one of its applications in measure theory.

Let X be a vector space over F (the field of real or complex numbers) with a topology τ such that the maps

$$f_1: X \times X \to X \quad \text{and} \quad f_2: F \times X \to X,$$

defined by

$$f_1(x, y) = x + y \quad \text{and} \quad f_2(\beta, x) = \beta x,$$

are continuous, where $X \times X$ and $F \times X$ have the usual product topologies. For convenience, 0 will denote both the zero vector and the scalar zero. The meaning will be clear from the context. The vector space X with topology τ is called a topological vector space or a TVS, in short. In a TVS, the translation map $x \to x + x_0$ is a homeomorphism, and thus, a local base† at 0 for τ gives automatically a local base at any other point. The following proposition gives properties that characterize a base for a topology for a TVS.

Proposition 6.28. Let X be a TVS. Then there is a local base $\mathscr{B}$ at 0 having the following properties:

(i) If G and H are in $\mathscr{B}$, then there is $A \in \mathscr{B}$ such that $A \subset G \cap H$.
(ii) If $G \in \mathscr{B}$ and $x \in G$, then there exists $H \in \mathscr{B}$ such that $x + H \subset G$.
(iii) If $G \in \mathscr{B}$, then there exists $H \in \mathscr{B}$ such that $H + H \subset G$.
(iv) If $x \in X$ and $G \in \mathscr{B}$, then there exists $\beta \in F$ such that $x \in \beta G$.
(v) If $G \in \mathscr{B}$ and $0 < |t| \leq 1$, then $tG \subset G$ and $tG \in \mathscr{B}$.

Conversely, a given collection $\mathscr{B}$ of subsets of a vector space X containing 0 and satisfying the above properties is a local base at 0 for a topology τ that makes X a TVS. The topology τ is Hausdorff if and only if

(vi) $\cap \{G: G \in \mathscr{B}\} = \{0\}$. ∎

Proof. Let X be a TVS with topology τ. Let $V \in \tau$ and $0 \in V$. Since $0.0 = 0$ and scalar multiplication is continuous, there exists $u > 0$ and

† A local base at 0 is a family of open subsets containing 0 such that each open set containing 0 contains a member of the family.

$W \in \tau$, $0 \in W$ such that $tW \subset V$, whenever $|t| \leq u$. If $U = uW$, then $0 \in U$, $U \in \tau$ and $tU \subset V$ for $|t| \leq 1$. (Note that we have used the fact that multiplication by a nonzero scalar is a homeomorphism so that $U \in \tau$.) Let $G = \cup \{tU: |t| \leq 1\}$. Then, $0 \in G \subset V$, $G \in \tau$ and $tG \subset G$ for $|t| \leq 1$. Let $\mathscr{B}$ be the collection of all such G. Then the reader can verify that properties (ii), (iii), and (iv) follow from the continuity of addition and scalar multiplication and the observations that $x + 0 = x$, $0 + 0 = 0$, and $0x = 0$. Property (i) follows from the definition of topology and property (v) from the definition of $\mathscr{B}$.

Conversely, let $\mathscr{B}$ be a collection of subsets of a vector space X containing 0 and satisfying properties (i)–(v). Define τ to be the collection of arbitrary union of sets from the family

$$\mathscr{F} = \{x + G: x \in X \text{ and } G \in \mathscr{B}\}.$$

Note that if $z \in x + G \in \mathscr{F}$, then by (ii) there exists $U \in \mathscr{B}$ such that $z + U \subset x + G$. Thus, if $z \in (x + G) \cap (y + H)$, G and H in $\mathscr{B}$, then there exists [by (i) and (ii)] $U_0 \in \mathscr{B}$ such that $z + U_0 \subset (x + G) \cap (y + H)$. Thus, $\mathscr{F}$ fulfills the conditions to be a base for a topology τ. We need to show that in this topology addition and scalar multiplication are continuous.

If $x + y = z$ and $V \in \mathscr{B}$, then by (iii) there exists $U \in \mathscr{B}$ such that $(x + U) + (y + U) \subset z + V$; thus, addition is continuous in τ. To prove the continuity of $(\beta, x) \to \beta x$, let us take first $\beta = 0$. Given $U \in \mathscr{B}$, there exists $V \in \mathscr{B}$ such that $V + V \subset U$. By (iv) and (v), there exists $0 < u < 1$ such that for $|t| < u$, $tx \in V$ and therefore, $t(x + V) \subset V + tV \subset V + V \subset U$. Now let $\beta \neq 0$. Let $W \in \mathscr{B}$. By (iii), there exists $W' \in \mathscr{B}$ satisfying (v) such that $W' + W' \subset W$. Let $W_0 = (1/1 + |\beta|)W'$. Then $W_0 \in \mathscr{B}$ and there exists $0 < u < 1$ such that for $|t| \leq u$, $tx \in W'$ and $(\beta + t)W_0 \subset W'$. Thus, for $|t| < u$, we have

$$(\beta + t)(x + W_0) = (\beta + t)x + (\beta + t)W_0 \subset \beta x + W' + W' \subset \beta x + W.$$

Finally, we consider the Hausdorff property of τ. Suppose that (vi) holds and $x \neq y$. Then there exist W_1 and W_2 in $\mathscr{B}$ such that $x - y \notin W_2$ and $W_1 + W_1 \subset W_2$. Since $-W_1 = W_1$, $(x + W_1) \cap (y + W_1) = \varnothing$. Thus, τ is T_2. The converse is immediate, since $x \neq 0$ and the T_2 property imply that there exists $V' \in \mathscr{B}$ such that $x \notin V'$. ∎

For further discussions on topological vector spaces, we need a definition.

Definition 6.17. Let X be a vector space over F and $A \subset X$. The set A is called balanced if $tA \subset A$ for $|t| \leq 1$. The set A is called absorbing if given $x \in X$, there exists $u > 0$ such that $tx \in A$, whenever $|t| < u$. ▮

Notice that in a TVS X, if V is a balanced open set and if given $x \in X$, there exists some nonzero $t_0 \in F$ such that $t_0 x \in V$, then for $|t| \leq |t_0|$, $tx = (t/t_0)t_0 x \in (t/t_0)V \subset V$, and consequently, V is absorbing. It follows from Proposition 6.28 that in a TVS, 0 has a local base consisting of balanced and absorbing sets.

Let us now consider two important examples of topological vector spaces which will also lead us to the study of a special important class of these spaces, called locally convex spaces. First, let $\mathscr{F}$ be a family of linear functionals on a vector space X. Then, the weakest topology on X which makes each member in $\mathscr{F}$ continuous has a local base at 0 consisting of sets of the form

$$\{x: |f_i(x)| < r, \ i = 1, 2, \ldots, n\} \tag{6.23}$$

where $r > 0$ and each f_i belongs to $\mathscr{F}$. This base has the properties (i)–(v) of Proposition 6.28 and therefore, induces a topology that makes X a TVS. Note that the sets in (6.23) are convex [that is, for all $t \in (0, 1)$, $tx + (1 - t)y$ are in the set whenever x and y are]. This topology is called the weak topology on X induced by $\mathscr{F}$. As for the second example, let $C_1(S)$ be the complex-valued continuous functions on some Hausdorff topological space S. Let us define the seminorm† p_K on $C_1(S)$ for each compact subset $K \subset S$ by

$$p_K(x) = \sup\{|x(s)| : s \in K\}.$$

Then the collection of sets of the form

$$\{x \in C_1(S): p_{K_i}(x) < r, \ i = 1, 2, \ldots, n\} \tag{6.24}$$

induces a Hausdorff topology on $C_1(S)$ making it a TVS. The sets in (6.24) are again convex. This topology is called the topology of uniform convergence on compact sets.

Definition 6.18. A TVS X with topology τ is called locally convex if τ has a local base at 0 consisting of convex sets. ▮

† A seminorm n on a vector space X over F is a nonnegative real function such that for x, y in X and $\beta \in F$, $n(\beta x) = |\beta| n(x)$ and $n(x + y) \leq n(x) + n(y)$. Thus, a seminorm is a norm if and only if $n(x) = 0$ implies that $x = 0$.

A normed linear space with its norm topology is locally convex. It follows from the proof of Proposition 6.28 that in a locally convex space (that is, a locally convex TVS), 0 has a local base consisting of balanced, convex, and absorbing sets. The reason is that in this proof the convex hull of the set G, when V is convex and open, is open, balanced, absorbing, and contained in V. In the second example above, the family of seminorms $\{p_K: K \text{ is a compact subset of } S\}$ induced a topology [meaning the topology induced by the sets in (6.24)] making $C_1(S)$ a locally convex space. The same is the case for all locally convex spaces, as the following theorem shows.

Theorem 6.26. Let X be a TVS. Then X is locally convex if and only if the topology τ of X has a local base at 0 given by the sets

$$N_{\beta_1,\beta_2,\ldots,\beta_n;r} = \{x: p_{\beta_i}(x) < r,\ i = 1, 2, \ldots, n\},$$

where (p_β), $\beta \in A$ (A an indexed set), is a family of seminorms and r is a positive real number. ∎

Proof. The "if" part is obvious since the sets $N_{\beta_1,\beta_2,\ldots,\beta_n;r}$ are convex. For the "only if" part, let $\mathscr{B}$ be the local base at 0 consisting of balanced, convex, and absorbing sets. For each $U \in \mathscr{B}$, let us define the Minkowski functional $p_U : X \to [0, \infty)$ by

$$p_U(x) = \inf\{\beta: \beta > 0 \text{ and } x \in \beta U\}. \tag{6.25}$$

Then p_U can be verified to have the following properties:

(i) $p_U(\beta x) = |\beta| p_U(x)$ for $\beta \in F$. (6.26)

(ii) $p_U(x + y) \leq p_U(x) + p_U(y)$. (6.27)

(iii) $\{x: p_U(x) < r\} \subset rU$ and belongs to τ, $r > 0$. (6.28)

Thus, p_U is a seminorm by (i) and (ii). It is clear from (iii) that the topology induced by the family of seminorms $\{p_U: U \in \mathscr{B}\}$ is the same as the original locally convex topology. ∎

It is always important to know if a particular topology is metrizable, since metric spaces have many nice properties. The next theorem considers this question for a locally convex space.

Theorem 6.27. Let X be a locally convex space. Then the following are equivalent:

(i) The topology of X is given by a pseudometric.
(ii) 0 has a countable local base.
(iii) The topology of X is generated by an at most countable family of seminorms.

Moreover, the topology of X is given by a seminorm if and only if the family of seminorms determining the topology of X has a finite subfamily (i.e., finitely many seminorms from the same family) determining the same topology. ▮

Proof. (i) $\Rightarrow$ (ii) by Remark 1.79. Also, (ii) $\Rightarrow$ (iii), since if $\mathscr{B}$ is a countable local base at 0 consisting of convex, balanced and absorbing sets, then the seminorms $\{p_U: U \in \mathscr{B}\}$ [defined as in (6.25)] will generate the same topology as $\mathscr{B}$.

Now let us assume that (p_n) is a countable family of seminorms generating the topology of X. Define

$$p(x, y) = \sum_{n=1}^{\infty} \frac{(1/2^n)p_n(x-y)}{1 + p_n(x-y)}. \tag{6.29}$$

Then p is a pseudometric. Notice that for x, y in X,

$$\frac{p_n(x+y)}{1+p_n(x+y)} \leq \frac{p_n(x)}{1+p_n(x)} + \frac{p_n(y)}{1+p_n(y)}.$$

For $\beta > 0$, $a > \beta$ if and only if $a/(1+a) > \beta/(1+\beta)$, and therefore, it follows that $\lim_{i\to\infty} p(x_i, y) = 0$ if and only if for each n, $\lim_{i\to\infty} p_n(x_i - y) = 0$. Thus, p induces the same topology on X as that induced by $(p_n)_{n=1}^{\infty}$.

For the second part of the theorem, we observe that the topology induced by the seminorms $\{p_1, p_2, \ldots, p_n\}$ is the same as that induced by the seminorm $\sum_{i=1}^{n} p_i$. Now suppose that the topology of X is generated by the family of seminorms $\{p_\beta: \beta \in A\}$, and that it is also generated by the seminorm p. Then there exists $u > 0$ and $\beta_1, \beta_2, \ldots, \beta_m$ in A such that

$$\{x: p_{\beta_i}(x) < u,\ i = 1, 2, \ldots, m\} \subset \{x: p(x) < 1\}. \tag{6.30}$$

We claim that, for all x in X,

$$p(x) \leq \frac{1}{u} \sum_{i=1}^{m} p_{\beta_i}(x). \tag{6.31}$$

If the claim is false, there exists z in X such that

$$p(z) > r > \frac{1}{u} \sum_{i=1}^{m} p_{\beta_i}(z), \tag{6.32}$$

for some real number r. Then $p_{\beta_i}(z/r) < u$, $i = 1, 2, \ldots, m$. But $p(z/r) > 1$. This contradicts (6.30) and the claim (6.31) follows. It follows that the seminorms $\{p_{\beta_i}: i = 1, 2, \ldots, m\}$ generate the same topology as does p. ▮

We now turn to the three basic principles of functional analysis discussed in Section 6.3 in the context of topological vector spaces. We need a definition.

Definition 6.19. A complete metrizable TVS is called a Fréchet space. ▮

In Fréchet spaces, the open mapping theorem, the closed graph theorem, and the principle of uniform boundedness hold in the following forms.

Theorem 6.28. Let X and Y be Fréchet spaces and $T: X \to Y$ be a closed linear onto map. Then T is open. ▮

(As before, the map T above is called closed if its graph is a closed subset of $X \times Y$ with usual product topology.)

Theorem 6.29. A closed linear map from a Fréchet space into another is continuous. ▮

Theorem 6.30. Let X and Y be two complete metrizable locally convex topological vector spaces. Let $\mathscr{T}$ be a family of continuous linear maps from X into Y such that for every continuous seminorm p on Y and every x in X, the set $\{p(T(x)): T \in \mathscr{T}\}$ is bounded. Then for each such p, there is a constant $C > 0$ and a continuous seminorm q on X such that for all x in X and $T \in \mathscr{T}$,

$$p(T(x)) \leq Cq(x). \quad ▮$$

The proofs of Theorems 6.28, 6.29 and 6.30 are analogous to those of Theorems 6.8, 6.9, and 6.11, respectively, and will not be given here. However, we prove below two other forms of uniform boundedness principles in topological vector spaces. We need another definition.

Definition 6.20. A set A in a TVS X is called bounded if given any open neighborhood U of 0, there exists $u > 0$ such that $uA \subset U$. ∎

Note that in a normed linear space, a set is bounded if and only if it is bounded in the norm.

Theorem 6.31. Let X be a complete metrizable TVS and Y be a TVS. Let $\mathscr{F}$ be a pointwise bounded family of continuous linear maps from X into Y. Then $\mathscr{F}$ is equicontinuous.† ∎

Proof. Let $V \subset Y$ be an open neighborhood of 0. Then there is an open and balanced neighborhood U of 0 such that $U + U + U + U \subset V$. Let W be the set defined by

$$W = \{x \in X: T(x) \in \bar{U} \text{ for all } T \text{ in } \mathscr{F}\}.$$

Then W is closed and balanced. Let $z \in X$. The set $\{T(z): T \in \mathscr{F}\}$ is a bounded subset of Y and therefore, there exists $\beta > 0$ such that

$$\beta\{T(z): T \in \mathscr{F}\} \subset U.$$

Hence, $\beta z \in W$ and therefore, W is absorbing. Thus, $X = \bigcup_{n=1}^{\infty} nW$. Since X is a complete metric space, there is a positive integer n such that nW is not nowhere dense. It follows that W contains an open set G. Let $x, y \in W$ and $T \in \mathscr{F}$. Then $T(x - y) \in \bar{U} - \bar{U} \subset U + U + U + U \subset V$, since $\bar{U} \subset U + U$. Thus, the set $H = G - G$ is an open neighborhood of 0 and for all $T \in \mathscr{F}$, $T(H) \subset V$. ∎

Theorem 6.32. Let X be a complete metrizable TVS and Y be a TVS. Suppose that (T_n) is a sequence of continuous linear maps from X into Y such that $\lim_{n\to\infty} T_n(x)$ exists for all $x \in X$. If $T(x) = \lim_{n\to\infty} T_n(x)$, then T is continuous. ∎

Proof. By Theorem 6.31, the family $(T_n)_{n=1}^{\infty}$ is equicontinuous. Let W be an open subset of Y containing 0. Let V be an open neighborhood of 0 in Y such that $V + V \subset W$. By Theorem 6.31, there is an open neighborhood U of 0 in X such that for each n, $T_n(U) \subset V$. Then for any x in U, $T(x) = T(x) - T_n(x) + T_n(x) \in V + V \subset W$ for sufficiently large n, since $T(x) - T_n(x) \to 0$ as $n \to \infty$. ∎

† $\mathscr{F}$ is called equicontinuous if given any open set $V \subset Y$ and $0 \in V$, there exists an open set $U \subset X$, $0 \in U$, and $T(U) \subset V$ for every $T \in \mathscr{F}$.

An important feature of locally convex spaces is that in such spaces there exist many continuous linear functionals, and consequently many useful results hold in such spaces. One such result is the famous Krein–Milman theorem. In what follows, we will present first some useful results concerning the existence of continuous linear functionals on locally convex spaces, and then, to conclude this section we will present the Krein–Milman theorem and one of its many applications.

We start with an extension of the Hahn–Banach theorem to locally convex spaces.

Theorem 6.33. Let X be a locally convex space and $Y \subset X$ be a linear subspace. Let f be a continuous linear functional on Y. Then there is a continuous linear functional g on X such that for $y \in Y$, $g(y) = f(y)$. ∎

Proof. Let $(p_\beta)_{\beta \in A}$ be a family of seminorms determining the topology of X. The relative topology on Y is given by the restrictions of these seminorms to Y. Thus, there exists $u > 0$ and $\beta_1, \beta_2, \ldots, \beta_n$ in A such that

$$\{y \in Y: p_{\beta_i}(y) < u, i = 1, 2, \ldots, n\} \subset \{y \in Y: | f(y) | < 1\}.$$

Let $p = (1/u) \sum_{i=1}^{n} p_{\beta_i}$. Then the above set inclusion implies that for each $y \in Y$, $| f(y) | \leq p(y)$. By Theorems 6.4 and 6.5, there exists a linear functional g on X such that g extends f and for all $x \in X$, $| g(x) | \leq p(x)$. Clearly g is continuous. ∎

Definition 6.21. Let X be a vector space over F. If f is a nonzero real linear functional on X regarded as a vector space over the reals, then the set $\{x \in X: f(x) = r\}$, where r is a real number, is called a hyperplane. ∎

Thus, a subset L in X, a vector space over the reals, is a hyperplane if and only if $L = x + W$, where $x \in X$ and W is a maximal linear subspace (that is, a proper subspace W such that the subspace spanned by $W \cup \{z\}$, $z \in X - W$, is X). (See Problem 6.2.20.)

Definition 6.22. In a TVS X, the sets U and V are said to be separated by a hyperplane if for some continuous real-valued functional f on X (regarded as a vector space over the reals) and some real number r,

$$U \subset \{x \in X: f(x) \leq r\} \text{ and } V \subset \{x \in X: f(x) \geq r\}.$$

In the case

$$U \subset \{x \in X: f(x) < r\} \quad \text{and} \quad V \subset \{x \in X: f(x) > r\},$$

we say that U and V are strictly separated. ∎

The above two definitions will now allow us to present a theorem which shows, among other things, that in a locally convex Hausdorff space for any two distinct points x and y there always exists a continuous linear functional f such that $f(x) \neq f(y)$.

Theorem 6.34. Let U and V be disjoint convex sets in a locally convex space X. Then the following assertions hold:

(a) If U is open, then U and V can be separated by a hyperplane.
(b) If U and V are both open, then they can be strictly separated by a hyperplane.
(c) If U is compact and V is closed, then they can be strictly separated by a hyperplane. ∎

Proof. We regard X as a locally convex space over the reals. Let U be open, $x \in U$ and $y \in V$. Let $W = W_0 - W_0$, where $W_0 = U - V + \{y - x\}$. Then W is open (and therefore, absorbing), convex, and symmetric. Since $0 \in W$, W is also balanced. Consider the Minkowski functional p_W as defined in (6.25). Let $Y = \{a(y - x): a \in R\}$ and $f(a(y - x)) = a$ for a in R. Then f is a linear functional on the subspace Y and for each $z \in Y$, $f(z) \leq p_W(z)$ since $y - x \notin W$ and $p_W(y - x) \geq 1$. By Theorem 6.4, f can be extended to a linear functional g on X such that for each $z \in X$, $g(z) \leq p_W(z)$. Since W is an open neighborhood of 0 and for each $z \in W$, $p_W(z) \leq 1$, it is clear that $W \subset g^{-1}[-1, 1]$ and g is continuous. Notice that for $x' \in U$ and $y' \in V$, $x' - y' + y - x \in W_0 \subset W$, and therefore, since $g(y - x) = 1$ and $g(x' - y' + y - x) \leq 1$, we have $g(x') \leq g(y')$. Thus,

$$\sup\{g(x'): x' \in U\} \leq \inf\{g(y'): y' \in V\}, \tag{6.33}$$

and the proof of part (a) is complete.

To prove (b), let U and V be both open. Consider the nonzero functional g above. Then $g(U)$ is open. To see this, let $x \in U$. If $g(x) = 0$, then since g is nonzero, there exists $y \in U_0 \subset U$ such that $g(y) \neq 0$, where U_0 is balanced. It follows that the interval $(-g(y), g(y)) \subset g(U_0) \subset g(U)$. If $g(x) \neq 0$, then there exists $\beta > 0$ such that $\{(1 + t)x: |t| \leq \beta\} \subset U$ so that the interval $((1 - \beta)g(x), (1 + \beta)g(x)) \subset g(U)$. This proves that $g(U)$ is open. Similarly, $g(V)$ is also open. By (6.33), $g(U) \cap g(V)$ is a singleton, if nonempty. Since $g(U) \cap g(V)$ is open, it must be empty.

To prove (c), let U be compact and V be closed. Then for each $x \in U$, there is a convex open set G_x such that $0 \in G_x$ and $(x + G_x) \cap V = \varnothing$. Since U is compact, there exist $x_1, x_2, \ldots, x_n$ in U such that $U \subset \bigcup_{i=1}^n (x_i + \frac{1}{2}G_{x_i})$. If $G = \bigcap_{i=1}^n \frac{1}{2}G_{x_i}$, then $(U + G) \cap V = \varnothing$. This means that $(U + \frac{1}{2}G) \cap (V - \frac{1}{2}G) = \varnothing$. Since $U + \frac{1}{2}G$ and $V - \frac{1}{2}G$ are disjoint open convex sets containing U and V respectively, the proof of (b) now applies. ∎

We now present the celebrated Krein–Milman theorem† and give J. L. Kelly's proof of it. The Krein–Milman theorem is an abstract analog for compact convex sets in locally convex spaces of the fact that in two dimensions any point inside or on a triangle determined by three vertices can be expressed as a convex linear combination of these vertices. For its proof, we need a definition and a proposition.

Definition 6.23. Let X be a TVS over F and $K \subset X$. A nonempty subset $A \subset K$ is called an extremal subset of K if the following holds: If x and y are both points of K and the point $\beta x + (1 - \beta)y$ is in A for some β in $(0, 1)$, then x and y must be points of A. An extremal subset of K consisting of exactly one point is called an extreme point of K. ∎

The vertices of a triangle in R^2 are the extreme points of the triangle, whereas its sides are examples of its other extremal subsets. Note that an open triangle in R^2 or an open solid sphere in R^3 has no extreme points, whereas for a closed solid sphere, its surface is an extremal subset and every point on it is an extreme point. It follows immediately from the definition that if in a TVS X, $B \subset A \subset K$, B is an extremal subset of A and A is an extremal subset of K, then B is an extremal subset of K.

Proposition 6.29. Let X be a locally convex Hausdorff space over F and $K \subset X$ be a nonempty compact subset. Then K has extreme points. ∎

Proof. Let $\mathscr{F}$ be the family of closed extremal subsets of K. Then $\mathscr{F}$ is nonempty since $K \in \mathscr{F}$. Partially order $\mathscr{F}$ by inclusion. Consider a linearly ordered subfamily $\{A_\beta: A_\beta \in \mathscr{F}, \beta \in B\}$. Then the set $\cap \{A_\beta: \beta \in B\}$ is nonempty (and closed) since K is compact and the A_β's have finite intersection property. (See Remark 1.31.) This intersection is also an ex-

† M. Krein and D. Milman, *Studia Math.* **9**, (1940).

tremal subset of K. Thus, it is an upper bound for the subfamily. By Zorn's Lemma, $\mathscr{F}$ contains a maximal element A. Suppose that A has two distinct points z and w. By Theorem 6.34, there exists a nonzero real-valued continuous linear functional f on X (as a TVS over the reals) such that $f(z) < f(w)$. This means that the set D defined by $D = \{x \in A: f(x) = \sup_{y \in A} f(y)\}$ is a nonempty closed proper subset of A. (Note that A is compact and therefore f attains its supremum in A.) Also, it follows from the definition of D that D is an extremal subset of A, and therefore, an extremal subset of K. This contradicts the maximality of A. Hence, A must be a singleton, and therefore, is an extreme point of K. ∎

We need one final definition in this section.

Definition 6.24. Let X be a TVS over F and $E \subset X$. Then the convex hull of E is defined as

$$\mathrm{co}(E) = \left\{ \sum_{i=1}^{n} a_i x_i : a_i \geq 0, \sum_{i=1}^{n} a_i = 1; x_i \in E, i = 1, 2, \ldots, n, 1 \leq n < \infty \right\}.$$

The closure of $\mathrm{co}(E)$ is called the closed convex hull of E, and denoted by $\overline{\mathrm{co}}(E)$. ∎

We may remark that $\mathrm{co}(E) = \cap \{A: A \supset E, A \text{ convex}\}$ and $\overline{\mathrm{co}}(E) = \cap \{A: A \supset E, A \text{ closed convex}\}$, since $\mathrm{co}(E)$ ($\overline{\mathrm{co}}(E)$, respectively) is a convex (closed convex, respectively) set containing E.

Theorem 6.35. *The Krein–Milman Theorem.* Let X be a locally convex Hausdorff space and K a compact subset of X. Let E be the set of its extreme points. Then $K \subset \overline{\mathrm{co}}(E)$. If K is also convex, $K = \overline{\mathrm{co}}(E)$. ∎

Proof. Let $x \in K - \overline{\mathrm{co}}(E)$. By Theorem 6.34, there is a nonzero real continuous linear functional f on X (as a TVS over the reals) and real numbers r, s such that

$$f(x) \leq r < s \leq \inf\{f(y): y \in \overline{\mathrm{co}}(E)\}.$$

Let $K_0 = \{z \in K: f(z) = \inf_{y \in K} f(y)\}$. Then K_0 is a closed extremal subset of K and $K_0 \cap E = \varnothing$. Since an extreme point of K_0 is an extreme point of K, K_0 has no extreme points, and this contradicts Proposition 6.29. ∎

Finally, we give an application of the Krein–Milman theorem to measure theory. We present below J. Lindenstrauss' proof† of the Liapounoff convexity theorem using Theorem 6.35.

Theorem 6.36. Let $\mu_1, \mu_2, \ldots, \mu_n$ be finite nonatomic (positive) measures on some measurable space $(X, \mathscr{A})$. Then the set of points in R^n of the form $(\mu_1(A), \mu_2(A), \ldots, \mu_n(A))$, $A \in \mathscr{A}$, is a closed and convex subset of R^n. ∎

Proof. We use induction on n. Before we give the proof for $n = 1$, it is helpful to carry out the induction step. (As we will see, the proof for $n = 1$ will be clear from that of the induction step.) Write $\mu = \mu_1 + \mu_2 + \cdots + \mu_n$, and let $W = \{g \in L_\infty(\mu): 0 \leq g \leq 1\}$. Define $T: W \to R^n$ by $T(g) = (\int g\, d\mu_1, \int g\, d\mu_2, \ldots, \int g\, d\mu_n)$. Let W be given the weak*-topology of $L_\infty(\mu)$ as the dual of $L_1(\mu)$. Then W is compact by Theorem 6.16, T is continuous, and $T(W)$ is compact and convex. We claim that $T(W) = \{(\mu_1(A), \mu_2(A), \ldots, \mu_n(A)): A \in \mathscr{A}\}$. To prove this, let $\beta = (\beta_1, \beta_2, \ldots, \beta_n) \in T(W)$. Then the set $W_0 = T^{-1}(\beta)$ is a weak*-compact convex subset of $W \subset L_\infty(\mu)$. Since $L_\infty(\mu)$, with the weak*-topology, is a locally convex Hausdorff space, it follows by Theorem 6.35 that W_0 has extreme points. Let g be an extreme point of W_0. Suppose that there exists $u > 0$ and a subset $A \in \mathscr{A}$ such that $\mu_1(A) > 0$ and $u \leq g(x) \leq 1 - u$ for $x \in A$. We will show that this will contradict the fact that g is an extreme point of W_0, thus proving that g is of the form χ_H, $H \in \mathscr{A}$.

Since μ_1 is nonatomic, there exists $B \in \mathscr{A}$, $B \subset A$ such that $\mu_1(B) > 0$ and $\mu_1(A - B) > 0$. By induction hypothesis, replacing X by B, we see that there exists $C \subset B$, $C \in \mathscr{A}$, such that

$$(\mu_2(C), \ldots, \mu_n(C)) = \tfrac{1}{2}(\mu_2(B), \ldots, \mu_n(B)) + \tfrac{1}{2}(\mu_2(\varnothing), \ldots, \mu_n(\varnothing)).$$

Similarly, there exists $D \in \mathscr{A}$ such that $D \subset A - B$ and

$$(\mu_2(D), \ldots, \mu_n(D)) = \tfrac{1}{2}(\mu_2(A-B), \ldots, \mu_n(A-B)) + \tfrac{1}{2}(\mu_2(\varnothing), \ldots, \mu_n(\varnothing)).$$

Now we choose real numbers s and t such that

$$|s| \leq u, \qquad |t| \leq u, \qquad s^2 + t^2 > 0,$$

and

$$s(\mu_1(B) - 2\mu_1(C)) = t(\mu_1(A - B) - 2\mu_1(D)).$$

† J. Lindenstrauss, *J. Math. Mech.* **15** (6), (1966).

Let $h = s(2\chi_C - \chi_B) + t(\chi_{A-B} - 2\chi_D)$. Then $h \not\equiv 0$ in $L_\infty(\mu)$, since $\mu(C) > 0$ and on C, $h = s$. [If $s = 0$, then $h = -t$ on D, $\mu(D) > 0$ and $t \neq 0$.] Notice that

$$|h| \leq u \leq g \leq 1 - u \leq 1 - |h|, \qquad \text{on } A,$$

and $h \equiv 0$ on $X - A$. This means that $g + h$ and $g - h$ are both in W. Since $\int h \, d\mu_i = 0$ for $i = 1, 2, \ldots, n$, the functions $g + h$ and $g - h$ are both in W_0. This contradicts the assertion that g is an extreme point of W_0, since

$$g = (g + h)/2 + (g - h)/2.$$

This means that g is a measurable function of the form χ_H and $T(g) = \beta = (\mu_1(H), \mu_2(H), \ldots, \mu_n(H))$. This proves the claim and completes the induction step. The reader can now furnish the proof for $n = 1$ (following the induction step). ∎

Problems

× **6.6.1.** Let $T: X \to Y$ be a linear map from a TVS X into a locally convex TVS Y. Suppose that the topology of Y is given by a family $\{p_\beta: \beta \in B\}$ of seminorms. Prove that T is continuous if and only if $p_\beta \circ T$ is continuous for each $\beta \in B$.

6.6.2. Give examples of two distinct locally convex topologies on a vector space X with the same class of continuous linear functionals.

× **6.6.3.** Show that a locally convex TVS is Hausdorff if and only if the family of seminorms $\{p_\beta: \beta \in B\}$ generating its topology is total, i.e., $x \in X$ is 0 whenever $p_\beta(x) = 0$ for each $\beta \in B$.

× **6.6.4.** Prove that a maximal subspace of a TVS is either closed or dense.

× **6.6.5.** Prove that a TVS is regular. [Hint: Observe that for any subset B, $\bar{B} = \cap \{B + U: U \text{ is open and } 0 \in U\}$.]

× **6.6.6.** Let X be a TVS. Show that X is T_3 if and only if $\{0\}$ is closed in X if and only if given $x \neq 0$, there exists an open set U such that $0 \in U$ and $x \notin U$.

× **6.6.7.** *Bounded sets in a TVS.* A set S in a TVS X is called bounded if for every open set U containing 0, there exists $\varepsilon > 0$ such that $\varepsilon S \subset U$. Prove the following assertions:

(i) A compact set in a TVS is bounded.
(ii) The closure of a bounded set in a TVS is bounded.
(iii) In a locally convex TVS, the convex hull of a bounded set is bounded.

× **6.6.8.** Prove that if a TVS has a bounded convex open set containing 0, then its topology is given by a seminorm. (Hint: Use the seminorm p_V, where V is a bounded, balanced, and convex open set containing 0.)

× **6.6.9.** Let X be a locally convex TVS and $\{p_\beta: \beta \in B\}$ is a family of seminorms generating its topology. Prove that a subset $S \subset X$ is bounded if and only if $p_\beta(S)$ is bounded for each $\beta \in B$.

6.6.10. Consider the space l_p, $0 < p < 1$. Notice that $\| x \| = (\sum_{i=1}^\infty | x_i |^p)^{1/p}$ does not define a norm, since for $0 < p < 1$, $\| x + y \|$ may exceed $\| x \| + \| y \|$. However, if we define

$$d(x, y) = \sum_{i=1}^{\infty} | x_i - y_i |^p,$$

where $x = (x_i)$ and $y = (y_i)$ are in l_p, then (l_p, d) becomes a complete metric TVS. Show that a set in l_p is bounded as a bounded set in a TVS if and only if it is bounded in the metric d. Also show that the convex hull of $\{x: d(0, x) \leq 1\}$ is not bounded in d. [Hint: The sequence $(1/n, 1/n, \ldots, 1/n, 0, 0, \ldots) = x_n$, the first n entries of x_n being $1/n$, is not bounded in d. By Problem 6.6.7, (l_p, d), $0 < p < 1$, is not locally convex.]

6.6.11. Let X be a vector space. Suppose that d_1 and d_2 are two complete metrics such that (X, d_1) and (X, d_2) are topological vector spaces. If d_1 is stronger than d_2, then show that they are equivalent. (Hint: Use the open mapping theorem.)

6.6.12. Show that the product of an arbitrary family of topological vector spaces, equipped with the product topology, is a TVS.

6.6.13. Show that an infinite product of Hausdorff topological vector spaces (each of positive dimension) cannot have a bounded absorbing set. [Hint: Let $X = \prod_{\beta \in B} X_\beta$ and $\pi_\beta: X \to X_\beta$ be the projection on X_β. If $A \subset X$ is absorbing, then $\pi_\beta(A) = X_\beta$ for all but finitely many β.]

6.6.14. Let X be a vector space over F and p a seminorm on X. Prove the following assertions:

(i) The sets $A = \{x: p(x) \leq 1\}$ and $B = \{x: p(x) < 1\}$ are balanced and convex.
(ii) If p' is another seminorm on X and $\{x: p(x) \leq 1\} = \{x: p'(x) \leq 1\}$, then $p = p'$. {Hint: Let $\varepsilon > 0$ and $y = [p(x) + \varepsilon]^{-1}x$.}
(iii) If X is a TVS and p is continuous at 0, then B is the interior of A and $A = \bar{B}$.

× **6.6.15.** Let $T: X \to Y$ be a linear map from a TVS X into a TVS Y. Show that if $T(V)$ is bounded for some open set V containing 0, then T is continuous. Further show that if Y has an open and bounded set containing 0, then $T(V)$ is bounded for some open set V containing 0, whenever T is continuous.

6.6.16. Prove that the product of infinitely many normed linear spaces is not a normed linear space.

6.6.17. Let X be a TVS. Prove the following assertions:

(i) A seminorm p on X is continuous if and only if $\{x: p(x) < 1\}$ is open.

(ii) Suppose that a family $\mathscr{F}$ of seminorms generate the topology of X. Then a seminorm p on X is continuous if and only if there exists $M > 0$ and $p_1, p_2, \ldots, p_n$ in $\mathscr{F}$ such that

$$p(x) \leq M \max\{p_i(x): 1 \leq i \leq n\},$$

for all $x \in X$.

(iii) Let X be as in (ii). Then the family of continuous seminorms on X also generates the topology of X.

6.6.18. Let X be a TVS and M a linear subspace of X. Then the set $X/M = \{x + M: x \in X\}$ of cosets of M is a vector space over F if we define vector addition and scalar multiplication by

$$(x + M) + (y + M) = x + y + M, \qquad \beta(x + M) = \beta x + M.$$

Let $\pi: X \to X/M$ be defined by $\pi(x) = x + M$. Then π is linear and onto. Let X/M be given the quotient topology, i.e., the largest topology on X/M such that π is continuous. Prove the following assertions:

(i) X/M is a TVS.

(ii) X/M is Hausdorff if and only if M is closed.

(iii) π is open.

(iv) A local base $\mathscr{B}$ of 0 in X generates a local base $\{\pi(U): U \in \mathscr{B}\}$ of 0 in X/M.

6.6.19. Let X be a complete metrizable locally convex TVS and $\mathscr{F}$ be a pointwise bounded family of continuous seminorms on X. Show that there exists an open neighborhood U of 0 such that $U \subset \{x: p(x) \leq 1$ for each $p \in \mathscr{F}\}$. [Hint: Let $V = \{x: p(x) \leq 1$ for each $p \in \mathscr{F}\}$. Then V is closed, balanced, and convex. V is also absorbing since given $x \in X$, $x \in rV$, where $r > \sup\{p(x): p \in \mathscr{F}\}$. Therefore, V cannot be nowhere dense. Note that $\frac{1}{2}V - \frac{1}{2}V = V$.]

6.6.20. Prove that the product of an arbitrary number of locally convex spaces is locally convex.

6.6.21. Let X be a TVS and Y, Z linear subspaces of X such that $X = Y + Z$ and $Y \cap Z = \{0\}$. Note that each x in X has a unique representation $x = y + z$, where $y \in Y$ and $z \in Z$. Prove that X is the topological direct sum of Y and Z (that is, the mappings $(y, z) \to y + z$ and $x = y + z \to (y, z)$, where $y \in Y$ and $z \in Z$, are continuous when $Y \times Z$ is given the product topology) if and only if the mapping $y + z \to y$ is continuous, and in this case, Z is called the topological supplement of Y.

6.6.22. If X is a Hausdorff TVS over F with finite dimension d, then show that every vector space isomorphism $f\colon F^d \to X$ is also a homeomorphism.

× **6.6.23.** Let X be a Hausdorff TVS over F. Show that X is finite dimensional if and only if X is locally compact.

6.6.24. Let X be a locally convex TVS over F. Consider the product space $F' = \prod_{\beta \in B} F_\beta$, where each F_β is F. Show that every continuous linear mapping $f\colon Y \to F'$, where Y is a linear subspace of X, can be extended to a continuous linear mapping from X into F'. (Hint: Let $\pi_\beta\colon F' \to F_\beta$ be the coordinate projection so that $\pi_\beta \circ f\colon Y \to F$ is a continuous linear functional on Y.)

6.6.25. Let Y be a linear subspace of a TVS X. Prove that Y possesses a topological supplement (see Problem 6.6.21) if and only if there is a continuous linear map $p\colon X \to X$ such that $p \circ p = p$ and $p(X) = Y$.

6.6.26. Let X be a locally convex Hausdorff TVS over F. Prove that every finite-dimensional linear subspace Y of X has a topological supplement. (Hint: Let $f\colon Y \to F^n$ be a linear homeomorphism. Use Problem 6.6.24 to extend f to a continuous linear map f_0 on X. Then $p = f^{-1} \circ f_0$: $X \to X$ is a continuous linear map with range Y such that $p \circ p = p$.)

× **6.6.27.** Let $f\colon X \to F$ be a nonzero linear functional on a TVS X over F. Prove that f is continuous if and only if $f^{-1}(0)$ is closed if and only if $f^{-1}(0)$ is not dense.

6.6.28. Let X be a TVS over the reals and U, V be subsets such that U has nonempty interior. Suppose that U and V are separated by a nonzero linear functional f on X. Prove that f is continuous. [Hint: Use Problem 6.6.27. If $U \subset \{x: f(x) \leq r\}$ and $V \subset \{x: f(x) \geq r\}$, then $f^{-1}(r) \cap (\text{Int } U) = \varnothing$.]

6.6.29. Do Problems 6.2.24 and 6.2.25 when X is a TVS over the reals.

6.6.30. Prove that the set of extreme points of a compact convex set in R^2 is closed. Show that this is not always true in R^3.

★ **6.6.31.** Show that in a uniformly convex normed linear space the set of extreme points of the closed ball $\{x: \| x - x_0 \| \leq r\}$ is the set $\{x: \| x - x_0 \| = r\}$, which is its surface.

6.6.32. Let X be a vector space over the reals and let $\mathscr{B}$ be defined by

$$\mathscr{B} = \{V \subset X: 0 \in V, V \text{ is convex such that for any } x \in X, \text{ there exists } \beta > 0 \text{ such that } \beta x \in V\}.$$

Prove the following assertions:

(i) $\mathscr{B}$ is a local base for a topology on X, making X a locally convex TVS.

(ii) The topology induced by $\mathscr{B}$ is stronger than any other locally convex topology on X.

6.6.33. Recall that c is the space of all real sequences (x_n) such that $\lim x_n$ exists, endowed with the norm $\| x_n \| = \sup | x_n |$, and c_0 is the subspace of sequences converging to zero. Show that the unit ball of c_0 does not have extreme points, whereas the unit ball of c has exactly two extreme points, namely, $(1, 1, 1, \ldots)$ and $(-1, -1, -1, \ldots)$.

6.6.34. Use the Krein–Milman theorem to show that neither c nor c_0 can be isometrically isomorphic to the dual of any normed linear space.

6.6.35. Let $1 < p < \infty$. Show that the set of extreme points of the closed unit ball of $L_p[0, 1]$ is $\{f: \| f \|_p = 1\}$. (See Problem 6.6.31.)

6.6.36. Show that the closed unit ball of $L_1[0, 1]$ has no extreme points. [Hint: If f has zero norm, then f is not an extreme point. Since $f = \frac{1}{2}(2 - \| f \|_1)f + \frac{1}{2} \| f \|_1 f$, the norm of an extreme point must be 1. Let $\| f \|_1 = 1$ and $f = f\chi_A + f\chi_B$, where $m(A) > 0$, $m(B) > 0$, and $A \cup B = \{x: f(x) \neq 0\}$. Notice that $f = \beta f_1 + (1 - \beta) f_2$, where $\beta = \| f\chi_A \|_1$, $f_1 = f\chi_A / \| f\chi_A \|_1$ and $f_2 = f\chi_B / \| f\chi_B \|_1$.]

6.6.37. Prove that the set of extreme points of the closed unit ball of $L_\infty[0, 1]$ is the set $\{f: | f(x) | = 1 \text{ a.e. in } [0, 1]\}$.

6.6.38. Find the extreme points of the closed unit ball of l_p, $1 \leq p \leq \infty$.

★ **6.6.39.** Find the extreme points of the closed unit ball of $C(S)$, the Banach space of real continuous functions with supremum norm, where S is a compact Hausdorff space.

6.6.40. *Totally Bounded Subsets in a TVS.* Let X be a Hausdorff TVS. A subset A is called totally bounded if for every open neighborhood V of 0, there exists a finite subset $\{x_1, x_2, \ldots, x_n\}$ of A such that $A \subset \bigcup_{i=1}^n (x_i$

$+ V$). Prove the following assertions:

(i) A totally bounded subset of X is bounded.

(ii) Let Y be a TVS, Z a proper closed subspace of Y, and U a bounded open neighborhood of 0. Let V be a balanced open neighborhood of 0 such that $V + V \subset U$. Then U is not contained in $Z + V$. {Hint: Suppose that $U \subset Z + V$. Then $V + V \subset U \subset Z + V$ implies that $2^n V \subset Z + V$ for each positive integer n. Then $V \subset \bigcap_{m \geq 1} [Z + (1/m)V] = Z$.}

(iii) Let $0 \in U \subset A$, where U is open and A is a totally bounded subset of X. Then X is finite dimensional. [Hint: Let X be infinite dimensional. Use (ii) to construct an infinite sequence (x_n) in A such that $x_{n+1} \notin V + W_n$, where V is a balanced open subset, $0 \in V$, $V + V \subset A$, and W_n is the subspace spanned by $\{x_1, x_2, \ldots, x_n\}$. Then $x_n - x_m \notin V$ if $n \neq m$. Let V' be a balanced open neighborhood of 0 such that $V' + V' \subset V$. There exist $y_1, y_2, \ldots, y_k$ such that $A \subset \bigcup_{i=1}^k (y_i + V')$. But then there are m, n such that $x_n - x_m \in V' + V' \subset V$.]

(iv) Every subset of a totally bounded subset A of X is totally bounded.

(v) Let $A_1, A_2, \ldots, A_n$ be a finite number of compact convex subsets of X. Then the convex hull of $A = \bigcup_{i=1}^n A_i$ is also compact. [Hint: Let $\beta_j \geq 0$, $\sum_{j=1}^n \beta_j = 1$, and $x_i \in A_i$ for $1 \leq i \leq n$. Consider the mapping $(\beta_1, \ldots, \beta_n, x_1, \ldots, x_n) \to \sum_{i=1}^n \beta_i x_i$, which is continuous as a map from a compact space into a Hausdorff space.]

(vi) Let X be a locally convex Hausdorff TVS and A a totally bounded subset. Then the convex hull of A is also totally bounded. [Hint: Let $0 \in V$, V open and convex. Let $A \subset \bigcup_{i=1}^n (x_i + V)$. If $y \in \text{co}(A)$, then there exists $z \in \text{co}\{x_1, x_2, \ldots, x_n\} \equiv K$, say, such that $y - z \in V$. Thus, $\text{co}(A) \subset K + V$, and by (v), K is compact.]

(vii) Let A and X be as in (vi). Then the balanced hull of A, i.e., the set $\{\beta x : x \in A \text{ and } |\beta| \leq 1\}$ is also totally bounded.

6.7. The Kakutani Fixed Point Theorem and the Haar Measure on a Compact Group

The theory of Haar measure is an important branch of measure theory and constitutes an extremely useful generalization of the theory of Lebesgue measure.

The Haar measure is a translation-invariant measure on a locally compact topological group [i.e., an algebraic group with locally compact topology where the mappings $(x, y) \to x \cdot y$ and $x \to x^{-1}$ are continuous]. The foundations of the theory of topological groups were laid around 1926–1927 by O. Schreier and F. Leja. To study the structure of certain topological groups, D. Hilbert in 1900 posed the following problem (now famous as Hilbert's fifth problem—fifth in the list of 23 problems he posed at the International Congress of Mathematics): Is every topological group that is locally Euclidean (i.e., every point has an open neighborhood homeomorphic to an open subset of R^n) necessarily a Lie group [i.e., a manifold that is a group and where the mappings $(x, y) \to xy$ and $x \to x^{-1}$ are analytic]? In 1933 A. Haar took a fundamental step towards the solution of this problem. He established the existence of a translation-invariant measure (now known as the Haar measure) on a second countable locally compact topological group. Soon after, in the same year, von Neumann utilized Haar's result and solved Hilbert's fifth problem in the affirmative for compact locally Euclidean groups. He also proved the uniqueness of the Haar measure in 1934, and later on in 1940 A. Weil extended Haar's result to all locally compact topological groups. Hilbert's fifth problem was solved completely (in 1952) in the affirmative by A. Gleason, D. Montgomery, and L. Zippin.

In this section we will present a fixed point theorem of S. Kakutani[†] and utilize it to prove the existence of the Haar measure on a compact topological group.

First we need a definition.

Definition 6.25. A family $\mathscr{F}$ of linear operators on a topological vector space X is called *equicontinuous* on a subset K of X if for every open set V with $0 \in V$ there is an open U with $0 \in U$ such that if $k_1, k_2 \in K$ and $k_1 - k_2 \in U$, then $\mathscr{F}(k_1 - k_2) \subset V$, i.e., $T(k_1 - k_2) \in V$ for every $T \in \mathscr{F}$. ▮

Theorem 6.37. *The Kakutani Fixed Point Theorem.* Let K be a nonempty compact convex subset of a locally convex Hausdorff space X, and let $\mathscr{F}$ be a group of linear operators on X. Suppose $\mathscr{F}$ is equicontinuous on K and $\mathscr{F}(K) \subset K$.[‡] Then there is $x \in X$ such that $T(x) = x$ for every $T \in \mathscr{F}$. ▮

Proof. By Zorn's Lemma there is a minimal nonempty compact convex set K_1 such that $\mathscr{F}(K_1) \subset K_1$. If K_1 is a singleton, there is nothing to

[†] S. Kakutani, *Proc. Imp. Acad. Tokyo* **14**, 242–245 (1938).

[‡] $\mathscr{F}(K) = \cup \{T(K): T \in \mathscr{F}\}$.

prove. If K_1 is not a singleton, we will reach a contradiction to the minimality of K_1, and the theorem will follow.

Suppose there are $z_1 \neq z_2$, $z_1, z_2 \in K_1$. Let $y = z_1 - z_2$. Then there are open sets U, V with $0 \in U \cap V$ such that V is convex and balanced,

$$y \notin \bar{V}, \tag{6.34}$$

and

$$\mathscr{F}(k_1 - k_2) \subset V \text{ if } k_1, k_2 \in K_1 \text{ and } k_1 - k_2 \in U. \tag{6.35}$$

Let $U_0 = \text{co}(\mathscr{F}(U))$. Then U_0 is open. Using the group property of $\mathscr{F}$, the reader can verify that

$$\mathscr{F}(U_0) = U_0 \quad \text{and} \quad \mathscr{F}(\bar{U}_0) = \bar{U}_0.$$

Since $\mathscr{F}$ contains the identity operator, $U \subset U_0$ and $\inf\{t\colon t > 0$ and $K_1 - K_1 \subset tU_0\} = t_0 < \infty$. Let $W = t_0 U_0$. Then it is clear that, for each ε in $(0, 1)$,

$$K_1 - K_1 \not\subset (1 - \varepsilon)\bar{W} \tag{6.36}$$

and

$$K_1 - K_1 \subset (1 + \varepsilon)W. \tag{6.37}$$

Since $0 \in W$, $K_1 \subset \bigcup_{k \in K_1}(k + \frac{1}{2}W)$; and therefore by the compactness of K_1, there exist $k_i \in K_1$ with $1 \leq i \leq n$ such that

$$K_1 \subset \bigcup_{i=1}^{n} (k_i + \tfrac{1}{2}W). \tag{6.38}$$

Now we claim that

$$K_2 \equiv K_1 \cap \left\{ \bigcap_{k \in K_1} \left[k + \left(1 - \frac{1}{4n}\right)\bar{W} \right] \right\}$$

is a *nonempty* compact convex set satisfying

(i) $\mathscr{F}(K_2) \subset K_2$,

(ii) $K_2 \neq K_1$.

This claim, once verified, will prove the theorem since K_2 will contradict the minimality of K_1. We prove the "nonempty" assertion last.

To prove (i), let $z \in K_2$ and $k \in K_1$. Since for $T \in \mathscr{F}$, $T^{-1}(K_1) \subset K_1$, then $k = T(k_0)$ for some $k_0 \in K_1$. Also $z \in k_0 + (1 - 1/4n)\bar{W}$ or $T(z) \in k + (1 - 1/4n) \cdot \bar{W}$, thus proving (i). Now to prove (ii), by (6.36) there are $x, y \in K_1$ such that $x - y \notin (1 - 1/4n)\bar{W}$. This means that $x \notin y +$

$(1 - 1/4n)\overline{W}$ or $x \notin K_2$, thus proving (ii). To prove $K_2 \neq \varnothing$, we show that $p = (1/n)\sum_{i=1}^{n} k_i$ [k_i's as in (6.38)] belongs to K_2. Clearly $p \in K_1$, since K_1 is convex. Let $y \in K_1$. Then by (6.38)

$$y \in k_j + \tfrac{1}{2}W, \tag{6.39}$$

for some j, where $1 \leq j \leq n$. Also, by (6.37), for each $i \neq j$ where $1 \leq i \leq n$,

$$y \in k_i + (1 + 1/4n)W. \tag{6.40}$$

By (6.39) and (6.40), we have

$$p \in \frac{1}{n}\left\{(y - \tfrac{1}{2}W) + (n-1)\left[y - \left(1 + \frac{1}{4n}\right)W\right]\right\}$$
$$= y - \left(1 - \frac{1}{4n} - \frac{1}{4n^2}\right)W,$$

or $p \in y + (1 - 1/4n)W$ (since $W = -W$), proving that $p \in K_2$. The proof of the theorem is now complete. ∎

Before we go into the existence of the Haar measure, let us consider a few examples of topological groups. First, a definition and a few basic remarks.

Definition 6.26. A *topological group* is a group G with a topology such that the mappings $(x, y) \to xy$ from $G \times G$ into G and $x \to x^{-1}$ from G into G are continuous. Clearly, the continuity requirements in a topological group are equivalent to the requirement that $(x, y) \to xy^{-1}$ is continuous. ∎

Remarks

6.30. In a topological group the mappings $x \to zx$, $x \to xz$, and $x \to x^{-1}$ are all homeomorphisms.

6.31. Suppose a topological group G has the property (T_0): if $x \neq y$ and $x, y \in G$, then there is an open set V containing one of them but excluding the other. Note that $(T_1) \Rightarrow (T_0)$ (see Chapter 1, Section 1.4). But curiously enough, the following are equivalent in G:

(a) G is a T_0-space.
(b) G is a T_1-space.
(c) G is a T_2-space.
(d) $\cap\{U\colon U \text{ open and } e \in U\} = \{e\}$, where e is the identity of G.

Proof. (a) $\Rightarrow$ (b). Let $x \neq y$ and $x \in V$ open and $y \notin V$. Then $e \in x^{-1}V = W$ [open by (i)]. Now if $U = W \cap W^{-1}$, then $y \in yU$ (open). Now if $x \in yU$; then $x^{-1} \in Uy^{-1}$ (since $U = U^{-1}$), $x^{-1} \in x^{-1}Vy^{-1}$, or $y \in V$ (which is a contradiction). Hence $x \in yU$. Hence (b) is true.

(b) $\Rightarrow$ (c). Let $x \neq y$. Then by (b), $y \in V = G - \{x\}$ (open). Hence $e \in y^{-1}V$ (open). Since G is a topological group, there is open W with $e \in W$ and $WW^{-1} \subset y^{-1}V$. Let $U = G - \overline{yW}$. If $x \in \overline{yW}$, then $xW \cap yW \neq \varnothing$ and so $x \in yWW^{-1} \subset V$, which a contradiction. Hence $x \notin \overline{yW}$ and $x \in U$. Hence (c).

The proofs of (c) $\Rightarrow$ (d) $\Rightarrow$ (a) are easy and are left to the reader. ∎

6.32. A Hausdorff topological group is completely regular. A first countable topological group is metrizable. The proofs of these facts are somewhat involved and are omitted. The reader might consult Pontrjagin's text.†

6.33. Every real-valued continuous function f with *compact support* defined on a topological group G is *uniformly continuous* [i.e., given $\varepsilon > 0$, there exists an open set V with $e \in V$ such that $|f(x) - f(y)| < \varepsilon$ whenever $x^{-1}y \in V$].

Proof. Suppose K is the compact support of f. By using the compactness of K and the continuity of $(x, y) \to xy$, we can find for $\varepsilon > 0$ an open set V_1 with $e \in V_1$ such that

$$|f(x) - f(xy)| < \varepsilon, \tag{6.41}$$

whenever $x \in K$ and $y \in V_1$. Now for each $x \in A$, where

$$A = \{y: |f(y)| \geq \varepsilon\} \subset \text{the interior of } K,$$

let V_x be an open set with $e \in V_x$ such that $xV_x \subset K$. Let $e \in W_x$ (open) such that $W_xW_x \subset V_x$. By the compactness of A there exist $x_1, x_2, \ldots, x_n \in A$ such that $A \subset \bigcup_{i=1}^n x_iW_{x_i}$ or $AW \subset K$, where $W = \bigcup_{i=1}^n W_{x_i}$. Therefore if $x \notin K$ and $y \in W^{-1}$, then

$$|f(x) - f(xy)| = |f(xy)| < \varepsilon. \tag{6.42}$$

From (6.41) and (6.42), taking $V = V_1 \cap W^{-1}$, we have for every $x \in G$ and $y \in V$,

$$|f(x) - f(xy)| < \varepsilon.$$

The proof of Remark 6.33 is now clear. ∎

† L. Pontrjagin, *Topological Groups*, Princeton University Press, Princeton, New Jersey (1939).

6.34. Let G be a compact topological group and $C(G)$ be the real-valued continuous functions on G with "sup" norm. Let

$$\mathscr{F}_f = \left\{ \sum_{i=1}^{n} a_i f(s_i x) : 0 \leq a_i \leq 1, \ \sum_{i=1}^{n} a_i = 1, \ s_i \in G, \text{ and } n \in N \right\}.$$

In other words, $\mathscr{F}_f$ is the convex hull of the set of all left translates of f [in $C(G)$]. Then $\bar{\mathscr{F}}_f$ is a compact subset of $C(G)$.

Proof. Suffice it to show that $\mathscr{F}_f$ is equicontinuous, since then the compactness of $\bar{\mathscr{F}}_f$ will follow by Arzela–Ascoli's Theorem (Chapter 1, Section 1.6). Since G is compact, by Remark 6.33, there is an open V with $e \in V$ and $x^{-1}y \in V$ such that $| f(sx) - f(sy) | < \varepsilon$ for every $s \in G$. The equicontinuity of $\mathscr{F}_f$ now follows. ∎

Examples

6.18. $R - \{0\}$ with the usual relative topology of R is a locally compact topological group under multiplication. R itself is also so under addition, and the Lebesgue measure is a translation-invariant measure on R.

6.19. The complex numbers of absolute value 1 is a compact topological group under multiplication and usual topology of R^2.

6.20. Let S be the set of all matrices of the form $\begin{pmatrix} x & y \\ 0 & 1 \end{pmatrix}$, where $x > 0$ and y is any real number. Then S is a locally compact topological group with the usual matrix multiplication and with the topology induced by the usual topology of R^2 (in an obvious manner).

Theorem 6.38. There exists a unique regular probability measure μ [i.e., $\mu(G) = 1$] on every compact Hausdorff topological group G such that for every $s \in G$,

(i) $\displaystyle\int f_s \, d\mu = \int f \, d\mu,$

(ii) $\displaystyle\int f^s \, d\mu = \int f \, d\mu,$

(iii) $\displaystyle\int f_{-1} \, d\mu = \int f \, d\mu,$

where $f \in C(G)$, $f_s(x) = f(sx)$, $f^s(x) = f(xs)$, and $f_{-1}(x) = f(x^{-1})$. This μ is called the *Haar measure* on G. ∎

Proof. If L_s: $f \to f_s$, then for $s \in G$, L_s is an isometry from $C(G)$ (with "sup" norm) into itself. Therefore, $\{L_s : s \in G\}$ is an *equicontinuous* group of linear operators on the Banach space $C(G)$. From Remark 6.34, $\bar{\mathscr{F}}_f$ is a compact convex subset of $C(G)$, and for $s \in G$, $L_s(\bar{\mathscr{F}}_f) \subset \bar{\mathscr{F}}_f$. By Theorem 6.37, there exists $f_0 \in \bar{\mathscr{F}}_f$ such that $f_0(xs) = f_0(x)$ for every x, $s \in G$; and therefore $f_0(x) = f_0(e)$ or f_0 is a constant c *which can be uniformly approximated by convex combinations of left translates of f*. If c' is a similar constant corresponding to the right translates of f, and c'' is a similar constant corresponding to the left translates of f_{-1}, we claim that $c = c' = c''$. To prove this, let $\varepsilon > 0$. Then there are $0 \leq a_i,\ b_j,\ c_k \leq 1$ with $\sum_{i=1}^n a_i = \sum_{j=1}^m b_j = \sum_{k=1}^p c_k = 1$ such that

$$\left| c - \sum_{i=1}^{n} a_i f(s_i x) \right| < \varepsilon, \tag{6.43}$$

$$\left| c' - \sum_{j=1}^{m} b_j f(x t_j) \right| < \varepsilon, \tag{6.44}$$

$$\left| c'' - \sum_{k=1}^{p} c_k f_{-1}(w_k x) \right| < \varepsilon, \tag{6.45}$$

for all $x \in G$ and some (s_i), (t_j), and $(w_k) \in G$. In the inequality (6.43), writing t_j for x, multiplying by b_j, and then summing over j, we have

$$\left| c - \sum_{i\ j} a_i b_j f(s_i t_j) \right| < \varepsilon. \tag{6.46}$$

Similarly, inequalities (6.44) and (6.45) can be written as

$$\left| c' - \sum_{i,j} a_i b_j f(s_i t_j) \right| < \varepsilon, \tag{6.47}$$

and

$$\left| c'' - \sum_{i,k} a_i c_k f(s_i w_k^{-1}) \right| < \varepsilon. \tag{6.48}$$

From inequalities (6.46) and (6.47), $c = c'$; also by writing (6.43) in a form similar to (6.48), we get $c = c''$. This argument also shows that there can be *only one* constant obtained as above corresponding to the translates of f or f_{-1}. If we call this constant $I(f)$, then we have

(A) $I(1) = 1$,
(B) $I(af) = aI(f)$, for all reals a,
(C) $I(f) \geq 0$, for $f \geq 0$,
(D) $I(f_s) = I(f^s) = I(f_{-1})$, $s \in G$,
(E) $I(f + g) = I(f) + I(g)$.

We establish only (E). Let $\varepsilon > 0$. Then, as in (6.43), we get

$$\left| I(f) - \sum_{i=1}^{n} a_i f(s_i x) \right| < \varepsilon, \qquad x \in G, \tag{6.49}$$

where $\sum_{i=1}^{n} a_i = 1$, $0 \leq a_i$, and $s_i \in G$. Let $h(x) = \sum_{i=1}^{n} a_i g(s_i x)$. Then $h \in \mathscr{P}_g$ and $\mathscr{P}_h \subset \mathscr{P}_g$; therefore, since each of these sets contains a unique constant function, $I(g) = I(h)$. Hence there are $b_j > 0$, $\sum_{j=1}^{m} b_j = 1$, and $t_j \in G$ such that

$$\left| I(g) - \sum_{j=1}^{m} b_j h(t_j x) \right| < \varepsilon. \tag{6.50}$$

By replacing h with g, we have

$$\left| I(g) - \sum_{i,j} a_i b_j g(s_i t_j x) \right| < \varepsilon. \tag{6.51}$$

Writing (6.49) as

$$\left| I(f) - \sum_{i,j} a_i b_j f(s_i t_j x) \right| < \varepsilon, \tag{6.52}$$

we have by inequalities (6.51) and (6.52)

$$\left| I(f) + I(g) - \sum_{i,j} a_i b_j (f + g)(s_i t_j x) \right| < 2\varepsilon, \tag{6.53}$$

thus proving (E).

By the Riesz Representation Theorem (Theorem 5.10), there is a unique regular Borel measure μ satisfying $I(f) = \int f \, d\mu$ for all $f \in C(G)$. Since $I(1) = 1$, $\mu(G) = 1$. All the properties of μ in the theorem now follow from property (D) above. ∎

Problems

6.7.1. Suppose μ is the Haar measure in a compact group G. Show that

(i) $\mu(V) > 0$ for every open set $V(\neq \varnothing)$, and

(ii) $\mu(Bx) = \mu(xB) = \mu(B^{-1})$ for every Borel set B and $x \in G$.

6.7.2. Suppose μ is a weakly Borel measure (possibly infinite) on a Hausdorff topological group G such that

(i) $\mu(xK) = \mu(K)$ for all compact sets K and $x \in G$, and

(ii) $0 < \mu(V) < \infty$ for some open set V with compact closure.

Prove that (a) G is locally compact, and (b) G is compact if $\mu(G) < \infty$.

(This means that it is impossible to have any meaningful translation-invariant measure on a non-locally-compact topological group.)

6.7.3. Let H be the component containing the identity of a topological group G. Prove that H is a subgroup of G such that $x^{-1}Hx = H$ for all $x \in G$.

6.7.4. Prove that every open subgroup of a topological group is closed.

6.7.5. Suppose G is a group with first countable topology such that

(i) $(x, y) \rightarrow xy$ is separately continuous, and

(ii) for any two compact sets A and B the set $AB^{-1} = \bigcup_{x \in B} \{y: yx \in A\}$ is compact.

Prove that G is a topological group.

6.7.6. Suppose that G is a group with a metric topology with property (i) of Problem 6.7.5 and the property $d(x, y) = d(xz, yz)$ for any x, y, and $z \in G$. Then prove that G is a topological group. [Here it is relevant to mention a beautiful result of R. Ellis: Suppose G is a group with locally compact Hausdorff topology such that $(x, y) \rightarrow xy$ is separately continuous. Then G is a topological group. For a proof, see his paper.[†]]

6.7.7. Let G be a Hausdorff topological group and let μ be a weakly Borel measure such that

(i) $\mu(Kx) = \mu(K)$ for every compact set K and $x \in G$,

(ii) $0 < \mu(V) < \infty$ for some open set V, and

(iii) $\mu(\{y\}) > 0$ for some $y \in G$.

Prove that G is discrete.

6.7.8. *Subsemigroups with Nonempty Interiors in a Compact Topological Group.* Let H be a subsemigroup of a compact topological group G such that H has a nonempty interior. Show that H is a compact subgroup of G. [Hint: Let S be the interior of H. Then $\bar{S}$ is a subsemigroup. If μ is the Haar measure of G, then for $x \in \bar{S}$ and any open set V, $\mu(x^{-1}V \cap \bar{S}) = \mu(V \cap \bar{S}) = \mu(Vx^{-1} \cap \bar{S})$. This means that $\bar{S}x = x\bar{S} = \bar{S}$, or $\bar{S}$ is a compact subgroup. Observe now that $S = \bar{S}$, since for $y \in \bar{S}$, $yS^{-1} \cap S \neq \varnothing$.]

6.7.9. Use Problem 6.7.8 to prove that every locally compact subsemigroup of positive Haar measure in a compact topological group is a compact subgroup. [Note: This result remains true without the requirement of "positive Haar measure." Actually, using an important result of K. Numakura (that a cancellative semigroup with compact topology and

[†] R. Ellis, *Duke Math. J.* **24**, 119–125 (1957).

jointly continuous multiplication is a topological group), it is very easy to do Problems 6.7.8 and 6.7.9. The reader should try this.]

6.7.10. *A Fixed Point Theorem for Affine Maps.* Let T be a continuous map from a compact convex subset K of a normed linear space X into itself. Suppose that T is affine, i.e., for $0 \leq a \leq 1$ and $x, y \in K$, $T(ax + (1-a)y) = aT(x) + (1-a)T(y)$. Then T has a fixed point. [Hint: For $x \in K$ and $x_n = (1/n)\sum_{k=0}^{n-1} T^k(x)$, observe that $\| T(x_n) - x_n \| \to 0$ as $n \to \infty$.]

6.7.11. *The Markov–Kakutani Fixed Point Theorem.* Let $\mathscr{F}$ be a family of continuous maps from a compact convex subset K of a normed linear space X into itself such that for any T and S in this family, T is affine and $T(S(x)) = S(T(x))$ for all $x \in K$. Then $\mathscr{F}$ has a common fixed point. [Hint: For $T \in \mathscr{F}$, the set $K_T = \{x \in K: T(x) = x\}$ is nonempty by Problem 6.7.10. If $S \in \mathscr{F}$, then $S: K_T \to K_T$ and there exists $x \in K_T$ such that $S(x) = x$. An induction argument shows that every finite subfamily of $\mathscr{F}$ has a common fixed point. Now use the finite intersection property of the compact sets of common fixed points of the finite subfamilies of $\mathscr{F}$.] This theorem was first proven by A. Markov in 1936. In 1938, S. Kakutani gave an alternative proof of this theorem, along with an extension in the noncommutative case.

Apply the above fixed point theorem to show the existence of Banach limits on l_∞. For a discussion of Banach limits, see Problem 6.2.16.

6.7.12. *The Schauder Fixed Point Theorem.* It T is a continuous map from a compact convex subset of a Banach space into itself, then T has a fixed point.

The proof of this theorem is difficult and depends on the Brouwer theorem in Chapter 1. For a proof, the reader might consult [11]. Around 1930 J. Schauder first proved this theorem. In 1935, A. Tychonoff proved this theorem in the more general context of locally convex Hausdorff spaces.

This theorem is useful for various applications in differential and integral equations. Using this theorem, the existence part of the initial-value problem considered in Chapter 1, Theorem 1.24 can be proven easily assuming only the continuity of the function f and requiring no Lipschitz condition. Demonstrate this.

6.7.13. (*Johnson*). Let $Y = [0, \Omega]$ with the order topology and ν be the nonregular measure on Y as in Problem 5.3.13 or Example 5.11. Let A be a discrete group with two elements and $X = \times\{X_\lambda: \lambda \in [0, \Omega)\}$, where $X_\lambda \equiv A \ \forall \lambda$, with the product topology. Let μ be the Haar measure on the compact group X (with coordinatewise multiplication). Prove that $M = \{(x, y): x \in X, y \in Y \text{ and } x_\lambda = e \text{ if } \lambda \geq y\}$ is a compact set such that $\nu(M_x)$ is not measurable with respect to the completion of μ.

7

Hilbert Spaces

In this chapter we will study aspects of the theory of Hilbert spaces. Roughly we may say that a Hilbert space is a Banach space whose norm is defined in a particular manner. We shall give a characterization in terms of the norm of those Banach spaces that are actually Hilbert spaces. This well-known result (Proposition 7.2) is due to Jordan and von Neumann.

Infinite-dimensional Hilbert spaces are natural generalizations of the finite-dimensional spaces R^n and C^n with the usual "Euclidean norms." Their study was initiated in the early 1900's by Hilbert, who studied the particular spaces l_2 and L_2. The abstract axiomatization of Hilbert space was later given by von Neumann in the separable case in the 1920's,† and in general by Löwig‡ and Rellich.§ Many others have made significant contributions.

Our aim in this chapter is to study Hilbert spaces starting with very basic properties of the structure of Hilbert spaces and ending with a brief exposition of some essential data concerning the spectral theory of self-adjoint operators. Primarily our aim is to prove the spectral theorem for bounded self-adjoint operators—an important tool in the further study of bounded linear operators in Hilbert space theory in that self-adjoint operators are represented as a sum (integral) of projection operators.

† J. von Neumann, Allgemeine Eigenwerttheorie Hermitescher Functionaloperen, *Math. Ann.* **102**, 49–131 (1929–1930); Mathematische Begründung der Quantenmechanik, *Nachr. Ges. Wiss. Göttingen Math.-Phys. Kl.*, 1–57 (1927).

‡ H. Löwig, Komplexe euklidische Räume von beliebiger endlicher oder unendlicher Dimensionzahl, *Acta Sci. Math.* (*Szeged.*) **7**, 1–33 (1934).

§ F. Rellich, Spectraltheorie in nichtseparabeln Räumen, *Math. Ann.* **110**, 342–356 (1935).

7.1. The Geometry of Hilbert Space

In this section V and W will denote vector spaces over the field F of real or complex numbers. $\bar{\alpha}$ will denote the complex conjugate of the complex number α.

Definition 7.1. A *sesquilinear form* B on $V \times W$ is a mapping B: $V \times W \to F$ such that, for all α and β in F, x and y in V, and w and z in W,

(i) $B(\alpha x + \beta y, z) = \alpha B(x, z) + \beta B(y, z)$

and

(ii) $B(x, \alpha w + \beta z) = \bar{\alpha} B(x, w) + \bar{\beta} B(x, z)$. ∎

In case $V = W$, a sesquilinear form on $V \times W$ is referred to as a sesquilinear form *on* V. A sesquilinear form on V is called *Hermitian* if $B(x, y) = \overline{B(y, x)}$ for all x and y in V. Since $B(x, x)$ is necessarily a real number if B is Hermitian, we say that a Hermitian form on V is *positive* if $B(x, x) \geq 0$ for all x in V and *strictly positive* if $B(x, x) > 0$ when $x \neq 0$.

Sometimes when F is the field of real numbers so that $\bar{\alpha} = \alpha$ for all scalars, a sesquilinear form on V is called a *bilinear form* and a Hermitian form is called a *symmetric form* since $B(x, y) = B(y, x)$ for all x and y in V. Since the development to follow is true—unless specifically indicated—for real and complex vector spaces, we will continue to use the terms sesquilinear and Hermitian regardless of whether $F = R$.

The following proposition gives in a nutshell some facts we will find extremely useful regarding sesquilinear forms on V.

Proposition 7.1. Let B be a sesquilinear form on V.

(i) *Polarization Identity*. If V is a complex vector space, then for all x and y in V

$$B(x, y) = \tfrac{1}{4}[B(x + y, x + y) - B(x - y, x - y) + iB(x + iy, x + iy) - iB(x - iy, x - iy)]. \tag{7.1}$$

If V is a real vector space, then for all x and y

$$B(x, y) = \tfrac{1}{4}[B(x + y, x + y) - B(x - y, x - y)], \tag{7.2}$$

provided B is Hermitian.

(ii) *Parallelogram Law.* For all x and y in V

$$B(x + y, x + y) + B(x - y, x - y) = 2B(x, x) + 2B(y, y). \quad (7.3)$$

(iii) *Cauchy–Schwarz Inequality.* If B is a positive Hermitian sesquilinear form on V, then for all x and y in V

$$| B(x, y) |^2 \leq B(x, x)B(y, y). \quad (7.4)$$

(iv) If B is a positive Hermitian sesquilinear form on V, then for all x and y in V,

$$[B(x + y, x + y)]^{1/2} \leq [B(x, x)]^{1/2} + [B(y, y)]^{1/2}. \quad ▮$$

Proof. Statements (i) and (ii) are verified by direct computation and the verifications are left to the reader. Assuming momentarily that (iii) has been verified, we can easily prove (iv). Indeed using (iii)

$$\begin{aligned} B(x + y, x + y) &= B(x, x) + B(x, y) + B(y, x) + B(y, y) \\ &= B(x, x) + 2\text{Re}[B(x, y)] + B(y, y) \\ &\leq B(x, x) + 2 \, | B(x, y) | + B(y, y) \\ &\leq B(x, x) + 2[B(x, x)]^{1/2}[B(y, y)]^{1/2} + B(y, y) \\ &= \{[B(x, x)]^{1/2} + [B(y, y)]^{1/2}\}^2. \end{aligned}$$

It remains therefore to establish (iii). For all real numbers r and for $\alpha \in F$ with $| \alpha | = 1$,

$$\begin{aligned} 0 \leq B(r\alpha x + y, r\alpha x + y) &= r^2 B(x, x) + r\alpha B(x, y) + r\bar{\alpha} B(y, x) + B(y, y) \\ &= r^2 B(x, x) + 2r \, \text{Re}[\alpha B(x, y)] + B(y, y). \end{aligned} \quad (7.5)$$

Since equation (7.5) holds for all real numbers r, the quadratic function $f(r) = B(x, x)r^2 + 2\text{Re}[\alpha B(x, y)]r + B(y, y)$ has at most one distinct real root. Hence its discriminant must be nonpositive, that is,

$$\{\text{Re}[\alpha B(x, y)]\}^2 \leq B(x, x)B(y, y), \quad (7.6)$$

for all α with $| \alpha | = 1$. Choose α so that $\alpha B(x, y) = | B(x, y) |$. Then inequality (7.6) yields

$$| B(x, y) |^2 \leq B(x, x)B(y, y). \quad ▮$$

Remark 7.1. Clearly if B is a strictly positive sesquilinear form on V

and $B(x, y) = B(x, z)$ for all x in V, then $y = z$. Indeed

$$B(y - z, y - z) = B(y - z, y) - B(y - z, z) = 0.$$

Other interesting and useful facts are given in the following corollary of Proposition 7.1 (i).

Corollary 7.1. Assume V is a complex vector space.

(i) If $B: V \times V \to C$ and $B': V \times V \to C$ are sesquilinear forms such that $B(x, x) = B'(x, x)$ for all x, then $B = B'$.

(ii) A sesquilinear form $B: V \times V \to C$ is Hermitian if and only if $B(x, x)$ is real for all x. ∎

Proof. The proof of (i) is readily seen by examining equation (7.1). To prove (ii) note that $B(x, x)$ is real if B is Hermitian since $B(x, x) = \overline{B(x, x)}$. Conversely, if $B(x, x)$ is real for all x, the sesquilinear form $B'(x, y) = \overline{B(y, x)}$ is such that $B'(x, x) = B(x, x)$ for all x. By (i), $B = B'$ or B is Hermitian. ∎

With the information given in Proposition 7.1 we are in a good position to begin our study of Hilbert and pre-Hilbert spaces.

Definition 7.2. A *pre-Hilbert space* P over the field F is a vector space P over F together with a strictly positive Hermitian sesquilinear form on P. ∎

The sesquilinear form in a pre-Hilbert space is often called an *inner product* and a pre-Hilbert space is accordingly called an *inner product space.* The image in F of the ordered pair (x, y) in $P \times P$ by the inner product B on P will be denoted by $(x \mid y)$ instead of $B(x, y)$. In what follows, P, unless otherwise mentioned, will denote a pre-Hilbert space.

Examples. Here are some simple yet important examples of inner product spaces:

7.1. For any positive integer n the space $C^n(R^n)$ of ordered n-tuples $x = (x_1, \ldots, x_n)$ of complex (real) numbers with inner product given by

$$(x \mid y) = \sum_{i=1}^{n} x_i \overline{y_i}.$$

7.2. The space l_2 of all complex (real) sequences $x = (x_i)_N$ such that $\sum_{i=1}^{\infty} |x_i|^2 < \infty$ with inner product given by

$$(x \mid y) = \sum_{i=1}^{\infty} x_i \overline{y_i}.$$

7.3. For any measure space $(X, \mathscr{A}, \mu)$, the space $L_2(\mu)$ of all measurable functions f for which $\int |f|^2 \, d\mu < \infty$ with inner product given by

$$(f \mid g) = \int f\bar{g} \, d\mu.$$

(Note that Examples 7.1 and 7.2 are special cases of Example 7.3 if X is chosen to be $\{1, 2, \ldots, n\}$ and N, respectively—each with the counting measure.)

7.4. The vector space of continuous functions f on an interval $[a, b]$ with inner product

$$(f \mid g) = \int_a^b f(t)\overline{g(t)} \, dt.$$

Any pre-Hilbert space P is a normed linear space by virtue of the following definition: If $x \in P$, define the norm of x by

$$\| x \| = (x \mid x)^{1/2}. \tag{7.7}$$

Since an inner product is a positive sesquilinear form, we have

$$\| x \| \geq 0 \text{ and } \| x \| = 0, \quad \text{if and only if } x = 0.$$

Also,

$$\| \alpha x \| = | \alpha | \, \| x \|, \quad \text{since } \| \alpha x \|^2 = (\alpha x \mid \alpha x) = \alpha\bar{\alpha}(x \mid x) = | \alpha |^2 \, \| x \|^2.$$

Finally, Proposition 7.1 (iv) becomes the triangle inequality

$$\| x + y \| \leq \| x \| + \| y \|. \tag{7.8}$$

One should also note that in any inner product space the *Parallelogram Law* and the *Cauchy–Schwarz Inequality*, respectively, now have the following forms:

$$\| x + y \|^2 + \| x - y \|^2 = 2 \| x \|^2 + 2 \| y \|^2, \tag{7.9}$$

$$| (x \mid y) | \leq \| x \| \, \| y \|. \tag{7.10}$$

Examining the geometrical meaning of the Parallelogram Law in R^2 demonstrates the aptness of its title: The sum of the squares of the diagonals in a parallelogram is equal to the sum of the squares of the four sides.

Remark 7.2. If $x_n \to x$ and $y_n \to y$ in P, then $(x_n \mid y_n) \to (x \mid y)$ in F. This follows from the inequality

$$\begin{aligned} | (x \mid y) - (x_n \mid y_n) | &= | (x \mid y) - (x \mid y_n) + (x \mid y_n) - (x_n \mid y_n) | \\ &\leq \| x \| \, \| y - y_n \| + \| x - x_n \| \, \| y_n \|. \end{aligned}$$

Definition 7.3. A *Hilbert space* is a complete pre-Hilbert space with norm $\| x \| = (x \mid x)^{1/2}$. ∎

Not all pre-Hilbert spaces are Hilbert spaces. For example the subspace of l_2 (Example 7.2) consisting of finitely nonzero sequences $x = (x_i)_N$ [a sequence $x = (x_i)_N$ is finitely nonzero if there exists some positive integer M such that $x_i = 0$ for all $i > M$] is a pre-Hilbert space that is not complete. Also Example 7.4 is not complete. The completion (see Problem 7.1.3) of this space is the space $L_2([a, b])$. The verifications of these statements are left as exercises (Problem 7.1.4).

Briefly we can say that a Hilbert space is a Banach space with the norm defined by an inner product as in equation (7.7). When is a Banach space a Hilbert space? The Parallelogram Law gives us one characterization.† Precisely, we have the following characterization, whose proof we have outlined in the Problems (Problem 7.1.5).

Proposition 7.2. A Banach space is a Hilbert space with its norm given by an inner product if and only if its norm satisfies the parallelogram identity (7.9). ∎

Definition 7.4. Two vectors x and y in a pre-Hilbert space P are said to be *orthogonal* (or perpendicular), written $x \perp y$, if $(x \mid y) = 0$. If E and F are subsets of P, then E and F are said to be *orthogonal* (to each other), written $E \perp F$, if $x \perp y$ for each x in E and y in F. A subset E of P is said to be an *orthogonal set* if $x \perp y$ for each nonequal pair of vectors x and y in E. If in addition $\| x \| = 1$ for each x in E, then E is said to be *orthonormal.* ∎

Remark 7.3. Any orthogonal set E in a pre-Hilbert space P that does not contain the zero vector is linearly independent. Indeed, if $\{x_1, x_2, \ldots, x_n\}$ is a finite subset of E and $\alpha_1 x_1 + \alpha_2 x_2 + \cdots + \alpha_n x_n = 0$, then $\alpha_j \| x_j \|^2 = (\sum_{i=1}^n \alpha_i x_i \mid x_j) = 0$ so that $\alpha_j = 0$ for each j.

In the space l_2 the countable set of vectors like $(0, 0, \ldots, 1, 0, 0, \ldots)$ where 1 is the ith coordinate for $i = 1, 2, \ldots$ is an orthonormal set. In R^3, the set $\{(1, 1, 0), (0, -1, 0)\}$ is independent but not orthogonal. In $L_2([0, 2\pi])$ the set $\{e^{int} \mid n = 0, \pm 1, \pm 2, \ldots\}$ is orthogonal, but not orthonormal.

Definition 7.5. A family of vectors $(x_i)_{i \in I}$ in a normed linear space is

† For other characterizations, see [22].

called *summable* to x, written $\sum_I x_i = x$, if for each $\varepsilon > 0$ there exists a finite subset $F(\varepsilon)$ of I such that if J is a finite subset of I containing $F(\varepsilon)$ then $\| \sum_{i\in J} x_i - x \| < \varepsilon$. ∎

It can be shown by the reader that in a Banach space a family $(x_i)_I$ is summable (to some x) if and only if for each number $\varepsilon > 0$ there exists a finite subset $F(\varepsilon)$ of I such that if J is a finite subset of I with $J \cap F(\varepsilon) = \varnothing$, then $\| \sum_{i\in J} x_i \| < \varepsilon$. From this criterion it follows that if $(x_i)_I$ is summable, the set of indices i for which $x_i \neq 0$ is at most countable. Indeed for each positive integer n, let $F(1/n)$ be the finite subset of I such that $\| \sum_J x_i \| < 1/n$ if $J \cap F(1/n) = \varnothing$ and J is finite. If $x \notin \bigcup_{n=1}^{\infty} F(1/n)$, a countable set, $\| x \| < 1/n$ for all n.

It is easy to verify the following rules in any pre-Hilbert space:

(i) If $\sum_I x_i = x$, then $\sum_I \alpha x_i = \alpha x$ for any scalar α.

(ii) If $\sum_I x_i = x$ and $\sum_I y_i = y$, then $\sum_I x_i + y_i = x + y$. (7.11)

(iii) If $\sum_I x_i = x$, then $\sum_I (x_i \mid y) = (x \mid y)$ and $\sum_I (y \mid x_i) = (y \mid x)$ for every vector y.

To verify (i) for instance, let $\varepsilon > 0$ be arbitrary. Then there is a finite subset $F(\varepsilon)$ such that if J is finite and $J \supset F(\varepsilon)$, then $\| \sum_J \alpha x_i - \alpha x \| = | \alpha | \, \| \sum_J x_i - x \| < | \alpha | \, \varepsilon$. Hence $\sum_I \alpha x_i = \alpha x$.

Proposition 7.3. *Pythagorean Theorem.*

(i) If $\{x_1, x_2, \ldots, x_n\}$ is any orthogonal family of vectors in a Hilbert space H, then

$$\left\| \sum_{i=1}^{n} x_i \right\|^2 = \sum_{i=1}^{n} \| x_i \|^2.$$

(ii) Any orthogonal family $(x_i)_I$ of vectors in H is summable if and only if $(\| x_i \|^2)_I$ is summable. If $x = \sum_I x_i$, then $\| x \|^2 = \sum_I \| x_i \|^2$. ∎

Proof. An inductive argument proves (i). To prove (ii) note that $(x_i)_I$ is summable if and only if for each $\varepsilon > 0$ there exists a finite subset $F(\varepsilon)$ of I such that if J is a finite subset of I with $J \cap F(\varepsilon) = \varnothing$, then

$$\sum_J \| x_j \|^2 = \| \sum_J x_j \|^2 < \varepsilon^2. \tag{7.12}$$

By virtue of the equality in (7.12), this condition is also necessary and sufficient for $(\| x_i \|^2)_I$ to be summable.

If $x = \sum_I x_i$, then by equation (7.11) we have

$$\| x \|^2 = (x \mid x) = \left(\sum_I x_i \,\middle|\, x\right)$$

$$= \sum_I (x_i \mid x) = \sum_{i \in I} \left(x_i \,\middle|\, \sum_{j \in I} x_j\right) = \sum_{i \in I} \sum_{j \in I} (x_i \mid x_j) = \sum_I \| x_i \|^2. \quad \blacksquare$$

Proposition 7.4. *Bessel's Inequality.* Let P be a pre-Hilbert space.

(i) If $\{x_1, \ldots, x_n\}$ is any finite family of orthonormal vectors in P and x is any vector in P, then

$$\sum_{i=1}^{n} | (x \mid x_i) |^2 \leq \| x \|^2.$$

(ii) If $(x_i)_I$ is any orthonormal family of vectors in P and x is any vector in P, then

$$\sum_I | (x \mid x_i) |^2 \leq \| x \|^2. \quad \blacksquare$$

Proof. (i)

$$0 \leq \left\| x - \sum_{i=1}^{n} (x \mid x_i) x_i \right\|^2 = \| x \|^2 - \sum_{i=1}^{n} \overline{(x \mid x_i)}(x \mid x_i)$$

$$- \sum_{i=1}^{n} (x \mid x_i)(x_i \mid x) + \sum_{i=1}^{n} (x \mid x_i)\overline{(x \mid x_i)} = \| x \|^2 - \sum_{i=1}^{n} | (x \mid x_i) |^2,$$

from which (i) follows.

(ii) By (i), $\sum_{i \in F} | (x \mid x_i) |^2 \leq \| x \|^2$ for any finite subset F of I.[†] Hence by the definition of summability the inequality must hold for I. $\blacksquare$

Using the concept of summability of an arbitrary family of vectors or scalars, we can give an example of a class of Hilbert spaces which we shall see later represents all Hilbert spaces.

Example 7.5. For any nonempty set I let $C^I(R^I)$ be the vector space of all complex- (real-) valued functions on I, that is, the set of all families of elements $(x_i)_{i \in I}$, where x_i is a scalar. Let $l_2(I)$ be the vector subspace of $C^I(R^I)$ of all families $(x_i)_{i \in I}$ such that $\sum_I | x_i |^2$ is summable (written $\sum_I | x_i |^2 < \infty$). $l_2(I)$ is a Hilbert space with inner product given by $(x \mid y)$

[†] Note that $\{i \in I: | (x \mid x_i) | > 0\}$ is at most countable.

$= \sum_I x_i \bar{y}_i$ for $x = (x_i)_I$ and $y = (y_i)_I$. Using the Hölder Inequality (Proposition 3.13 in Chapter 3), the reader can verify that this does in fact define an inner product, and in particular that $(x_i \bar{y}_i)_I$ is summable. We here establish the completeness of $l_2(I)$. [The completeness also follows from that of L_2 (see Theorem 3.12 in Chapter 3), where the measure is the counting measure; but we here give a different proof.]

To this end, let $x^k = (x_i^k)_{i \in I}$, $k = 1, 2, \ldots$ be a Cauchy sequence in $l_2(I)$. Since for each i

$$\| x_i^n - x_i^m \|^2 \leq \sum_{i \in I} | x_i^n - x_i^m |^2 = \| x^n - x^m \|^2,$$

$(x_i^k)_{k=1,2,\ldots}$ is a Cauchy sequence of scalars for each i in I. Hence there exists for each i in I an x_i such that $\lim_{k \to \infty} x_i^k = x_i$. Let $x = (x_i)_{i \in I}$. We wish to show $x \in l_2(I)$ and $x^k \to x$ in $l_2(I)$.

Let J be any finite subset of I and $\varepsilon > 0$ be arbitrary. Then there exists $N > 0$ such that if $n, m > N$, then

$$\sum_{i \in J} | x_i^n - x_i^m |^2 \leq \sum_{i \in I} | x_i^n - x_i^m |^2 \leq \varepsilon^2.$$

Letting $n \to \infty$, then for $m \geq N$

$$\sum_{i \in J} | x_i - x_i^m |^2 \leq \varepsilon^2.$$

Since J is an arbitrary finite subset of I, this means $\sum_I | x_i - x_i^m |^2 \leq \varepsilon^2$ and $\| (x_i - x_i^m)_I \| \to 0$ as $m \to \infty$. In particular $(x_i - x_i^m)_I$ is in $l_2(I)$, so that $(x_i)_I = (x_i - x_i^m)_I + (x_i^m)_I$ is in $l_2(I)$.

Problems

× **7.1.1.** (i) If x $(\neq 0)$ and y are any vectors in a pre-Hilbert space, prove $| (x \mid y) | = \| x \| \, \| y \|$ if and only if $y = \lambda x$ for some $\lambda \in F$. [Hint: Look at the proof of Proposition 7.1 (iii).]

(ii) If x and y are nonzero vectors in a pre-Hilbert space, prove $\| x + y \| = \| x \| + \| y \|$ if and only if $y = \lambda x$ for some $\lambda > 0$.

(iii) Prove in a pre-Hilbert space, $\| x - z \| = \| x - y \| + \| y - z \|$ if and only if $y = \alpha x + (1 - \alpha) z$ for some α in $[0, 1]$.

7.1.2. (i) Prove that if (x_n) is an orthogonal sequence of vectors in a pre-Hilbert space such that $\sum_{i=1}^{\infty} \| x_i \|^2 < \infty$, then the sequence $(\sum_{i=1}^{n} x_i)_{n \in N}$ is a Cauchy sequence.

(ii) Give an example where the conclusion of (i) may fail if (x_n) is not orthogonal.

× **7.1.3.** If P is a pre-Hilbert space with inner product $(x \mid y)$, prove that there is a Hilbert space H with inner product $B(x, y)$ and a linear map T: $P \to H$ such that $B(Tx, Ty) = (x \mid y)$ for all x and y in P and $T(P)$ is dense in H. Prove that if (H', B') is another Hilbert space satisfying these criteria, then H and H' are linearly isometric—that is, there is a linear map S from H onto H' such that $B'(Sx, Sy) = B(x, y)$ for all x and y in H. H is called the *completion* of P.

× **7.1.4.** (i) Prove that the subspace of l_2 (see Example 7.2) consisting of finitely nonzero sequences is not complete, but its completion is l_2.

(ii) Prove that the space of Example 7.4 is not complete, but its completion is $L_2([a, b])$. {Hint: Look at the sequence $f_n(t) = 0$ if $a \leq t \leq (a + b)/2$; $= n[t - (a + b)/2]$ if $(a + b)/2 \leq t \leq (a + b)/2 + 1/n$; $= 1$ otherwise.}

× **7.1.5.** (i) Prove Proposition 7.2 by showing that each Banach space whose norm satisfies the Parallelogram Law, equation (7.9), is a Hilbert space. [Hint: If B is a real Banach space, define $(x \mid y)$ as in equation (7.2) by $\frac{1}{4}\{\| x + y \|^2 - \| x - y \|^2\}$ while if B is a complex Banach space define $(x \mid y)$ as in equation (7.1) by $\frac{1}{4}\{\| x + y \|^2 - \| x - y \|^2 + i \| x + iy \|^2 - i \| x - iy \|^2\}$. In the real case show $(x \mid y) + (z \mid y) = (x + z \mid y)$, $(x_n \mid y) \to (x \mid y)$ if $x_n \to x$ in B, and conclude $(\alpha x \mid y) = \alpha(x \mid y)$ for all real α. Note that in the complex case $\operatorname{Im}(x \mid y) = \operatorname{Re}(x \mid iy)$.]

(ii) Prove $L_1[0, 1]$ is not a Hilbert space by showing that the Parallelogram Law is not satisfied.

7.1.6. Show that the result in Problem 7.1.5 (i) can be extended as follows: Let V be a real vector space and $\| \cdot \|$: $V \to R$ be a function satisfying the parallelogram law [equation (7.9)] and the following property: For every $x \in V$, the function $\alpha \to \| \alpha x \|$ on R is continuous at 0. Then $(x \mid y) = \frac{1}{4}\{\| x + y \|^2 - \| x - y \|^2\}$ defines a nonnegative Hermitian bilinear form on V. (This extension is due to D. Fearnley-Sander and J. Symons.)

7.1.7. State and prove a complex version of the result outlined in Problem 7.1.6.

7.1.8. Define on l_2 the norm

$$\| f \| = \left(\sum_n | f(n) |^2\right)^{1/2} + \sup_{n \in N} | f(n) |.$$

Show that this norm is equivalent to the usual norm in l_2 but it does not come from an inner product.

7.2. Subspaces, Bases, and Characterizations of Hilbert Spaces

We will now turn our attention to subspaces of pre-Hilbert and Hilbert spaces. It is clear that any vector subspace of a pre-Hilbert space is a pre-Hilbert space with the restricted inner product.

Crucial to the study of the structure of Hilbert spaces and subspaces is the following result, not valid in every normed linear space.

Theorem 7.1. Let S be a complete and convex ($x, y \in S$ implies $\alpha x + (1-\alpha)y \in S$ for all $\alpha \in [0, 1]$) subset of a pre-Hilbert space P. Given any vector x in P there exists one and only one vector $y_0 \in S$ such that $\| x - y_0 \| \leq \| x - y \|$ for all y in S. ∎

(In regard to this theorem see Problems 7.2.1 and 7.2.2.)

Proof. Let y_i be a sequence of vectors in S such that $\| x - y_i \|$ converges to δ, the inf of $\{\| x - y \|: y \text{ in } S\}$. We will show that y_i is a Cauchy sequence in S converging to the desired vector y_0. Using the Parallelogram Law, equation (7.9) of Section 7.1,

$$\| (y_i - x) + (x - y_j) \|^2 + \| (y_i - x) - (x - y_j) \|^2 = 2 \| y_i - x \|^2 + 2 \| x - y_j \|^2$$

or

$$\| y_i - y_j \|^2 = 2 \| y_i - x \|^2 + 2 \| x - y_j \|^2 - 4 \| \tfrac{1}{2}(y_i + y_j) - x \|^2.$$

Since $\frac{1}{2}(y_i + y_j) \in S$,

$$\| \tfrac{1}{2}(y_i + y_j) - x \|^2 \geq \delta^2.$$

Hence,

$$\| y_i - y_j \|^2 \leq 2 \| y_i - x \|^2 + 2 \| x - y_j \|^2 - 4\delta^2. \tag{7.13}$$

As $i, j \to \infty$, the right-hand side of equation (7.13) goes to zero so that $(y_i)_N$ is a Cauchy sequence in S. Since S is complete, y_i converges to some y_0 in S. Since $\| y_i - x \| \to \| y_0 - x \|$, $\| y_0 - x \| = \delta$.

If $y_0' \in S$ also satisfies $\| y_0' - x \| = \delta$, then using the Parallelogram Law again

$$\begin{aligned} \| y_0 - y_0' \|^2 &= 2 \| y_0 - x \|^2 + 2 \| x - y_0' \|^2 - 4 \| \tfrac{1}{2}(y_0 + y_0') - x \|^2 \\ &\leq 2 \| y_0 - x \|^2 + 2 \| x - y_0' \|^2 - 4\delta^2 = 4\delta^2 - 4\delta^2 = 0. \end{aligned}$$ ∎

Lemma 7.1. If S is a proper complete subspace of pre-Hilbert space P, then there exists x in $P - S$ such that $\{x\} \perp S$. ▮

Proof. By Theorem 7.1 for any vector z in $P - S$ there exists a unique vector $y_0(z)$ in S such that $\| z - y_0(z) \| \leq \|z - y\|$ for all y in S. Let $x = z - y_0(z)$. We will show $x \in P - S$ and $\{x\} \perp S$. Clearly $x \notin S$ since $z \notin S$. Since for every scalar α, $y_0(z) + \alpha y \in S$ for every y in S,

$$\| \alpha y - x \|^2 = \| [y_0(z) + \alpha y] - z \|^2 \geq \| z - y_0(z) \|^2 = \| x \|^2.$$

Hence

$$0 \leq \| x - \alpha y \|^2 - \| x \|^2 = -\alpha(y \mid x) - \bar{\alpha}(x \mid y) + \alpha\bar{\alpha} \| y \|^2.$$

Letting $\alpha = -\beta(x \mid y)$ for any *real* β we get

$$\begin{aligned} 0 &\leq \beta(x \mid y)(y \mid x) + \beta\overline{(x \mid y)}(x \mid y) + \beta^2(x \mid y)\overline{(x \mid y)} \| y \|^2 \\ &= 2\beta \mid (x \mid y) \mid^2 + \beta^2 \mid (x \mid y) \mid^2 \| y \|^2 \\ &= \beta \mid (x \mid y) \mid^2 [2 + \beta \| y \|^2]. \end{aligned} \tag{7.14}$$

If β is chosen to be a negative number such that $\beta > -2/\| y \|^2$, (7.14) forces $\mid (x \mid y) \mid = 0$ or $x \perp y$. ▮

Definition 7.6. If S is any subset of a pre-Hilbert space P, the *annihilator* or *orthogonal complement* of S is the set

$$S^{\perp} = \{x \in P\colon x \perp y, \text{ for all } y \text{ in } S\}.$$ ▮

The orthogonal complement of any set S always contains the zero element of P. Clearly $S \cap S^{\perp} \subset \{0\}$ and $S \subset (S^{\perp})^{\perp}$. More can be said.

Lemma 7.2. If S is any subset of P, then $S^{\perp}$ is a closed subspace of P. ▮

The proof is easy using Remark 7.2.

Note that in Theorem 7.1 and Lemma 7.1 if P is a Hilbert space, the word "complete" may be replaced by "closed" since in any complete metric space a subset is closed if and only if it is complete. In particular, Lemma 7.2 assures us that $S^{\perp}$ is complete in a Hilbert space.

If M and N are subspaces of a pre-Hilbert space P, then $M + N$ is the subspace defined as $\{m + n\colon m \in M, n \in N\}$. If $M \perp N$, then each element of $M + N$ has a unique representation as $m + n$ with $m \in M$ and $n \in N$. Indeed, if $m + n = m' + n'$ with $m, m' \in M$ and $n, n' \in N$, then

$m - m' = n' - n$. Hence $m - m' \in M \cap N$, so that $(m - m' \mid m - m') = 0$. Hence $m = m'$. Similarly $n = n'$. In this case we write $M + N$ as $M \oplus N$.

Theorem 7.2. If M is a complete linear subspace of pre-Hilbert space P, then $P = M \oplus M^\perp$ and $M = (M^\perp)^\perp$. (See also Problem 7.2.6.) ∎

Proof. Let z be any vector in P. By Theorem 6.1, there exists a unique vector $y_0(z)$ in M such that

$$\| z - y_0(z) \| \leq \| z - y \|, \qquad \text{for all } y \text{ in } M.$$

As shown in the proof of Lemma 7.1, $x(z) = z - y_0(z)$ is in $M^\perp$. Hence $z = y_0(z) + x(z) \in M + M^\perp$. Since $M \perp M^\perp$, $P = M + M^\perp = M \oplus M^\perp$.

Clearly for any set S, $S \subset (S^\perp)^\perp$. If $z \in (M^\perp)^\perp$, then $z = y + x$ with $y \in M$ and $x \in M^\perp$. Since also $y \in (M^\perp)^\perp$, $x = z - y \in (M^\perp)^\perp$. But $(M^\perp) \cap (M^\perp)^\perp = 0$, so $x = 0$. Thus $z = y \in M$. Hence $(M^\perp)^\perp = M$. ∎

If S is a set in a pre-Hilbert space P, we denote by $[S]$ the smallest subspace of P containing S. It is easy to see that $[S]$ is the vector space of all finite linear combinations of elements of S.

Proposition 7.5. *The Gram–Schmidt Orthonormalization Process.* If $\{x_i: i = 1, 2, \ldots, N\}$ for $1 \leq N \leq \infty$ is a linearly independent set in a pre-Hilbert space P, then there is an orthonormal set $\{z_i: i = 2, \ldots, N\}$ such that $[\{z_i: i = 1, 2, \ldots, n\}] = [\{x_i: i = 1, 2, \ldots, n\}]$ for each $n = 1, 2, \ldots, N$. ∎

Proof. For $N = \infty$ we proceed by induction. Let $z_1 = \| x_1 \|^{-1} x_1$. Clearly $[x_1] = [z_1]$. Assume an orthonormal set $\{z_i: i = 1, 2, \ldots, n - 1\}$ exists such that $[\{z_i: i = 1, \ldots, n - 1\}] = [\{x_i: i = 1, \ldots, n - 1\}]$. Define $z_n = \| y_n \|^{-1} y_n$, where $y_n = x_n - \sum_{i=1}^{n-1} (x_n \mid z_i) z_i$. This equation and the inductive hypothesis guarantee that each z_i for $i = 1, 2, \ldots, n$ is a linear combination of the set $\{x_i: i = 1, \ldots, n\}$ and each x_i for $i = 1, 2, \ldots, n$ is a linear combination of the set $\{z_i: i = 1, 2, \ldots, n\}$. Also, for each $j = 1, 2, \ldots, n - 1$,

$$\begin{aligned}(z_n \mid z_j) = \| y_n \|^{-1} (y_n \mid z_j) &= \| y_n \|^{-1} \left\{ (x_n \mid z_j) - \sum_{i=1}^{n-1} (x_n \mid z_i)(z_i \mid z_j) \right\} \\ &= \| y_n \|^{-1} \{ (x_n \mid z_j) - (x_n \mid z_j)(z_j \mid z_j) \} \\ &= 0.\end{aligned}$$

The modifications for the case when N is finite are obvious. ∎

If $\{x_i: i = 1, 2, \ldots, N\}$ for some N in $1 \leq N \leq \infty$ is a vector space basis for a pre-Hilbert space P, then by Proposition 7.5, P has a basis $\{z_i: i = 1, 2, \ldots, N\}$ which is an orthonormal set. This basis $\{z_i: i = 1, 2, \ldots, N\}$ has the following property: If x in P is orthonormal to $\{z_i: i = 1, 2, \ldots, N\}$, then $x = 0$. Indeed if x is in P, $x = \sum_{i=1}^{N} \alpha_i z_i$ for some scalars α_i. However, for each $j = 1, 2, \ldots, N$,

$$\alpha_j = \alpha_j \| z_j \| = (\alpha_j z_j \mid z_j) = \left(\sum_{i=1}^{N} \alpha_i z_i \mid z_j \right) = (x \mid z_j) = 0,$$

so that $x = 0$. Because the set $\{z_i: i = 1, 2, \ldots, N\}$ has this property, it is an example of a "complete" orthonormal set. Precisely, we have the following definition.

Definition 7.7. An orthonormal set S in a pre-Hilbert space is *complete* if whenever $\{x\} \perp S$ then $x = 0$. A complete orthonormal set in a pre-Hilbert space is called a (pre-Hilbert space) *basis*. ∎

A word of caution and explanation is in order. The word "basis" is used with two different meanings. In one sense it means an algebraic or Hamel basis, that is, a linearly independent set in a vector space which spans the vector space. In the other sense it means a Hilbert space basis as defined in Definition 7.7.

Subsequently, whenever we speak of a basis we mean a pre-Hilbert space basis unless otherwise indicated.

Proposition 7.6. Let $(x_i)_I$ be an orthonormal family in a Hilbert space H. The following statements are equivalent.

(i) The family $(x_i)_I$ is a basis for H.

(ii) The family $(x_i)_I$ is a maximal orthonormal family in H.

(iii) If $x \in H$, then $x = \sum_I (x \mid x_i) x_i$ (the Fourier expansion of x).

(iv) If x and y are in H, then

$$(x \mid y) = \sum_I (x \mid x_i)(x_i \mid y) \qquad \text{(Parseval's identity)}.$$

(v) If x is in H, then $\| x \|^2 = \sum_I | (x \mid x_i) |^2$. ∎

Proof. (i) $\Rightarrow$ (ii). Suppose S is an orthonormal family containing

$(x_i)_I$ and $s \in S - (x_i)_I$. Then $s \perp x_i$ for each $i \in I$ implies $s = 0$. However, this contradicts the fact that $\| s \| = 1$.

(ii) $\Rightarrow$ (iii). By Bessel's Inequality (Proposition 7.4) the family

$$[| (x \mid x_i) |^2]_I$$

has a convergent sum since for any finite subset A of I,

$$\sum_{i \in A} | (x \mid x_i) |^2 \leq \| x \|^2.$$

Hence $\sum_I (x \mid x_i) x_i$ converges since for each $\varepsilon > 0$ there exists a finite subset $F(\varepsilon)$ of I so that if J is finite and $J \cap F(\varepsilon) = \emptyset$, then (using Proposition 7.3)

$$\| \sum_J (x \mid x_i) x_i \|^2 = \sum_J \| (x \mid x_i) x_i \|^2 = \sum_J | (x \mid x_i) |^2 < \varepsilon.$$

Let $y = \sum_I (x \mid x_i) x_i$. It remains to show $x = y$. For each $j \in I$,

$$(x - y \mid x_j) = \Big(x - \sum_I (x \mid x_i) x_i \,\Big|\, x_j\Big) = (x \mid x_j) - (x \mid x_j) = 0.$$

If $x - y \neq 0$, then $\{\| x - y \|^{-1}(x - y)\} \cup \{x_i\}_{i \in I}$ is an orthonormal family properly containing $\{x_i\}_{i \in I}$. Since this is impossible, $x = y$.

(iii) $\Rightarrow$ (iv).

$$(x \mid y) = \Big(\sum_{i \in I} (x \mid x_i) x_i \,\Big|\, \sum_{j \in I} (y \mid x_j) x_j\Big) = \sum_{i \in I} \Big((x \mid x_i) x_i \,\Big|\, \sum_{j \in I} (y \mid x_j) x_j\Big)$$
$$= \sum_{i \in I} (x \mid x_i)\Big(\sum_{j \in I} \overline{(y \mid x_j)(x_i \mid x_j)}\Big) = \sum_{i \in I} (x \mid x_i)(x_i \mid y).$$

(iv) $\Rightarrow$ (v). Take $x = y$ in (iv).

(v) $\Rightarrow$ (i). If $(x \mid x_i) = 0$ for each i, then $\| x \|^2 = \sum (x \mid x_i)^2 = 0$. ∎

Does each pre-Hilbert space have a basis? It is sufficient to ask whether each pre-Hilbert space has a maximal orthonormal set. Consider the collection of all orthonormal sets in a given pre-Hilbert space. Order this collection by set inclusion. Since the union of an increasing family of orthonormal sets is orthonormal, Zorn's Lemma guarantees the existence of a maximal orthonormal set. We have proved the following theorem.

Theorem 7.3. Every pre-Hilbert space has a basis. ∎

Proposition 7.7. Any two bases of a pre-Hilbert space H have the same cardinality. ∎

Proof. If H is finitely generated, the result follows from the theory of finite-dimensional vector spaces. So assume that $\{x_i\}_I$ and $\{y_j\}_J$ are two bases of infinite cardinality. Since for each $k \in I$, $x_k = \sum_J (x_k \mid y_j) y_j$ by Proposition 6.6 (iii), the set J_k of those indices j for which $(x_k \mid y_j) \neq 0$ is at most countable. Since $\{x_i\}_I$ is a basis, no y_j can be orthogonal to each x_k. This means $J \subset \bigcup_{k \in I} J_k$. Hence

$$\text{card } J \leq \aleph_0 \cdot \text{card } I = \text{card } I.$$

A symmetrical argument gives card $I \leq$ card J. ∎

Proposition 7.7 gives meaning to the following definition.

Definition 7.8. The *dimension* of a pre-Hilbert space H is the cardinality of any basis for H. ∎

It is an interesting fact that in a pre-Hilbert space H the "distance" between any two distinct elements x_i and x_j of a basis $\{x_i\}_I$ is $2^{1/2}$. Indeed,

$$\| x_i - x_j \|^2 = (x_i - x_j \mid x_i - x_j) = (x_i \mid x_i) + (x_j \mid x_j) = 2.$$

It follows that an open neighborhood $N(x_i, 2^{1/2}/2) = \{x \in H: \| x - x_i \| < 2^{1/2}/2\}$ of x_i contains no other element of the basis $\{x_i\}_I$ except x_i. In fact the collection of such neighborhoods is pairwise disjoint. If S is a dense subset of H, each $N(x_i, 2^{1/2}/2)$ for $i \in I$ must contain a point of S. This means that the cardinality of I is no greater than that of S. In other words if H is a separable Hilbert (metric) space so that S is countable, card $I \leq \aleph_0$. We have partly proved the following proposition.

Proposition 7.8. The dimension of a Hilbert space H is less than or equal to $\aleph_0$ if and only if H is separable. ∎

The converse is left to the reader.

A *linear isometry* T from a Hilbert space H into a Hilbert space K is a linear mapping from H into K such that $\| Tx \| = \| x \|$ for all x in H. If $T: H \to K$ is a linear isometry, then it follows from the polarization identity (Proposition 7.1) and the equation $(Tx \mid Tx) = \| Tx \|^2 = \| x \|^2 = (x \mid x)$ that $(Tx \mid Ty) = (x \mid y)$ for all x and y in H. Hence $T: H \to K$ is a linear isometry if and only if T is linear and $(Tx \mid Ty) = (x \mid y)$ for all x and y in H. Two Hilbert spaces H and K are said to be *linearly isometric* if there is a linear isometry from H *onto* K.†

† In this chapter we write Tx, rather than $T(x)$, as was done in Chapter 6, for ease of notation in the context of inner products.

Proposition 7.9. Two Hilbert spaces H and K are linearly isometric if and only if they have the same dimension. ▮

Proof. If H and K are linearly isometric, let $T: H \to K$ be an isometry from H onto K. If $\{h_i: i \in I\}$ is a basis of H, then the set $\{Th_i: i \in I\}$ is a basis in K of the same cardinality.

Conversely, if the dimension of K equals the dimension of H, let $\{h_i: i \in I\}$ and $\{k_i: i \in I\}$ be bases of H and K, respectively, indexed by the same set I. If $h = \sum_I (h \mid h_i) h_i$ is in H, define Th as $\sum_I (h \mid h_i) k_i$. Clearly T is a linear mapping from H onto K and $\| Th \|^2 = \sum | (h \mid h_i) |^2 = \| h \|^2$. ▮

Theorem 7.4

(i) A Hilbert space of finite dimension n is linearly isometric to $l_2(I)$, where $I = \{1, 2, \ldots, n\}$.
(ii) A separable Hilbert space of infinite dimension is linearly isometric to l_2.
(iii) A Hilbert space of dimension α is linearly isometric to $l_2(I)$, where I is a set of cardinality α. ▮

Proof. For any set I, a basis for $l_2(I)$ is the set $B = \{f_i \in l_2(I): i \in I$ and $f_i(j) = \delta_{ij}\}$, where δ_{ij} is the Kronecker delta. The dimension of $l_2(I)$ is then the cardinality of B, which is the cardinality of I. The result follows from Proposition 6.9. ▮

• **Remark 7.4.** *Applications of Proposition 7.6: Fourier Analysis in the Hilbert Space L_2.* Recall the definition of Fourier series from Chapter 6. Remark 6.5. The Fourier series of a function $f \in L_1[-\pi, \pi]$ is sometimes written as the trigonometric series

$$\frac{1}{2} a_0 + \sum_{n=1}^{\infty} [a_n \cos nt + b_n \sin nt], \tag{7.15}$$

where

$$a_n = \frac{1}{\pi} \int_{-\pi}^{\pi} f(s) \cos ns \, ds, \qquad b_n = \frac{1}{\pi} \int_{-\pi}^{\pi} f(s) \sin ns \, ds.$$

Note that the series (7.15) can be easily derived from the series

$$\sum_{k=-\infty}^{\infty} \hat{f}(k) e^{ikt}, \qquad \hat{f}(k) = \frac{1}{2\pi} \int_{-\pi}^{\pi} f(s) e^{-iks} \, ds,$$

by using the formula $e^{ins} = \cos ns + i \sin ns$. By the Riemann–Lebesgue theorem (see Problem 3.2.18), we have

$$\lim_{|n|\to\infty} \hat{f}(n) = 0, \qquad f \in L_1[-\pi, \pi]. \tag{7.16}$$

This fact also follows from Theorem A below. It can be proven (see Problem 7.2.8) that the family $\{(1/2\pi)^{1/2}e^{ikt}: k = 0, \pm 1, \pm 2, \ldots\}$ is a basis for $L_2[-\pi, \pi]$. Using this fact, Theorem A below follows immediately from Proposition 7.6.

Theorem A. Let $f \in L_2[-\pi, \pi]$. Then

$$\lim_{n\to\infty} \int_{-\pi}^{\pi} \left| f(t) - \sum_{k=-n}^{n} \hat{f}(k)e^{ikt} \right|^2 dt = 0 \tag{7.17}$$

and

$$2\pi \sum_{k=-\infty}^{\infty} |\hat{f}(k)|^2 = \int_{-\pi}^{\pi} |f(t)|^2 \, dt. \tag{7.18}$$

For $g \in L_2[-\pi, \pi]$,

$$\int_{-\pi}^{\pi} f(t)\overline{g(t)} \, dt = 2\pi \sum_{k=-\infty}^{\infty} \hat{f}(k) \, \overline{\hat{g}(k)} \tag{7.19}$$

and if $(a_k)_{k=-\infty}^{\infty}$ is a sequence of scalars such that

$$\sum_{k=-\infty}^{\infty} |a_k|^2 < \infty, \tag{7.20}$$

then there exists a unique $f \in L_2[-\pi, \pi]$ such that $f = \sum_{k=-\infty}^{\infty} a_k e^{ikt}$ and $a_k = \hat{f}(k)$. ▮

If we take $g = \chi_{[a,b]}$, $-\pi \le a < b < \pi$, in equation (7.19) above, we obtain the following striking result.

Theorem B. For $f \in L_2[-\pi, \pi]$,

$$\int_a^b f(t) \, dt = \sum_{k=-\infty}^{\infty} \int_a^b \hat{f}(k)e^{ikt} \, dt. \qquad ▮$$

This theorem is surprising since we have obtained above the integral of f by integrating the terms of the Fourier series. Note that usually term-by-term integration of an infinite series is only possible by an assumption of uniform convergence, and in the case of the Fourier series above we have not assumed even pointwise convergence of the series.

Now we present a useful relationship between the convolution products of functions in L_2 and their Fourier coefficients. Let $f, g \in L_2[-\pi, \pi]$. We continue these functions on R with period 2π. For convenience, we define $f * g$ as

$$f * g(t) = \frac{1}{2\pi} \int_{-\pi}^{\pi} f(t-s)g(s)\, ds. \tag{7.21}$$

Then $f * g$ is a continuous function on $[-\pi, \pi]$, since

$$2\pi \,| f * g(t_2) - f * g(t_1) | \leq \| f(t_2 - s) - f(t_1 - s) \|_2 \, \| g(s) \|_2 \to 0$$
$$\text{as } | t_1 - t_2 | \to 0.$$

(Note that this convergence to zero is trivially justified when f is continuous, and therefore it follows for $f \in L_2$ since the continuous functions are dense in L_2.)

Theorem C. The Fourier series for $f * g$ [as defined in equation (7.21) above] is absolutely convergent (uniformly in t) and given by

$$f * g(t) = \sum_{k=-\infty}^{\infty} \hat{f}(k)\hat{g}(k)e^{ikt}. \tag{7.22}$$

∎

Proof. By using the periodicity of the functions, substitution, and Fubini's Theorem one can easily verify that

$$\widehat{(f * g)}(k) = \hat{f}(k)\hat{g}(k);$$

and thus the series in equation (7.22) is the Fourier series of $f * g$. Since $(\hat{f}(n))$ and $(\hat{g}(n))$ are both in l_2 by equation (7.18), it follows that the series in equation (7.22) is absolutely convergent, uniformly in t, and thus has a continuous sum function, say $F(t)$. Since the series is uniformly convergent, term-by-term integration is possible and it follows that the series is the Fourier series of F. Now it is clear from equation (7.17) that $F = f * g$ almost everywhere. Since these functions are continuous, $F = f * g$. ∎

Problems

× **7.2.1.** Show that Theorem 7.1 may be false if S is not complete.

[Hint: Consider the set S of all real sequences (x_i) such that $\sum_{i=1}^{\infty} x_i = 1$ in the pre-Hilbert space of finitely nonzero sequences of real numbers. S contains no vector of minimum norm.]

7.2.2. (i) Show that the uniqueness conclusion of Theorem 7.1 is not valid in $R \times R$ with norm $\| (x, y) \| = \max\{\| x \|, \| y \|\}$.

(ii) Show that the existence conclusion of Theorem 7.1 is not valid in some normed linear spaces.

× **7.2.3.** In the space $L_2[0, 1]$, let $\{f_1, f_2, \ldots\}$ be the orthonormal set obtained from the set $\{1, x, x^2, \ldots\}$ by the Gram–Schmidt orthonormalization process. Using the fact that the set of polynomials is dense in $L_2[0, 1]$, prove that $\{f_1, f_2, \ldots\}$ is a basis of $L_2[0, 1]$.

× **7.2.4.** (i) Show that on any real or complex vector space an inner product can be defined.

(ii) Give an example of a vector space (necessarily infinite dimensional) that is a pre-Hilbert space under two different inner products but the completions of each of the pre-Hilbert spaces are not linearly isometric Hilbert spaces. [Hint: Consider l_2 and form a new inner product by considering a Hamel basis $(x_i)_{i \in I}$ such that $(x_i \mid x_j) = \delta_{ij}$.]

7.2.5. As in Example 7.4, let P be the pre-Hilbert space of continuous real-valued functions on $[-1, 1]$. Show that the set of odd functions O is orthogonal to the set of even functions E. Show $P = E \oplus O$. Is $E = O^\perp$?

× **7.2.6.** Let P be the subspace of Example 7.2 of finitely nonzero real sequences. Let M be the subspace of P of all sequences $x = (x_k)$ such that $\sum_{k=1} x_k/k = 0$. Prove that M is closed but $M \neq (M^\perp)^\perp$. Compare Theorem 7.2. (Hint: Show $M^\perp = \{0\}$ since $\{1, 0, 0, \ldots, -n, 0, \ldots\}$ is in M where $-n$ is the nth coordinate.)

7.2.7. Prove Theorem 7.3 for separable Hilbert spaces without using Zorn's Lemma (or an equivalent).

7.2.8. Prove that $S = \{e^{ikt}/(2\pi)^{1/2}: k = 0, \pm 1, \pm 2, \ldots\}$ is a basis for $L_2([0, 2\pi])$. {Hint: Show S is complete by showing $f = 0$ a.e. if $\int_0^{2\pi} f(t) e^{-ikt}\, dt = 0$ and $f \in L_2([0, 2\pi])$. To accomplish this let $F(t)$ be the periodic absolutely continuous function $F(t) = \int_0^t f(u)\, du$. By partial integration show that $\int_0^{2\pi} [F(t) - c] e^{-ikt}\, dt = 0$ for $k = \pm 1, \pm 2, \ldots$. Pick c so that this is true also if $k = 0$. Since $F - c$ has period 2π, approximate $F - c$ uniformly on $[0, 2\pi]$ by a trigonometric polynomial $T(t) = \sum_{-n}^{n} c_k e^{ikt}$

(see Corollary 1.2 in Chapter 1). Show that $\int_0^{2\pi} | F(t) - c |^2 \, dt$ is arbitrarily small. Hence $F(t) = c$ a.e.}

× **7.2.9.** Show that (iii)–(v) of Proposition 7.6 are equivalent in any pre-Hilbert space P and so are (i) and (ii).

7.2.10. Show that a maximal orthonormal set $\{x_\alpha\}$ in a *pre-Hilbert space* P need *not* be a basis for P, in the sense that it may not be possible to write each $x \in P$ uniquely as $\sum c_\alpha x_\alpha$. (Hint: Let $\{e_i\}$ be an orthonormal basis for a Hilbert space H and P be the linear subspace spanned by $\{\sum_{n=1}^{\infty} n^{-1} e_n, e_2, e_3, \ldots\}$. Then $\{e_2, e_3, \ldots\}$ is a maximal orthonormal set in P, though not a basis for P.)

★ **7.2.11.** Let H_1 and H_2 be Hilbert spaces over the same scalars, and let $T\colon H_1 \to H_2$ be a linear operator such that $T(B) = 0$, where B is an orthonormal basis for H_1. Suppose that $\overline{T(H_1)} = H_2$, G is the graph of T, and $H = H_1 \oplus H_2$ [the direct sum $\{(x, y)\colon x \in H_1, y \in H_2\}$ with inner product $((x_1, y_1) \mid (x_2, y_2)) = (x_1 \mid x_2) + (y_1 \mid y_2)$]. Then show that $\bar{G} = H$ and $\dim G = \dim H_1$. (This result is taken from S. Gudder.†)

★ **7.2.12.** Is the dimension of a pre-Hilbert space the same as that of its completion? [Hint: Use Problem 7.2.11 to find the answer in the negative. Take $H_1 = l_2$, $\dim H_2 = c$, B an orthonormal basis for H_1, and $D(\supset B)$ a Hamel basis for H_1. Then $\dim D = c$. Let $T\colon H_1 \to H_2$ so that $T(B) = 0$ and $T(D - B)$ is an orthonormal basis for H_2. Then $\dim G$ (the graph of T) $= \aleph_0$, but $\dim \bar{G} = c$.]

7.2.13. Use Problem 7.2.12 to show that the "necessity" part of Proposition 6.8 is not necessarily true in a pre-Hilbert space.

7.2.14. *Measure in a Hilbert Space.* Let $(X, \mathscr{B}, \mu)$ be a measure space where $\mathscr{B}$ is the smallest σ-algebra containing the open sets of a Hilbert space X. Suppose $\mu(G) > 0$ for every open set G, $\mu(B + x) = \mu(B)$ for every $B \in \mathscr{B}$, $x \in X$, and X is infinite dimensional. Then prove that $\mu(G) = \infty$ for every nonempty open set G. [Hint: Let (x_i) be an infinite orthonormal set in X. Let $G = \{x\colon \| x \| < r\}$ and $G_n = \{x\colon \| x - (r/2) \cdot x_n \| < r/4\}$. Then each $G_n \subset G$, and G_n's are pairwise disjoint with the same positive measure.]

7.2.15. Prove that the linear subspace spanned by the set $\{x^n e^{-x^2/2}\colon n = 0, 1, 2, \ldots\}$ is dense in $L_2(-\infty, \infty)$. [Hint: Assume for some

† S. Gudder, *Am. Math. Mon.* **81**(1), 29–36 (1974).

$f \in L_2(-\infty, \infty)$,

$$\int_{-\infty}^{\infty} f(t)e^{-t^2/2}t^n \, dt = 0, \quad \text{for } n = 0, 1, 2, \ldots.$$

Let

$$F(z) = \int_{-\infty}^{\infty} f(t)e^{-t^2/2}e^{itz} \, dt,$$

for complex numbers z. Show $F^{(k)}(z) = 0$ for $k = 1, 2, \ldots$ implying that $F(z)$ is identically zero. Therefore

$$\int_{-\infty}^{\infty} f(t)e^{-t^2/2}e^{itx} = 0,$$

if $-\infty < x < \infty$. Multiply this equality by e^{-ixy}, where y is real and integrate with respect to x from $-w$ to w to get

$$\int_{-\infty}^{\infty} f(t)e^{-t^2/2} \frac{\sin w(t-y)}{t-y} \, dt = 0,$$

for every w and y. Conclude that $f(t) = 0$ a.e.]

7.3. The Dual Space and Adjoint Operators

In this section we will consider the dual of a Hilbert space and characterize each element of the dual space. Leading up to our study of operators in the next section, we will also define the adjoint of an operator and give its basic properties.

The reader will recall from Chapter 6 that the dual space E^* of a normed linear space E is the Banach space of all bounded linear functionals on E. In a pre-Hilbert space P each element y in P gives rise to a special element y^* of P^* defined by

$$y^*(x) = (x \mid y), \qquad \text{for all } x \text{ in } P. \tag{7.23}$$

It is easy to verify using the Cauchy–Schwarz Inequality [Proposition 7.1 (iii)] that y^* is a bounded linear functional with $\| y^* \| = \| y \|$. Thus we can define a mapping $y \to y^*$ of P into P^* that is an isometry and conjugate linear, that is,

$$(y + z)^* = y^* + z^*, \qquad \text{for all } y, z \in P$$

$$(\alpha y)^* = \bar{\alpha} y^*, \qquad \text{for all scalars } \alpha \text{ and all } y \text{ in } P.$$

Since P^* is always complete, P will be complete—a Hilbert space—if this mapping is surjective. The converse is also true, as shown in the next theorem.

Theorem 7.5. Let H be a Hilbert space. For each continuous linear functional f in H^*, there exists a unique y in H such that $f(x) = (x \mid y)$ for all x in H. Thus the conjugate linear isometry $y \to y^*$ [given by equation (7.23)] of a pre-Hilbert space P into its dual P^* is surjective if and only if P is actually a Hilbert space. ∎

Proof. Let N be the null space $\{x \in H: f(x) = 0\}$ of f, a closed linear subspace of H. If $N = H$, $f = 0$ and $y = 0$. If $N \neq H$, since $(N^\perp)^\perp = N$ by Theorem 7.2, $N^\perp$ does not equal the zero space. Let $z \in N^\perp$ with $z \neq 0$. Since $N \cap N^\perp = \{0\}$, $f(z) \neq 0$. Replacing z by $[f(z)]^{-1}z$ we may assume $f(z) = 1$.

Now for $x \in H, f(x - f(x)z) = f(x) - f(x) = 0$ so that $x - f(x)z \in N$. Since $z \in N^\perp$,

$$0 = (x - f(x)z \mid z) = (x \mid z) - f(x)(z \mid z)$$

or

$$f(x) = \frac{(x \mid z)}{(z \mid z)} = \left(x \,\middle|\, \frac{z}{\| z \|^2}\right).$$

Letting $y = z/\| z \|^2$, $f(x) = (x \mid y)$, where y is independent of x.

If also w is in H so that $f(x) = (x \mid w)$ for all x in H, then in particular $f(w - y) = (w - y \mid w) = (w - y \mid y)$ or $w = y$. ∎

Remark 7.5. The dual H^* of a Hilbert space H is a Hilbert space with inner product given by

$$(x^* \mid y^*) = (y \mid x)_H, \tag{7.24}$$

for each x^* and y^* in H^* where $x \to x^*$ and $y \to y^*$ as in equation (7.23) under mapping $H \to H^*$. [Here $(\ \mid\ )_H$ denotes the inner product in H.]

The inner product given by equation (7.24) is compatible with the already existent norm on H^* as $\| y^* \|^2 = \| y \|^2 = (y \mid y)_H = (y^* \mid y^*)$. We may say therefore that H and H^* are "conjugate" isomorphic since the isometric surjection $y \to y^*$ from H to H^* is conjugate linear and $(x^* \mid y^*) = \overline{(x \mid y)_H}$.

Corollary 7.2. Each Hilbert space is reflexive. ∎

Proof. The composition of the conjugate isometric surjections $H \to H^*$ and $H^* \to (H^*)^*$ is easily seen to be the natural mapping $J: H \to H^{**}$ given by $J(h) = \hat{h}$, where $\hat{h}(h^*) = h^*(h)$ for all h^* in H^*. ∎

Definition 7.9. If X and Y are normed linear spaces, a sesquilinear form B on $X \times Y$ is said to be bounded if there exists a constant M such that $| B(x, y) | \leq M \| x \| \| y \|$ for all x in X and y in Y. The norm $\| B \|$ of B is given by

$$\| B \| = \inf\{M: | B(x, y) | \leq M \| x \| \| y \| \text{ for all } x \text{ in } X, y \text{ in } Y\}. \blacksquare$$

Proposition 7.10.

(i) If B is a bounded sesquilinear form on $X \times Y$, where X and Y are normed linear spaces, then

$$\begin{aligned} \| B \| &= \sup\{| B(x, y) |: \| x \| < 1 \text{ and } \| y \| < 1\} \\ &= \sup\{| B(x, y) |: \| x \| \leq 1 \text{ and } \| y \| \leq 1\} \\ &= \sup\{| B(x, y) |: \| x \| = 1 \text{ and } \| y \| = 1\}, \end{aligned}$$

and $| B(x, y) | \leq \| B \| \| x \| \| y \|$ for all x in X and y in Y.

(ii) If P is a pre-Hilbert space and B is a bounded Hermitian sesquilinear form on P then

$$\| B \| = \sup\{| B(x, x) |: \| x \| \leq 1\}.$$

[Problem 7.3.6 shows that a converse of (ii) is not always true.] ∎

Proof. The proof of part (i) is analogous to the proof of the corresponding equalities for norms of linear operators on normed linear spaces and is left to the reader.

To prove (ii) we first note that by assumption $B(x, x)$ is real since B is Hermitian. If $\| x \| \leq 1$, then by part (i), $| B(x, x) | \leq \| B \|$ so that

$S \leq \| B \|$ if S equals the $\sup\{| B(x, x) |: \| x \| \leq 1\}$. It suffices to show that for any x and y in P with $\| x \| \leq 1$ and $\| y \| \leq 1$ we have $| B(x, y) | \leq S$. We look at two cases.

Case 1. Suppose $B(x, y)$ is real. By Proposition 7.1 (i)[†] we have

$$B(x, y) = \tfrac{1}{4}\{B(x + y, x + y) - B(x - y, x - y)\}.$$

Using the Parallelogram Law,

$$\begin{aligned} | B(x, y) | &\leq \tfrac{1}{4}\{| B(x + y, x + y) | + | B(x - y, x - y) |\} \\ &\leq \tfrac{1}{4}\{S \| x + y \|^2 + S \| x - y \|^2\} \\ &= \tfrac{1}{4}S\{2\| x \|^2 + 2\| y \|^2\} \leq S. \end{aligned}$$

Case 2. In general, write $| B(x, y) | = \alpha B(x, y)$, where α is a complex number of norm 1. Then $B(\alpha x, y)$, equal to $\alpha B(x, y)$, is a real number. Hence by case 1,

$$| B(x, y) | = | B(\alpha x, y) | \leq S. \quad \blacksquare$$

If H and K are Hilbert spaces each bounded linear operator $T: H \to K$ generates a bounded sesquilinear form B_T on $H \times K$ by the formula

$$B_T(x, y) = (Tx \mid y)_K, \tag{7.25}$$

where $(\mid)_K$ is the inner product of K. It is easy to verify that B_T is sesquilinear and that $\| B_T \| \leq \| T \|$ using the Cauchy–Schwarz Inequality, Proposition 7.1. If $\| x \| \leq 1$ and $\| y \| \leq 1$, then by definition of $\| B_T \|$,

$$| B_T(x, y) | = | (Tx \mid y)_K | = | (y \mid Tx)_K | \leq \| B_T \|.$$

Fixing x and taking the supremum over $\| y \| \leq 1$, we get $\| Tx \| \leq \| B_T \|$ since $(y \mid Tx)_K$ for fixed x is a continuous linear functional on K of norm $\| Tx \|$. Now taking the supremum over $\| x \| \leq 1$, we get $\| T \| \leq \| B_T \|$. Hence $\| T \| = \| B_T \|$.

It is interesting that every bounded sesquilinear form B on $H \times K$ is equal to B_T for some bounded linear operator T from H to K. This is the content of the next theorem.

Theorem 7.6. If H and K are Hilbert spaces, then for each bounded sesquilinear form B on $H \times K$ there exists a unique bounded linear operator $T: H \to K$ such that $B(x, y) = (Tx \mid y)$ for all x and y. Moreover $\| B \| = \| T \|$. ∎

[†] The other two terms on the right side of Proposition 7.1 (i) cancel out, in this case, after simplification.

Proof. For each x in H, define f_x: $K \to F$ (the scalar field) by the formula

$$f_x(y) = \overline{B(x, y)}. \tag{7.26}$$

Then f_x is linear (easily checked) and

$$| f_x(y) | = | B(x, y) | \leq \| B \| \; \| x \| \; \| y \|.$$

So f_x is bounded with $\| f_x \| \leq \| B \| \; \| x \|$. By Theorem 7.5, there exists for each x in H a unique element z_x in K such that $\| f_x \| = \| z_x \|$ and $f_x(y) = (y \mid z_x)$ for all y in K. Define T: $H \to K$ by $Tx = z_x$. Then

$$\overline{B(x, y)} = f_x(y) = (y \mid z_x) = (y \mid Tx),$$

so that $(Tx \mid y) = B(x, y)$. We now assert that T is linear, bounded with norm $\| B \|$, and is the unique operator satisfying $B(x, y) = (Tx \mid y)$:

T is linear: for all y, $(T(\alpha_1 x_1 + \alpha_2 x_2) \mid y) = (\alpha_1 Tx_1 + \alpha_2 Tx_2 \mid y)$.

T is bounded: $\| Tx \| = \| f_x \| \leq \| B \| \; \| x \|$. (Clearly now $\| T \| = \| B \|$ by our remarks before the theorem.)

T is unique: If S is also a bounded linear operator satisfying $(Sx \mid y) = B(x, y)$, then the equality $(Sx - Tx \mid y) = 0$ for all y implies $Sx = Tx$ for all x. ∎

Given a bounded linear operator T: $H \to K$, we have shown there is a unique bounded sesquilinear form B_T on $H \times K$ satisfying the formula $B_T(x, y) = (Tx \mid y)$. Moreover, $\| B_T \| = \| T \|$. In the same manner

$$B^*(y, x) = \overline{B_T(x, y)} = \overline{(Tx \mid y)} = (y \mid Tx)$$

defines a bounded sesquilinear form B^* on $K \times H$ with norm $\| B^* \|$ satisfying

$$\| B^* \| = \| B_T \| = \| T \|.$$

However, by Theorem 7.6 we know that to B^* there corresponds a unique operator S: $K \to H$ satisfying $B^*(y, x) = (Sy \mid x)$ and with $\| S \| = \| B^* \|$. Hence corresponding to T: $H \to K$ there corresponds a (necessarily unique) bounded linear operator S: $K \to H$ satisfying

$$(Tx \mid y) = B_T(x, y) = \overline{B^*(y, x)} = \overline{(Sy \mid x)} = (x \mid Sy)$$

and

$$\| T \| = \| B_T \| = \| B^* \| = \| S \|.$$

The operator S is called the *adjoint* of T and is generally denoted by T^*. Formally we have proved the following result.

Theorem 7.7. To each bounded linear operator $T: H \to K$ there corresponds one and only one linear operator $T^*: K \to H$, called the adjoint of T satisfying $(Tx \mid y)_K = (x \mid T^*y)_H$ for all x in H and y in K. Moreover $\| T \| = \| T^* \|$. ∎

Problem 7.3.7 gives some examples of adjoints of operators.

Remark 7.6. In Chapter 6, the adjoint $T^*: Y^* \to X^*$ of a continuous linear operator $T: X \to Y$ was defined for normed space X and Y by the equation $T^*(f^*)(x) = f^*(T(x))$ for $f^* \in Y^*$ and $x \in X$. Denoting this adjoint momentarily by T' instead of T^*, we remark that T' is *not* the same operator defined in Theorem 7.7 when X and Y are Hilbert spaces. Indeed the domains are different—that of T' being the dual space Y^*. To emphasize this distinction the T^* of Theorem 7.7 is sometimes called the Hilbert space adjoint. The relationship between T^* and T' is examined in Problem 7.3.8.

We conclude this section with several easily verified results giving some properties of bounded linear operators and their adjoints on Hilbert spaces.

Proposition 7.11. For any bounded linear operators $T: H \to K$ and $S: H \to K$ and scalar α we have the following:

(i) $(y \mid Tx)_K = (T^*y \mid x)_H$ for all $x \in H$ and $y \in K$;

(ii) $(S + T)^* = (T^* + S^*)$;

(iii) $(\alpha T)^* = \bar{\alpha} T^*$;

(iv) $(T^*)^* = T$;

(v) $\| T^*T \| = \| TT^* \| = \| T \|^2$;

(vi) $T^*T = 0$ if and only if $T = 0$. ∎

Proof

(i) Left to the reader.

(ii) $$(x \mid (S + T)^*y)_H = ((S + T)x \mid y)_K = (Sx \mid y)_K + (Tx \mid y)_K$$
$$= (x \mid S^*y)_H + (x \mid T^*y)_H = (x \mid S^*y + T^*y)_H$$

for all x in H and y in K. Hence

$$(S + T)^*y = (S^* + T^*)y \qquad \text{for all } y.$$

(iv) $((T^*)^*x \mid y)_K = \overline{(y \mid (T^*)^*x)_K} = \overline{(T^*y \mid x)_K} = (Tx \mid y)_K$

for all x in H and y in K. Hence $(T^*)^* = T$.

(v) For $\| x \| \leq 1$,

$$\| Tx \|^2 = (Tx \mid Tx)_K = (T^*Tx \mid x)_H \leq \| T^*Tx \| \; \| x \| \leq \| T^*T \|.$$

Hence $\| T \|^2 \leq \| T^*T \|$. However, $\| T^*T \| \leq \| T^* \| \; \| T \| = \| T \|^2$ so $\| T^*T \| = \| T \|^2$. Replacing T by T^* and T^* by $(T^*)^*$ or T gives $\| TT^* \| = \| T^* \|^2 = \| T \|^2$.

(vi) follows from (v) and the proof of (iii) is similar to that of (ii). ∎

Proposition 7.12. If T: $H \to K$ and S: $K \to L$ are continuous linear operators (H, K, and L are Hilbert spaces), then $(ST)^* = T^*S^*$. ∎

Proof. For all x in H and y in L,

$$((ST)^*y \mid x)_H = (y \mid STx)_L = (S^*y \mid Tx)_K = (T^*S^*y \mid x)_H. \qquad ∎$$

Proposition 7.13. If T: $H \to K$ is a continuous linear mapping and $M \subset H$ and $N \subset K$ with $T(M) \subset N$, then $T^*(N^\perp) \subset M^\perp$. ∎

Proof. If $y \in N^\perp$ and $m \in M$,

$$(T^*y \mid m)_H = (y \mid Tm)_K = 0 \text{ so that } T^*y \in M^\perp. \qquad ∎$$

A stronger conclusion in Proposition 7.13 can be obtained if N is a closed subspace, as Proposition 7.14 shows.

Proposition 7.14. If T: $H \to K$ is a continuous linear mapping, M is a linear subspace of H, and N is a closed linear subspace of K, then $TM \subset N$ if and only if $T^*(N^\perp) \subset M^\perp$. ∎

Proof. The necessity is proved by Proposition 7.13. If $T^*(N^\perp) \subset M^\perp$, then also $T^{**}(M^{\perp\perp}) \subset N^{\perp\perp}$ by Proposition 7.13. Since $M^{\perp\perp} \supset M$, $N^{\perp\perp} = N$ by Theorem 7.2 and $T^{**} = T$, $T(M) \subset N$. ∎

Proposition 7.15. If T: $H \to K$ is a continuous linear mapping, then

(i) $\{x\colon Tx = 0\} = [T^*(K)]^\perp$;

(ii) $\{x: Tx = 0\}^{\perp} = \overline{T^*(K)}$;
(iii) $\{y: T^*y = 0\} = [T(H)]^{\perp}$;
(iv) $\{y: T^*y = 0\}^{\perp} = \overline{T(H)}$. ∎

Proof. For a fixed $x \in H$,

$$(Tx \mid y)_K = 0 \text{ for any } y \in K, \tag{7.27}$$

if and only if

$$(x \mid T^*y)_H = 0 \text{ for any } y \in K. \tag{7.28}$$

This proves (i). From (i),

$$\{x: Tx = 0\} = [T^*(K)]^{\perp} = [\overline{T^*(K)}]^{\perp}, \tag{7.29}$$

so that

$$\{x: Tx = 0\}^{\perp} = [\overline{T^*(K)}]^{\perp\perp} = \overline{T^*(K)}. \tag{7.30}$$

This proves (ii). The other two parts follow similarly if we work with T^* instead of T and note that $T^{**} = T$. ∎

Having introduced the idea of the adjoint in this section and having given some of its essential properties, we will use it in the next section to define special classes of operators. In particular we will show in Section 7.5 that each bounded linear operator $T: H \to H$ which equals its adjoint T^* has a neat representation.

Problems

× **7.3.1.** If P is a pre-Hilbert space such that $N^{\perp\perp} = N$ for every closed linear subspace N of P, show P is a Hilbert space by showing that every continuous linear functional y^* in P^* is of the form $y^*(x) = (x \mid y)$ for some y in P.

× **7.3.2.** Prove that if f is a linear functional on a Hilbert space H, then the null space $N = \{x: f(x) = 0\}$ is dense in H if f is not continuous.

× **7.3.3.** (i) Prove that in a Hilbert space H, a sequence (x_n) converges weakly to x in H if and only if the sequence $((x_n \mid y))$ converges to $(x \mid y)$ for each y in H.

(ii) Give an example of a sequence (x_n) in the infinite-dimensional Hilbert space H that converges weakly but not strongly in H. [Hint: Use (i) and Bessel's Inequality.]

7.3.4. (i) Let $P\times Q$ be the direct product of pre-Hilbert spaces P and Q. For $x = (p_1, q_1)$ and $y = (p_2, q_2)$ in $P\times Q$, define $(x \mid y)$ as $(p_1 \mid p_2)_P + (q_1 \mid q_2)_Q$. Show that $P\times Q$ becomes a pre-Hilbert space with this inner product and that $\| (x \mid y) \|^2 = \| x \|^2 + \| y \|^2$. Show that $P\times Q$ equals $P_0 \oplus Q_0$, where P_0 is the range of the mapping from P into $P\times Q$ given by $p \to (p, 0)$ and similarly Q_0 is the range of the mapping from Q into $P\times Q$. Show that $P\times Q$ is a Hilbert space if and only if P and Q are Hilbert spaces.

(ii) Generalize (i) to an arbitrary collection (P_α) of pre-Hilbert spaces.

7.3.5. If X and Y are normed linear spaces, prove that the set $B(X, Y)$ of bounded sesquilinear forms on $X\times Y$ is a normed linear space linearly isometric to $L(X, (\bar{Y})^*)$, where $\bar{Y}$ is the complex conjugate of Y—that is, the same space as Y except that addition $\oplus$ and scalar multiplication $\odot$ in $\bar{Y}$ is given by $\alpha \odot x \oplus \beta \odot y = \bar{\alpha} \cdot x + \bar{\beta} \cdot y$ for all scalars α, β and vectors x and y, where $+$ and $\cdot$ represent the operations in Y. Conclude $B(X, Y)$ is a Banach space. Compare Theorem 7.6.

7.3.6. (i) Let P be a complex pre-Hilbert space and define $B(x, y) = i(x \mid y)$. Prove B is a bounded sesquilinear form on P with $\| B \| = \sup\{| B(x, x) | : \| x \| \leq 1\}$ but B is not Hermitian. Compare Proposition 7.10(ii).

(ii) Let P be any pre-Hilbert space. Suppose $T: P \to P$ is a continuous linear mapping with $\| T \| = 1$ and $Tp = p$ for some nonzero p in P. Define $B_T(x, y) = (Tx \mid y)$. Show that B_T is sesquilinear with $\| B_T \| = \sup\{| B_T(x, x) | : \| x \| \leq 1\}$, but B_T is not necessarily Hermitian.

× **7.3.7.** (i) Let T be a mapping from a separable Hilbert space H into itself and let (e_i) be a basis of H. Then $Te_j = \sum_{i=1}\alpha_{ij}e_i$ as in Proposition 6.6. The collection (α_{ij}) for $i = 1, 2, \ldots$ and $j = 1, 2, \ldots$ is called the matrix of T with respect to the basis (e_i). Show $(Te_j \mid e_k) = \alpha_{kj}$ and that the matrix of T^* with respect to (e_i) is the collection of scalars (β_{ij}) for $i = 1, 2, \ldots,$ and $j = 1, 2, \ldots$ where $\beta_{ij} = \bar{\alpha}_{ji}$.

(ii) What is the adjoint of the shift operator $T: l_2 \to l_2$ given by $T(a_1, a_2, \ldots) = (0, a_1, a_2, \ldots)$?

× **7.3.8.** If H and K are Hilbert spaces, and $U: H \to H^*$ and $V: K \to K^*$ are the conjugate isometries discussed in Theorem 7.5, prove that $T^* = U^{-1}T'V$, where T^* is the Hilbert space adjoint of T, and T' is the Banach space adjoint from K^* to H^* given by $T'(k^*)h = k^*(Th)$.

7.3.9. Let H and K be Hilbert spaces and $\{h_i\}_{i\in I}$ and $\{k_j\}_{j\in J}$ be bases for H and K, respectively. If $T: H \to K$ is a bounded linear mapping, $\sum_I \| Th_i \|^2$ converges if and only if $\sum_J \| T^*k_j \|^2$ converges and in this case both sums equal $\sum_{i,j} | (Th_i \mid k_j) |^2$.

7.3.10. A bounded linear operator $T: H \to K$ is said to be a *Hilbert–Schmidt operator* if there exists a basis $\{h_i\}$ in H such that $\sum_I \| Th_i \|^2 < \infty$. Prove the following using Problem 7.3.9:

(i) If T is a Hilbert–Schmidt operator, show $\sum_I \| Tg_i \|^2 < \infty$ for every basis $\{g_i\}_I$ of H.

(ii) T is a Hilbert–Schmidt operator if and only if T^* is a Hilbert–Schmidt operator.

(iii) The class of Hilbert–Schmidt operators is a pre-Hilbert space with the inner product $(S \mid T) = \sum_I (Sh_i \mid Th_i)$ where $\{h_i\}_I$ is a basis of H.

(iv) The pre-Hilbert space of (iii) is complete. Prove this by showing that it is isomorphic to $l_2(I \times J)$ via $T \to (x_{ij})_{I \times J}$, where $x_{ij} = (Th_i \mid k_j)$ and where $\{h_i\}_I$ and $\{k_j\}_J$ are bases of H and K, respectively.

(v) Show that a Hilbert–Schmidt operator is compact.

(vi) If $(X, \mathscr{B}, \mu)$ is a σ-finite measure space and $\int\int | K(x, y) |^2 d\mu(x)\, d\mu(y) < \infty$, prove that $T: L_2(\mu) \to L_2(\mu)$ given by $(Tf)(x) = \int f(y) K(xy)\, d\mu(y)$ is a Hilbert–Schmidt operator.

7.3.11. Prove that the following statements are equivalent in a *pre-Hilbert* space P.

(i) P is complete.

(ii) If M is a closed subspace of P, then $M \oplus M^{\perp} = P$.

(iii) If M is a closed subspace of P, then $M = M^{\perp\perp}$.

(iv) If M is a proper closed subspace of P, then $M^{\perp} \neq \{0\}$.

(v) If f is a continuous linear functional on P, then there exists $x \in P$ such that $f(y) = (y \mid x)$ for all $y \in P$.

× **7.3.12.** Suppose that $S \subset R$ and that for each s in S there is a bounded linear operator $T(s)$ on H. An operator A on H is the *weak limit* of the function $T: S \to L(H, H)$ as $s \to s_0$, written $A = w\text{-}\lim_{s \to s_0} T(s)$, if for any pair h and k in H

$$(Ah \mid k) = \lim_{s \to s_0} \big(T(s)h \mid k\big).$$

An operator A on H is the *strong limit* of $T(s)$ as $s \to s_0$, written $A = s\text{-}\lim_{s \to s_0} T(s)$ if for all h in H

$$\lim_{s \to s_0} \| (T(s) - A)h \| \to 0.$$

An operator A on H is the *uniform limit* of $T(s)$ as $s \to s_0$ if

$$\lim_{s \to s_0} \| T(s) - A \| \to 0.$$

Show that if A is the uniform limit of $T(s)$ as $s \to s_0$, then A is the strong limit of $T(s)$ as $s \to s_0$. Show also that if A is the strong limit of $T(s)$ as $s \to s_0$, then A is the weak limit of $T(s)$ as $s \to s_0$. (Note here that s_0 may be $\pm\infty$ and S may be the natural numbers N.)

× **7.3.13.** Let A be a linear operator from a Hilbert space H into itself with $(x \mid Ay) = (Ax \mid y)$ for all x and $y \in H$. Prove that A is bounded. (Hint: Use the Closed Graph Theorem.)

7.3.14. If $T \in L(H, H)$, define $\omega(T) = \sup\{|(Tx \mid x)| : \|x\| = 1\}$. Prove that

(i) $\omega(cT) = |c| \, \omega(T)$ for all scalars c;
(ii) $\omega(T_1 + T_2) \leq \omega(T_1) + \omega(T_2)$ for $T_1, T_2 \in L(H, H)$;
(iii) $\omega(T) \leq \|T\| \leq 2\omega(T)$.

Conclude that ω defines a norm on $L(H, H)$ equivalent to the usual operator norm. [Hint: To prove $\|T\| \leq 2\omega(T)$, let $B(x, y) = (Tx \mid y)$ and prove $|B(x, y)| \leq 2\omega(T)$ for all $\|x\| = \|y\| = 1$.]

7.4. The Algebra of Operators. The Spectral Theorem and the Approximation Theorem for Compact Operators

In this section we will first examine some special classes of operators on a Hilbert space H—that is, special classes of operators from H into itself. These special classes of operators can all be defined by use of the adjoint. Secondly, we will prove a spectral theorem for compact operators which the spectral theorem of the next section will generalize. Finally, we will show how each completely continuous operator can be approximated by operators with finite-dimensional ranges.

Let us first give in capsule form the definition of all types of operators we will consider in this section.

Definition 7.10. Let $T: H \to H$ be a continuous linear operator with adjoint T^*.

(i) T is *isometric* if and only if $T^*T = I$, the identity of H.
(ii) T is *unitary* if and only if $T^*T = TT^* = I$.
(iii) T is *self-adjoint* (or Hermitian) if and only if $T = T^*$.
(iv) T is a *projection* if and only if $T^2 = T$ and $T^* = T$.
(v) T is *normal* if and only if $T^*T = TT^*$. ∎

Note that when we speak of T as being any one of the types of operators of our Definition 7.10, T is understood to be continuous and linear. The reader should consult Problem 7.4.1 for various examples of these types of operators.

Clearly, every unitary, every self-adjoint, and every projection operator is normal. The following four propositions give equivalences of isometric, unitary, self-adjoint, and normal operators, respectively. We first prove the following lemma.

Lemma 7.3. If H is a complex Hilbert space and $S, T: H \to H$ are bounded linear operators such that $(Sx \mid x) = (Tx \mid x)$ for all x, then $S = T$. ∎

Proof. The sesquilinear forms $B_S(x, y) = (Sx \mid y)$ and $B_T(x, y) = (Tx \mid y)$ are such that $B_S(x, x) = B_T(x, x)$ for all x in H. By the polarization identity [Proposition 7.1(i)], $B_S(x, y) = B_T(x, y)$ for all x and y in H. By Remark 7.1, $S = T$. ∎

Proposition 7.16. The following conditions on $T: H \to H$ are equivalent.

(i) T is isometric.

(ii) $(Tx \mid Ty) = (x \mid y)$ for all x and y in H.

(iii) $\| Tx \| = \| x \|$ for all x in H. ∎

Proof. (i) ⇒ (ii) since $(Tx \mid Ty) = (x \mid T^*Ty) = (x \mid y)$. Trivially (ii) implies (iii). If we assume (ii), then $(T^*Tx \mid y) = (x \mid y)$ for all x and y so that $T^*Tx = x$ for all x, or $T^*T = I$. Hence (ii) ⇒ (i). If H is a real Hilbert space, then by Proposition 7.1 (i)

$$(Tx \mid Ty) = \tfrac{1}{4}\{\| Tx + Ty \|^2 - \| Tx - Ty \|^2\}.$$

Clearly (iii) implies (ii) if H is real. Similarly if H is complex Proposition 6.1 (i) gives that (iii) ⇒ (ii). ∎

Proposition 7.17. The following conditions on $T: H \to H$ are equivalent.

(i) T is unitary.

(ii) T^* is unitary.

(iii) T and T^* are isometric.
(iv) T is isometric and T^* is injective.
(v) T is isometric and surjective.
(vi) T is bijective and $T^{-1} = T^*$. ∎

Proof. Since $T^{**} = T$, the equivalences of (i), (ii), and (iii) are trivial, as are the implications (iii) ⇒ (iv) and (vi) ⇒ (i). The proof is completed by showing (iv) ⇒ (v) and (v) ⇒ (vi).

(iv) ⇒ (v): Since T is isometric, $T(H)$ is closed. Hence by Proposition 6.15,

$$H = \{0\}^\perp = \{x: T^*x = 0\}^\perp = \overline{T(H)} = T(H).$$

(v) ⇒ (vi): Let $S = T^{-1}$. To show $T^* = S$. Since T is isometric,

$$T^* = T^*I = T^*(TS) = (T^*T)S = IS = S.$$ ∎

Proposition 7.18. Let $T: H \to H$ be a continuous linear operator. Let (i), (ii), (iii), and (iv) represent the following statements:

(i) T is self-adjoint.
(ii) $(Tx \mid y) = (x \mid Ty)$ for all x and y in H.
(iii) $(Tx \mid x) = (x \mid Tx)$ for all x in H.
(iv) $(Tx \mid x)$ is real for all x in H.

Then (i) is equivalent to (ii), (iii) is equivalent to (iv), and if H is a complex Hilbert space, all the statements are equivalent. ∎

Proof. All the implications are trivial except perhaps (iv) ⇒ (i) in the complex case. Suppose (iv) holds. Then

$$(T^*x \mid x) = (x \mid Tx) = (Tx \mid x)$$

for each x in H. By Lemma 7.3, $T = T^*$. ∎

Proposition 7.19. If H is a Hilbert space, then $T: H \to H$ is normal if and only if $\| T^*x \| = \| Tx \|$ for each x in H. ∎

Proof. If T is normal,

$$\| T^*x \|^2 = (T^*x \mid T^*x) = (TT^*x \mid x) = (T^*Tx \mid x) = (Tx \mid Tx) = \| Tx\|^2. \quad (7.31)$$

Conversely, if $\| T^*x \| = \| Tx \|$, equation (7.31) shows that T is normal using Lemma 7.3 in the case H is complex. However, in either case the Hermitian sesquilinear forms B and B', given by $B(x, y) = (Tx \mid Ty)$ and $B'(x, y) = (T^*x \mid T^*y)$, respectively, are such that

$$B(x, x) = (Tx \mid Tx) = \| Tx \|^2 = \| T^*x \|^2 = (T^*x \mid T^*x) = B'(x, x),$$

for all x in H. By the polarization identity, Proposition 7.1 (i), $B(x, y) = B'(x, y)$ for all x and y in H. Hence

$$(T^*Tx \mid y) = (Tx \mid Ty) = (T^*x \mid T^*y) = (TT^*x \mid y),$$

for all x and y in H. Hence $T^*T = TT^*$ by Remark 7.1. ∎

We will take a closer look at self-adjoint operators in what follows. Before doing so, let us examine a few pertinent theorems regarding some special self-adjoint operators—the projection operators.

Recall that if N is a closed linear subspace of H, then $H = N \oplus N^\perp$. We can define $P: H \to H$ by the following rule: If $x = y + z$ with $y \in N$ and $z \in N^\perp$, let $Px = y$. P is easily checked to be a bounded linear operator, but also satisfies the equations $P^2 = P$ and $P^* = P$. Indeed, $Px \in N$ implies $P(Px) = Px$ and

$$(Px_1 \mid x_2) = (y_1 \mid y_2 + z_2) = (y_1 + z_1 \mid y_2) = (x_1 \mid Px_2),$$

where $x_1 = y_1 + z_1$ and $x_2 = y_2 + z_2$ with $y_1, y_2 \in N$ and $z_1, z_2 \in N^\perp$. P is called the projection† operator associated with N and is denoted by P_N.

Proposition 7.20. If $T: H \to H$ is a projection operator, then there is one and only one closed linear subspace N such that $T = P_N$. In fact $N = T(H)$ and $N^\perp = \operatorname{Ker} T$. ∎

Proof. Let $N = \{x: (T - I)x = 0\} = \{x: Tx = x\}$, a closed subspace. If y is in the range $R(T)$ of T, then for some x, $y = Tx = T(Tx) = Ty$ so that y is in N. Conversely, if $y \in N$, $y = Ty \in R(T)$. Therefore $N = R(T)$. Now by Proposition 7.15

$$\{x: Tx = 0\} = [T^*(H)]^\perp = [T(H)]^\perp$$

† P is also called the orthogonal projection on N.

or $N^\perp = \{x: Tx = 0\}$. Let $x \in H$. Then $x = y + z$ with $y \in N$ and $z \in N^\perp$. Hence $Tx = T(y + z) = Ty + Tz = y$. Hence $T = P_N$.

If $T = P_N = P_M$ for some subspaces N and M then $N = P_N(H) = P_M(H) = M$. ∎

• **An Application of Proposition 7.20: The Mean Ergodic Theorem.** In Chapter **3**, we discussed measure-preserving mappings and the Individual Ergodic Theorem showing the existence of an a.e. pointwise limit for L_1 functions. Here we show how the subject of unitary operators comes up naturally in the context of ergodic theory. We present below the Mean Ergodic Theorem, proven by von Neumann in 1933, which will consider L_2 functions and convergence in the L_2 norm. We will use Proposition 6.20.

First, we notice that if Φ is a measure-preserving mapping on a measure space $(X, \mathscr{A}, \mu)$ and if

$$T(f)(x) = f(\Phi(x)), \qquad f \in L_2(\mu),$$

then T is a unitary† operator on L_2. Thus the Mean Ergodic Theorem in L_2 can be presented in the following Hilbert space setup.

The Mean Ergodic Theorem. Let T be a unitary operator on a Hilbert space H. Let P be the orthogonal projection onto the closed linear subspace $N = \{x: Tx = x\}$. Then, for any $x \in H$,

$$T_n x = \frac{1}{n} \sum_{k=0}^{n-1} T^k x \to Px \qquad \text{as } n \to \infty.$$ ∎

Proof. First, we notice that by Proposition 7.15,

$$[(I - T)(H)]^\perp = \operatorname{Ker}(I - T^*) = \{x: T^*x = x\} = \{x: T^{-1}x = x\}$$
$$= N(= P(H)).$$

Now if $x = y - Ty$, then since $\| T \| = 1$ [by Proposition 7.11(v)],

$$\| T_n x \| = \| (1/n)[y - T^n y] \| \leq (2/n) \| y \| \to 0, \qquad \text{as } n \to \infty.$$

By Theorem 7.2 and Proposition 7.20,

$$\operatorname{Ker} P = N^\perp = \overline{(I - T)(H)}.$$

† Note that $T^*(f)(x) = f(\Phi^{-1}(x))$.

Therefore, for $x \in N^{\perp}$, $\lim_{n\to\infty} T_n x$† $= 0 = Px$. Also, for $x \in N = P(H)$, $\lim_{n\to\infty} T_n x = Tx = x = Px$. Since $H = N \oplus N^{\perp}$, the theorem now follows easily. ∎

Having established a one-to-one correspondence between projections on H and closed linear subspaces of H, it is interesting to investigate the preservation properties of this correspondence. Before giving these properties we define a partial ordering on the class of self-adjoint operators.

Definition 7.11. A self-adjoint operator T is said to be positive in case $(Tx \mid x) \geq 0$ for all $x \in H$. If S and T are self-adjoint operators, $S \leq T$ if $T - S$ is positive, written $T - S \geq 0$. ∎

Proposition 7.21. Suppose M_1 and M_2 are closed linear subspaces corresponding, respectively, to projections P_1 and P_2 on H. Then the following hold:

(i) The assertions $M_1 \perp M_2$, $P_1P_2 = 0$, $P_2P_1 = 0$, $P_1(M_2) = \{0\}$, and $P_2(M_1) = \{0\}$ are equivalent.

(ii) P_1P_2 is a projection if and only if $P_1P_2 = P_2P_1$. If this condition is satisfied the range of P_1P_2 is $M_1 \cap M_2$.

(iii) The assertions $P_1 \leq P_2$, $\| P_1x \| \leq \| P_2x \|$ for all x, $M_1 \subset M_2$, $P_2P_1 = P_1$, and $P_1P_2 = P_1$ are equivalent. ∎

Proof. (i) $M_1 \perp M_2$ implies $M_2 \subset M_1^{\perp}$. As shown in the proof of Proposition 7.20, $M_2 = \{x: P_2x = x\} = R(P_2)$ and $M_1^{\perp} = \{x: P_1x = 0\}$. Since $M_2 \subset M_1^{\perp}$ and $P_2x \in M_2$ for any x in H, $P_1P_2 = 0$. Now suppose $P_1P_2 = 0$. Then if $x \in M_2$, $P_2x = x$ and therefore $P_1x = P_1P_2x = 0$. Hence $P_1(M_2) = \{0\}$. Now suppose $P_1(M_2) = \{0\}$. Then $M_2 \subset M_1^{\perp}$ and therefore $M_2 \perp M_1$. The other equivalences of (i) follow by symmetry.

(ii) P_1P_2 a projection implies

$$P_1P_2 = (P_1P_2)^* = P_2^*P_1^* = P_2P_1.$$

Conversely, $P_1P_2 = P_2P_1$ implies

$$(P_1P_2)^2 = P_1(P_2P_1)P_2 = P_1^2P_2^2 = P_1P_2$$

† Note that

$$\| T_nx \| \leq \| T_nz \| + \| T_n(x - z) \| \leq \| T_nz \| + \| x - z \|$$

for $z \in (I - T)(H)$.

and

$$(P_1P_2)^* = P_2{}^*P_1{}^* = P_2P_1 = P_1P_2.$$

Finally if P_1P_2 is a projection, then the range M of $P = P_1P_2$ is contained in M_1 and M_2 since P_1 and P_2 commute. Also if $x \in M_1 \cap M_2$, then $P_1x = x = P_2x$ or $Px = x$ so that $x \in M$. Hence $M = M_1 \cap M_2$.

(iii) If $P_1 \leq P_2$, then $\| P_1x \|^2 = (P_1x \mid x) \leq (P_2x \mid x) = \| P_2x \|^2$ for all x. If $\| P_1x \| \leq \| P_2x \|$ for all x, then for $x \in M_1$,

$$\| x \| = \| P_1x \| \leq \| P_2x \|,$$

so that $\| P_2x \| \geq \| x \|$. Since $\| x \|^2 = \| P_2x \|^2 + \| x - P_2x \|^2$, $P_2x = x$ and $x \in M_2$. If $M_1 \subset M_2$, then for all x, $P_1x \in M_2$. Hence $P_2P_1x = P_1x$. If $P_2P_1 = P_1$, then $P_1P_2 = (P_2P_1)^* = P_1^* = P_1$. If $P_1P_2 = P_1$, then

$$(P_1x \mid x) = \| P_1x \|^2 = \| P_1P_2x \|^2 \leq \| P_2x \|^{2\dagger} = (P_2x \mid x). \quad \blacksquare$$

We have therefore seen that the one-to-one correspondence between closed subspaces of H and projections on H preserves order: $M \subset N$ if and only if $P_M \leq P_N$. Thus the set of projections is a partially ordered set such that for every family of projections $\{P_i\}_I$ there exists a "greatest" projection "smaller than" each P_i and a "smallest" projection "greater than" each P_i.

Self-adjoint operators are also called *Hermitian* operators. This terminology is first of all consistent with that used for square complex matrices (a_{ij}) where $a_{ij} = \bar{a}_{ij}$. Indeed if T is a self-adjoint operator on the Hilbert space C^n, let (a_{ij}) $i = 1, \ldots, n$ and $j = 1, \ldots, n$ be the matrix that represents T relative to some orthonormal basis $\{c_1, \ldots, c_n\}$ in C^n—that is, $Tc_j = \sum_{i=1}^n a_{ij}c_i$. Problem 7.3.7 (i) shows that the matrix of T^* is (b_{ij}), where $b_{ij} = \bar{a}_{ji}$. Thus T is Hermitian if and only if (a_{ij}) is a Hermitian matrix. In particular we know from the theory of diagonalization of Hermitian matrices that T can be represented with respect to some orthonormal basis, say $\{x_1, \ldots, x_n\}$, by a diagonal matrix so that $Tx_i = \lambda_i x_i$, $i = 1, \ldots, n$, for some scalars $\lambda_1, \ldots, \lambda_n$. Hence if $x \in C^n$, $x = \sum_{i=1}^n (x \mid x_i)x_i$ by Proposition 7.6 so that

$$Tx = \sum_{i=1}^{n} \lambda_i (x \mid x_i)x_i. \tag{7.32}$$

† For a projection operator $Q \neq 0$, $\| Q \| = 1$ since $Q = Q^2$ and $\| Qx \|^2 \leq \| Qx \|^2 + \| x - Qx \|^2 = \| x \|^2$. See 7.4.19.

It is formula (7.32) we wish to generalize to compact self-adjoint operators in this section. We return to it in Theorem 7.8.

The term "Hermitian" for self-adjoint operators is also consistent with that used for sesquilinear forms. Indeed, if B_T is defined on $H \times H$ by $B_T(x, y) = (Tx \mid y)$, then $\overline{B_T(x, y)} = \overline{(Tx \mid y)} = (y \mid Tx) = (T^*y \mid x)$ so that $\overline{B_T(x, y)} = B_T(y, x)$ if and only if $T = T^*$, by Remark 7.1. Hence if T is self-adjoint, then by Proposition 7.10 (ii),

$$\| T \| = \| B_T \| = \sup\{| (Tx \mid x) |: \| x \| \leq 1\}.$$

These remarks enable us to prove the next results.

Lemma 7.4. Let T be a self-adjoint operator on H. Let $m_T = \inf\{(Tx \mid x): \| x \| = 1\}$ and $M_T = \sup\{(Tx \mid x): \| x \| = 1\}$. Then $m_T I \leq T \leq M_T I$ and $\| T \| = \max\{M_T, -m_T\}$. ∎

Proof. For all x in H,

$$m_T(x \mid x) = m_T\| x \|^2 \leq (Tx \mid x) \leq M_T\| x \|^2 = M_T(x \mid x),$$

which shows $m_T I \leq T \leq M_T I$. The rest is clear. ∎

Lemma 7.5. Let T be a positive self-adjoint operator on H. Then

$$| (Tx \mid y) |^2 \leq (Tx \mid x)(Ty \mid y),$$

for all $x, y \in H$. ∎

Proof. Apply the Cauchy–Schwarz inequality, Proposition 7.1 (iii), to the positive Hermitian sesquilinear form $B(x, y) = (Tx \mid y)$. ∎

Lemma 7.6. Suppose T is a self-adjoint operator, on a Hilbert space H.

(i) If μ is an eigenvalue of T, then μ is real and $m_T \leq \mu \leq M_T$,

$$m_T = \inf\{(Tx \mid x): \| x \| = 1\}, \quad M_T = \sup\{(Tx \mid x): \| x \| = 1\}.$$

Eigenvectors corresponding to distinct eigenvalues are orthogonal.

(ii) $\sigma(T) \subset [m_T, M_T]$ and the endpoints m_T and M_T are both in $\sigma(T)$. ∎

Proof. (i) Since $(Tx \mid x)$ is real and equals $\mu(x \mid x)$ with $(x \mid x) \geq 0$, μ must be real. Also if x is an eigenvector with $\| x \| = 1$,

$$\mu = \mu \| x \|^2 = (\mu x \mid x) = (Tx \mid x)$$

so that $m_T \leq \mu \leq M_T$.

Now suppose μ_0 is an eigenvalue with $\mu_0 \neq \mu$ and let x_0 and x be eigenvectors corresponding to μ_0 and μ, respectively. Since μ_0 is real,

$$\mu(x \mid x_0) = (Tx \mid x_0) = (x \mid Tx_0) = (x \mid \mu_0 x_0) = \mu_0(x \mid x_0).$$

Thus $(x \mid x_0) = 0$ since $\mu \neq \mu_0$.

(ii) Suppose $\lambda \in F - [m_T, M_T]$. Then $\delta = d(\lambda, [m_T, M_T]) > 0$ and for every x in H with $\| x \| = 1$,

$$\| (T - \lambda I)x \| \geq | ((T - \lambda I)x \mid x) | = | (Tx \mid x) - \lambda | \geq \delta. \quad (7.33)$$

Hence by Proposition 6.5 in Chapter 6, $T - \lambda I$ is a bijection onto the range $R(T - \lambda I)$ of $T - \lambda I$ with a bounded inverse on $R(T - \lambda I)$. In particular $R(T - \lambda I)$ is complete and closed. Suppose $R(T - \lambda I)^\perp \neq \{0\}$. Then for $y \in R(T - \lambda I)^\perp$ with $\| y \| = 1$, we have $((T - \lambda I)(y) \mid y) = 0$, which contradicts (7.33). Hence $R(T - \lambda I) = H$ and $\lambda \notin \sigma(T)$.

Finally we show that m_T and M_T are in $\sigma(T)$. Since by Lemma 7.4 $M_T I - T \geq 0$, we have, using Lemma 7.5 that for $\| x \| = 1$

$$\begin{aligned} \| (M_T I - T)x \|^4 &= | ((M_T I - T)x \mid (M_T I - T)x) |^2 \\ &\leq ((M_T I - T)x \mid x)((M_T I - T)^2 x \mid (M_T I - T)x) \\ &\leq \| M_T I - T \|^3 (M_T - (Tx \mid x)). \end{aligned}$$

By virtue of the definition of M_T,

$$\begin{aligned} \inf\{\| (M_T I - T)x \|^4 : \| x \| = 1\} &\leq \inf\{M_T - (Tx \mid x) : \| x \| = 1\} \\ &\quad \cdot \| M_T I - T \|^3 = 0. \end{aligned}$$

Thus $(M_T I - T)^{-1}$ does not exist as a bounded operator, so $M_T \in \sigma(T)$. Similarly $m_T \in \sigma(T)$. ∎

In particular if T is a nonzero compact, self-adjoint operator on H, then Lemmas 7.4, 7.6 and Theorem 6.25 show us that either $\| T \|$ or $- \| T \|$ is an eigenvalue of T. Hence there is a real number λ such that $| \lambda | = \| T \|$ and a vector x in H with $\| x \| = 1$ and $Tx = \lambda x$. We have proved the following result.

Lemma 7.7. Suppose T is a nonzero compact, self-adjoint operator on H. Then either $\| T \|$ or $-\| T \|$ is an eigenvalue of T and there is a corresponding eigenvector x such that $\| x \| = 1$ and $| (Tx \mid x) | = \| T \|$. ∎

Before proving the spectral theorem for compact self-adjoint operators, let us establish two results concerning self-adjoint operators which will be needed in the next section.

Lemma 7.8. If $T\colon H \to H$ is a self-adjoint operator, then T is positive if and only if $\sigma(T) \subset [0, \infty)$. ∎

Proof. Let $m_T = \inf\{(Tx \mid x)\colon \| x \| = 1\}$. If T is positive, $m_T \geq 0$ and so by Lemma 7.6, $\sigma(T) \subset [0, \infty)$. On the other hand, if $\sigma(T) \subset [0, \infty)$ then $m_T \geq 0$ by Lemma 7.6 and $(Tx \mid x) \geq 0$ for all x. ∎

Lemma 7.9. Suppose (T_n) is a bounded sequence of self-adjoint operators with $T_n \leq T_{n+1}$ (as in Definition 7.11) for $n = 1, 2, \ldots$. Then there is a self-adjoint operator T such that T_n converges *strongly* to T (that is, $T_n x \to Tx$ for each x in H). ∎

Proof. By assumption there exists a positive number M such that $\| T_n \| \leq M$ or $-MI \leq T_n \leq MI$. If $\frac{1}{2}M^{-1}(T_n + MI)$ converges strongly to a self-adjoint operator S, then T_n converges strongly to the self-adjoint operator $2MS - MI$. Thus, replacing T_n by $\frac{1}{2}M^{-1}(T_n + MI)$, we may suppose $0 \leq T_n \leq I$ for $n = 1, 2, \ldots$.

If m and n are positive integers with $m > n$, then $0 \leq T_{mn} \leq I$ if $T_{mn} = T_m - T_n$, and $\| T_{mn} \| \leq 1$ by Lemma 7.4. Using Lemma 7.5 and Proposition 7.1 (iii)

$$\begin{aligned} \| T_m x - T_n x \|^4 &= \| T_{mn} x \|^4 = \| (T_{mn} x \mid T_{mn} x) \|^2 \\ &\leq (T_{mn} x \mid x)(T_{mn}^2 x \mid T_{mn} x) \\ &\leq (T_{mn} x \mid x) \, \| T_{mn}^2 x \| \, \| T_{mn} x \| \\ &\leq \| x \|^2 [(T_m x \mid x) - (T_n x \mid x)]. \end{aligned} \tag{7.34}$$

Since $0 \leq T_n \leq T_{n+1} \leq I$, $(T_n x \mid x)$ is a bounded increasing sequence of real numbers which converges so that $(T_n x)$ is a Cauchy sequence by (7.34). Let $Tx = \lim_{n\to\infty} T_n x$. It is easy to check that the operator T thus defined is a bounded self-adjoint operator. ∎

We have now come to a main objective of this section.

Theorem 7.8. Suppose $T\colon H \to H$ is a nonzero, compact, self-adjoint operator.

(i) If $T(H)$ is finite-dimensional, then the nonzero eigenvalues of T form a real finite sequence $\lambda_1, \lambda_2, \ldots, \lambda_n$ and $Tx = \sum_{i=1}^{n} \lambda_i(x \mid x_i)x_i$ for all x in H, where $x_1, x_2, \ldots, x_n$ is a corresponding orthonormal set of eigenvectors.

(ii) If T does not have a finite-dimensional range, then the nonzero eigenvalues of T form a real infinite sequence $\lambda_1, \lambda_2, \ldots$ with $|\lambda_i| \to 0$ such that $Tx = \sum_{i=1}^{\infty} \lambda_i(x \mid x_i)x_i$ for each x in H where $x_1, x_2, \ldots$ is an orthonormal set of eigenvectors with x_i corresponding to λ_i. ∎

Proof. Suppose first that T is not of finite rank. We show first that for each n, there exists a nonzero closed subspace X_n of H, an eigenvalue λ_n of T with $|\lambda_n| = \| T|_{X_n} \|$, and a corresponding eigenvector x_n for λ_n with $\| x_n \| = 1$ such that

$$\begin{aligned} &|\lambda_n| \geq |\lambda_{n+1}|, \\ &X_{n+1} = \{x : (x \mid x_i) = 0, \text{ for } i = 1, 2, \ldots, n\}, \\ &x_n \in X_n, \quad \text{for } n = 1, 2, \ldots. \end{aligned} \tag{7.35}$$

Let $X_1 = H$ and $T_1 = T$. By Lemma 7.7, there exists a vector x_1 and a scalar λ_1 such that $\| x_1 \| = 1$, $|\lambda_1| = \| T_1 \|$, and $Tx_1 = \lambda_1 x_1$.

Proceeding inductively, suppose that for each $i = 1, 2, \ldots, n-1$, X_i, λ_i, and x_i have been chosen satisfying equation (7.35). Define

$$X_n = \{x : (x \mid x_i) = 0, \text{ for } i = 1, 2, \ldots, n-1\}.$$

$X_n \neq 0$, for if $X_n = 0$ then the set $\{x_1, x_2, \ldots, x_{n-1}\}$ forms a basis for H, which contradicts the fact that T is of infinite rank. Clearly, $X_n \subset X_{n-1}$ and $T: X_n \to X_n$ since for $x \in X_n$

$$(Tx \mid x_i) = (x \mid Tx_i) = (x \mid \lambda_i x_i) = 0,$$

for $i = 1, 2, \ldots, n-1$. Letting $T_n = T|_{X_n}$, T_n is compact and self-adjoint. Now if $T_n = 0$, then letting $y_{n-1} = x - \sum_{i=1}^{n-1}(x \mid x_i)x_i$ we see that $y_{n-1} \in X_n$ so that $Ty_{n-1} = 0$ or $Tx = \sum_{i=1}^{n-1}\lambda_i(x \mid x_i)x_i$ for each x in H. This is impossible since T has infinite rank. Hence T_n is nonzero. By Lemma 7.7 again, there is a vector $x_n \in X_n$ and a scalar λ_n such that $\| x_n \| = 1$, $|\lambda_n| = \| T_n \|$, and $Tx_n = \lambda_n x_n$. Since $X_n \subseteq X_{n-1}$,

$$|\lambda_n| = \| T_n \| \leq \| T_{n-1} \| = |\lambda_{n-1}|.$$

The sequence (λ_i) of eigenvalues thus formed converges to zero. If not, $x_n = T(x_n/\lambda_n)$ would have a convergent subsequence x_{n_k} since (x_n/λ_n)

is bounded and T is compact. However, since the x_{n_k} are orthonormal, $\| x_{n_k} - x_{n_{k'}} \|^2 = 2$ if $k \neq k'$ and the sequence (x_{n_k}) can never converge.

Setting again $y_n = x - \sum_{k=1}^{n} (x \mid x_k)x_k$, we have

$$\| y_n \|^2 = \| x \|^2 - \sum_{k=1}^{n} | (x \mid x_k) |^2 \leq \| x \|^2.$$

Since $y_n \in X_{n+1}$ and $| \lambda_{n+1} | = \| T |_{X_{n+1}} \|$,

$$\| Ty_n \| \leq | \lambda_{n+1} | \; \| y_n \| \leq | \lambda_{n+1} | \; \| x \|,$$

which means $Ty_n \to 0$. Hence

$$Tx = \sum_{k=1}^{\infty} \lambda_k (x \mid x_k) x_k.$$

Finally, if λ is a nonzero eigenvalue excluded from (λ_k), then $Tx = \lambda x$ for some vector x with $\| x \| = 1$ and $(x \mid x_k) = 0$ for all k. Hence $\lambda x = Tx = 0$, which is a contradiction. We have proved (ii).

Suppose now T has finite rank. Clearly T has only a finite set of distinct eigenvalues, for if $\lambda_1, \lambda_2, \ldots$ is an infinite set of distinct nonzero eigenvalues with $x_1, x_2, \ldots$ a corresponding orthogonal set of eigenvectors, then $T(H)$ is infinite dimensional as each x_k is in $T(H)$ since $x_k = T(\lambda_k^{-1} x_k)$.

Just as in expression (7.35), using the same procedure, let (λ_k) be the finite set of all eigenvalues with corresponding eigenvectors (x_k) such that $| \lambda_{k+1} | \leq | \lambda_k | = \| T |_{X_k} \|$. If $m + 1$ is the smallest integer such that $\lambda_{m+1} = 0$, then $T |_{X_{m+1}} = 0$, and, as before, $Tx = \sum_{i=1}^{m} \lambda_i (x \mid x_i) x_i$. In the other case, there is a smallest integer m such that $X_{m+1} = \{0\}$; then $\{x_1, x_2, \ldots, x_m\}$ is a basis for H. ∎

At this juncture we can prove another significant result using the theorem just proved. In short, this result says that each Hilbert space satisfies the approximation property (see Chapter 6). More precisely, we have the following theorem.

Theorem 7.9. Let $T: H \to H$ be a compact operator. Then there is a sequence T_n of finite-rank linear operators converging to T in the norm (uniformly). ∎

Proof. First assume H is separable. Then H contains a complete orthonormal sequence (x_k). Define T_n by

$$T_n x = \sum_{k=1}^{n} (x \mid x_k) x_k.$$

Clearly (by Proposition 7.6) $\| T_n \| \leq 1$, $\| I - T_n \| \leq 1$, and $T_n x \to x$ as $n \to \infty$ for each x in H. If the conclusion of the theorem is false, then there exists a $\delta_0 > 0$ such that $\| T - A \| > \delta_0$ for all finite-rank linear operators A. Since $T_n T$ has finite rank, there exists for each n a y_n in H with $\| y_n \| = 1$ and $\| (T - T_n T) y_n \| \geq \delta_0/2$. Since the sequence (y_n) is bounded, there exists a subsequence (y_{n_i}) of (y_n) such that Ty_{n_i} converges to some z in H. However,

$$\| (T - T_{n_i} T) y_{n_i} \| = \| (I - T_{n_i}) T y_{n_i} \| \leq \| (I - T_{n_i})(T y_{n_i} - z) \| + \| (I - T_{n_i}) z \|, \tag{7.36}$$

and the left side of (7.36) goes to zero as $i \to \infty$. Since this is a contradiction, the theorem is proved when H is separable.

Now let H be any Hilbert space. Let $S = T^*T$. Then $S^* = S$, so S is self-adjoint and compact.[†] By Theorem 7.8, $Sx = \sum_{k=1}^{\infty} \lambda_k (x \mid x_k) x_k$ for each x in H, where (x_k) is some orthonormal sequence of eigenvectors with corresponding eigenvalues (λ_k). Let H_0 be the subspace of H spanned by $\{T^n x_k \colon n = 0, 1, \ldots, \text{ and } k = 1, 2, \ldots\}$. Then $\bar{H}_0$ is a separable Hilbert space and $T\colon \bar{H}_0 \to \bar{H}_0$. However, we know by what we have just proved that there is a sequence $\{F_n\}$ of finite-rank operators converging to $T_{\bar{H}_0}$ in norm. Now $H = \bar{H}_0 \oplus \bar{H}_0^{\perp}$. Therefore for each x in H, $x = y + z$ with $y \in \bar{H}_0$ and $z \in \bar{H}_0^{\perp}$. Since $z \perp x_k$ for all $k = 1, 2, \ldots$, $Sz = 0$. Therefore $T^*Tz = 0$ so that $Tz = 0$ as $(Tz \mid Tz) = (T^*Tz, z) = 0$.

Define G_n on H by $G_n x = F_n y$ for $n = 1, 2, \ldots$. Then G_n are finite-rank operators, and since $\| x \|^2 = \| y \|^2 + \| z \|^2$,

$$\| T - G_n \| = \sup_{\|x\|=1} \| Tx - G_n x \| = \sup_{\|x\|=1} \| Ty - F_n y \| \leq \| T_{\bar{H}_0} - F_n \|. \quad \blacksquare$$

• Application of Theorem 7.8: Fredholm Integral Equations and the Sturm–Liouville Problem

Let us first consider here solving what is known as a Fredholm integral equation with a Hilbert–Schmidt-type kernel. It is an equation of the form

$$f(s) - \lambda \int_a^b K(s, t) f(t)\, dt = g(s), \tag{7.37}$$

where $K(s, t)$ is a complex-valued measurable function on $[a, b] \times [a, b]$

† By Proposition 6.21 and Theorem 6.21.

such that

$$\| K \|_2 = \int_a^b \int_a^b | K(s, t) |^2 \, ds \, dt < \infty,$$

$f \in L_2[a, b]$ and $g \in L_2[a, b]$. The function K is called the *kernel* of equation (7.37). λ is a nonzero complex parameter. Solving equation (7.37) means finding a function f in $L_2[a, b]$ that satisfies equation (7.37).

As shown in Remark 6.29 of Chapter 6, the operator T given by

$$Tf(s) = \int_a^b K(s, t) f(t) \, dt$$

is a compact bounded linear operator from $L_2[a, b]$ to $L_2[a, b]$ with norm $\| T \|$ satisfying $\| T \| \leq \| K \|_2$. Moreover T is a self-adjoint operator if and only if $K(s, t) = \overline{K(t, s)}$ a.e. with respect to (s, t). In this case the kernel K is called symmetric.

If we assume that the kernel K is symmetric, the equation (7.37) takes the form

$$f = \lambda Tf + g$$

or

$$Tx - \mu x = y, \tag{7.38}$$

where $\mu = 1/\lambda$, $x = f$, $y = -(1/\lambda)g$, and T is a compact self-adjoint operator. Clearly $y = 0$ if and only if $g = 0$. For $g = 0$, the values of λ for which equation (7.37) has a nontrivial solution are called *characteristic values* of equation (7.37). For $y = 0$, the values of μ for which equation (7.38) has a nonzero solution are the eigenvalues of T. Clearly characteristic values and nonzero eigenvalues are reciprocally related.

We can consider the equation (7.38) as an equation in any Hilbert space H where T is a self-adjoint compact operator on H and x and y are elements of H. Since T is compact, the nonzero values of μ are either *regular* values of T or eigenvalues of T; that is, $T - \mu I$ either has a bounded inverse on H or μ is an eigenvalue of T. If $(T - \mu I)^{-1}$ exists, then equation (7.38) has a unique solution for each y in H, and in particular 0 is the only solution of $Tx - \mu x = 0$.

If μ is an eigenvalue of T, let H_μ be the eigenspace of T, the subspace of H consisting of all eigenvectors corresponding to μ. Let $H_\mu^\perp$ be the orthogonal complement of H_μ. Note that $T(H_\mu) \subset H_\mu$ and $T(H_\mu^\perp) \subset H_\mu^\perp$. If we consider T restricted to $H_\mu^\perp$, then μ is not an eigenvalue for this restricted operator. Consequently as an operator from $H_\mu^\perp$ to $H_\mu^\perp$,

$T - \mu I$ has a bounded inverse on $H_\mu^\perp$ and equation (7.38) has a unique solution for any y in $H_\mu^\perp$. Since any x in H can be written uniquely as $x_1 + x_2$ with $x_1 \in H_\mu$ and $x_2 \in H_\mu^\perp$ and $(T - \mu I)x_1 = 0$ we have

$$(T - \mu I)x = (T - \mu I)x_2 \in H_\mu^\perp.$$

Hence equation (7.38) has a solution only for y in $H_\mu^\perp$. If x_2 is any solution of (7.38) and x_1 is any element of H_μ, then $x_1 + x_2$ is also a solution. Consequently, as in the case when μ was not an eigenvalue, we can say that (7.38) has a unique solution for y in H if and only if the homogeneous equation $(T - \mu I)x = 0$ has a unique solution. Moreover if $\mu \neq 0$ is an eigenvalue, then (7.38) has a solution if and only if y is in the subspace $H_\mu^\perp$ of vectors orthogonal to the space of eigenvectors.

Theorem 7.8 furnishes a technique for solving equation (7.38) and of course equation (7.37). First of all, suppose μ is not an eigenvalue. By Theorem 7.8, for each x in H

$$Tx = \sum_i \mu_i(x \mid x_i)x_i.$$

From (7.38), $Tx = y + \mu x$ or

$$x = \frac{1}{\mu}(Tx - y) = \frac{1}{\mu}\sum_i \mu_i(x \mid x_i)x_i - y. \tag{7.39}$$

For any n, by taking scalar products of both sides of this equation, we have

$$\begin{aligned}\mu_n(x \mid x_n) &= \frac{\mu_n}{\mu}\left(\left[\sum_i \mu_i(x \mid x_i)x_i - y\right] \mid x_n\right)\\ &= \frac{\mu_n}{\mu}\left[\mu_n(x \mid x_n) - (y \mid x_n)\right],\end{aligned}$$

from which

$$\mu_n(x \mid x_n) = \frac{(\mu_n/\mu)(y \mid x_n)}{\mu_n/\mu - 1} = \frac{\mu_n(y \mid x_n)}{\mu_n - \mu}. \tag{7.40}$$

Hence from equation (7.39) the solution x is given by

$$x = \frac{1}{\mu}\sum_i \frac{\mu_i(y \mid x_i)}{\mu_i - \mu}x_i - y.$$

Secondly if μ is an eigenvalue, then equation (7.39) is still valid and as before we can solve for $\mu_n(x \mid x_n)$ as in equation (7.40) if $\mu \neq \mu_n$. If $\mu = \mu_n$, the coefficient of eigenvector x_n can be taken as arbitrary since this term represents a vector in H_μ. Consequently, a solution of equation (7.38) is given by

$$x = (1/\mu)[\sum c_i x_i - y],$$

where $c_n = \mu_n(y \mid x_n)/(\mu_n - \mu)$ if $\mu_n \neq \mu$ and c_n is arbitrary if $\mu_n = \mu$.

These results are summarized in the following theorem.

Theorem A. Suppose T is a compact self-adjoint operator on H. If $\mu \neq 0$ is not an eigenvalue of T and y is an element of H, then the solution of $(T - \mu I)x = y$ exists and is the unique vector

$$x = \frac{1}{\mu}\left[\sum_i \frac{\mu_i(y \mid x_i)x_i}{\mu_i - \mu} - y\right],$$

where the μ_i are the eigenvalues of T and the x_i are corresponding eigenvectors. If $\mu \neq 0$ is an eigenvalue of T and $y \in H_\mu{}^\perp$, then a solution of $(T - \mu I)x = y$ exists and is given by

$$x = \frac{1}{\mu}\left(\sum_i c_i x_i - y\right),$$

where $c_i = \mu_i(y \mid x_i)/(\mu_i - \mu)$ if $\mu_i \neq \mu$ and c_i is arbitrary if $\mu_i = \mu$. ∎

An equation which is solved by means of the theory for solving Fredholm integral equations is the so-called Stürm–Liouville equation. It is an equation of the form

$$-[\psi(x)f'(x)]' + \varphi(x)f(x) - \mu f(x) + g(x) = 0, \tag{7.41}$$

where $\psi(x)$, $\psi'(x)$, and $\varphi(x)$ are real continuous functions on $[a, b]$; $g(x)$ is a complex-valued continuous function on $[a, b]$, and $\psi(x) > 0$ for x in $[a, b]$. Equations of this sort arise in the study of vibrating strings and membranes, transmission lines, and resonance in a cavity. Solving the Sturm–Liouville problem means finding a function f in $L_2[a, b]$ satisfying equation (7.41) such that

$$a_1 f(a) + a_2 f'(a) = 0, \tag{7.42}$$

$$b_1 f(b) + b_2 f'(b) = 0, \tag{7.43}$$

$$f'' \text{ exists and is continuous on } [a, b], \tag{7.44}$$

where a_1, a_2, b_1 and b_2 are real constants such that $|a_1| + |a_2| > 0$ and $|b_1| + |b_2| > 0$.

Together with equation (7.41) the associated homogeneous equation (7.45) in which $g(x) = 0$ can also be considered:

$$-[\psi(x)f'(x)]' + \varphi(x)f(x) - \mu f(x) = 0. \tag{7.45}$$

In consideration of this equation, two possibilities arise. Either $\mu = 0$ admits the existence of a nontrivial solution to equation (7.45) that satisfies equations (7.42)–(7.44) or $\mu = 0$ admits only the trivial solution $f = 0$. Alternatively, we say, respectively, that either 0 is an eigenvalue of equation (7.45) or 0 is not an eigenvalue. We deal with each of these cases below.

Let us observe first that if μ_1 and μ_2 are two distinct eigenvalues of equation (7.45), that is, two scalars that admit nontrivial solutions f_1 and f_2, respectively, of equation (7.45), then f_1 and f_2 are orthogonal elements of $L_2[a, b]$. Indeed, since

$$[\psi(x)f_1'(x)]' - \varphi(x)f_1(x) + \mu_1 f_1(x) = 0,$$
$$[\psi(x)\overline{f_2'(x)}]' - \varphi(x)\overline{f_2(x)} + \bar{\mu}_2\overline{f_2(x)} = 0,$$

it follows that upon multiplying the first equation by $\bar{f}_2$ and the second by f_1 and then subtracting,

$$\{\psi(x)[\bar{f}_2(x)f_1'(x) - f_1(x)\overline{f_2'(x)}]\}' = (\bar{\mu}_2 - \mu_1)f_1\bar{f}_2. \tag{7.46}$$

Since $a_1 f_1(a) + a_2 f_1'(a) = 0$, $a_1\overline{f_2(a)} + a_2\overline{f_2'(a)} = 0$, and $|a_1| + |a_2| > 0$, the determinant $f_1(a)\overline{f_2'(a)} - f_1'(a)\overline{f_2(a)}$ is zero. Similarly $\overline{f_2(b)}f_1'(b) - f_1(b)\overline{f_2'(b)} = 0$. Upon integrating both sides of equation (7.46), we find therefore that $\int_a^b f_1(x)\overline{f_2(x)}\,dx = 0$.

Since $L_2[a, b]$ is separable, it contains at most a countable set of orthogonal elements. Consequently there are at most countably many distinct eigenvalues of equation (7.45) and uncountably many real numbers that are not eigenvalues of equation (7.45).

Let us first consider now the case where 0 is an eigenvalue of equation (7.45) and show that this case can be reduced to the case where 0 is not an eigenvalue. Suppose λ_0 is a real number that is not an eigenvalue of equation (7.45), that is,

$$-[\psi(x)f'(x)]' + \varphi(x)f(x) - \lambda_0 f(x) = 0$$

has only the trivial solution that satisfies equations (7.42) and (7.43). Letting $\gamma(x) = \varphi(x) - \lambda_0$, this equation has the form

$$[-\psi(x)f'(x)]' + \gamma(x)f(x) = 0,$$

and equation (7.41) has the form

$$-[\psi(x)f'(x)]' + \gamma(x)f(x) - (\mu - \lambda_0)f(x) + g(x) = 0. \qquad (7.47)$$

By assumption the homogeneous counterpart of equation (7.47), i.e., with $g(x) = 0$, has only the trivial solution when $\mu - \lambda_0 = 0$, which is precisely the assumption made when the case in which 0 is not an eigenvalue of equation (7.45) is considered.

Let us then assume the latter case: $\mu = 0$ admits only the trivial solution $f = 0$ to equation (7.45) that satisfies equations (7.42)–(7.44). Let $\mu = 0$. From the theory of elementary differential equations, a nontrivial real solution u_1 of equation (7.45) exists that satisfies equation (7.42) and a nontrivial solution u_2 of equation (7.45) exists that satisfies equation (7.43). Moreover, the Wronskian of u_1 and u_2 is given by $c/\psi(x)$ for each x in $[a, b]$, where c is a constant. Solutions u_1 and u_2 are linearly independent, for if one were a constant multiple of the other, each would satisfy equation (7.45) and conditions (7.42)–(7.44). This means that $c \neq 0$. By the proper choice of u_1 and u_2 we can assume that $c = -1$, whereby u_1 and u_2 satisfy the equation

$$u_1(x)u_2'(x) - u_1'(x)u_2(x) = -1/\psi(x), \qquad x \in [a, b]. \qquad (7.48)$$

Using the method of variation of parameters from elementary differential equations, a solution of equation (7.41) with $\mu = 0$ is given by

$$u(x) = c_1(x)u_1(x) + c_2(x)u_2(x),$$

where c_1 and c_2 are functions on $[a, b]$ satisfying

$$c_1'(x)u_1(x) + c_2'(x)u_2(x) = 0,$$

and

$$c_1'(x)u_1'(x) + c_2'(x)u_2'(x) = -g(x)/\psi(x).$$

From equation (7.48), the solution of these equations is given by

$$c_1'(x) = -u_2(x)g(x) \qquad \text{and} \qquad c_2'(x) = u_1(x)g(x)$$

or

$$c_1(x) = \int_x^b u_2(y)g(y)\,dy \qquad \text{and} \qquad c_2(x) = \int_a^x u_1(y)g(y)\,dy.$$

Hence

$$\begin{aligned} u(x) &= u_1(x)\int_x^b u_2(y)g(y)\,dy + u_2(x)\int_a^x u_1(y)g(y)\,dy \\ &= \int_a^b G(x, y)g(y)\,dy, \end{aligned}$$

where

$$G(x, y) = \begin{cases} u_1(x)u_2(y), & \text{if } x \leq y, \\ u_1(y)u_2(x), & \text{if } y \leq x. \end{cases} \tag{7.49}$$

Observe that the function u satisfies the boundary conditions (7.42) and (7.43). Indeed,

$$u(a) = u_1(a) \int_a^b u_2(y)g(y)\,dy$$

and

$$u'(a) = u_1'(a) \int_a^b u_2(y)g(y)\,dy,$$

so that

$$a_1u(a) + a_2u'(a) = [a_1u_1(a) + a_2u_1'(a)] \int_a^b u_2(y)g(y)\,dy = 0.$$

Similarly $b_1u(b) + b_2u'(b) = 0$. Observe also that u is the only solution of equation (7.41) satisfying equations (7.42) and (7.43), for if v is any other solution, then $u - v = 0$ as $u - v$ satisfies equation (7.45) with $\mu = 0$.

The function $G(x, y)$ is called *Green's function.* It is easily seen to be continuous on $[a, b] \times [a, b]$ and to satisfy the symmetry property $G(x, y) = G(y, x)$. The solution of equation (7.41) is seen therefore to be the operator value of the compact self-adjoint linear operator G: $L_2[a, b] \to L_2[a, b]$ given by

$$(Gg)(x) = \int_a^b G(x, y)g(y)\,dy, \tag{7.50}$$

corresponding to the continuous function g in $L_2[a, b]$. Specifically we have the following theorem relating the Stürm–Liouville equation to an integral equation.

Theorem B. (i) Assume $\mu = 0$ is not an eigenvalue of equation (7.45). Then there is a real continuous function G defined on $[a, b] \times [a, b]$ and satisfying $G(x, y) = G(y, x)$ such that $f(x)$ is a solution of

$$-[\psi(x)f'(x)]' + \varphi(x)f(x) + g(x) = 0$$

satisfying equations (7.42) and (7.43) if and only if

$$f(x) = \int_a^b G(x, y)g(y)\,dy.$$

In particular, there is a real symmetric function G on $[a, b]\times[a, b]$ such that f is a solution of

$$[\psi(x)f'(x)]' - \varphi(x)f(x) = \lambda f(x) \tag{7.51}$$

satisfying equations (7.42) and (7.43) if and only if f is a solution of the integral equation

$$f(x) = \lambda \int_a^b G(x, y)f(y)\,dy. \tag{7.52}$$

(ii) Assume 0 is an eigenvalue of equation (7.45). If λ_0 is not an eigenvalue, there is a real symmetric function $G(x, y, \lambda_0)$ such that $f(x)$ is a solution of

$$-[\psi(x)f'(x)]' + \varphi(x)f(x) - \lambda_0 f(x) + g(x) = 0$$

if and only if

$$f(x) = \int_a^b G(x, y, \lambda_0)g(y)\,dy.$$

In particular, $G(x, y, \lambda_0)$ exists such that f is a solution of

$$-[\psi(x)f'(x)]' + \varphi(x)f(x) = \lambda f(x)$$

if and only if f is a solution of

$$f(x) = (\lambda_0 - \lambda)\int_a^b G(x, y, \lambda_0)f(y)\,dy. \qquad \blacksquare$$

In the foregoing discussion of the Stürm–Liouville problem, the first two terms of equation (7.41) can be considered as the values of an operator S in $L_2[a, b]$ whose values are given by

$$Sf(x) = -[\psi(x)f'(x)]' + \varphi(x)f(x).$$

S is actually an unbounded self-adjoint operator on the domain of functions in $L_2[a, b]$ satisfying equations (7.42), (7.43), and (7.44). (This type of operator and its eigenvalues are studied in Chapter 8.) Equation (7.45) thus has the form

$$(S - \mu)f = 0. \tag{7.53}$$

We have seen in Theorem B that if 0 is not an eigenvalue then f is a solution of equation (7.53) if and only if f is a solution of the equation

$$(G - 1/\mu)f = 0, \tag{7.54}$$

where G is a compact operator on $L_2[a, b]$. Hence we see that the values of μ for which equation (7.53) has a nontrivial solution (the eigenvalues of S) are the reciprocals of the nonzero eigenvalues of the integral operator G. In other words, the eigenvalues of S are the characteristic values of the Fredholm integral equation (7.52). When zero is not an eigenvalue of G, the set of eigenvalues of G form a bounded real sequence converging to zero by Theorem 7.8, whereby the eigenvalues of S then form a real sequence $(\mu_i)_N$ such that $|\mu_i| \to \infty$.

Problems

× **7.4.1.** In this problem A, S, T, U, and Q are the continuous linear operators on l_2 given by the following rules:

$$A(x_1, x_2, \ldots) = (\alpha_1 x_1, \alpha_2 x_2, \ldots),$$

where (α_i) is a bounded sequence of scalars;

$$S(x_1, x_2, \ldots) = (x_2, x_3, \ldots);$$
$$T(x_1, x_2, \ldots) = (0, x_1, x_2, x_3, \ldots);$$
$$U(x_1, x_2, \ldots) = (x_{i_1}, x_{i_2}, x_{i_3}, \ldots),$$

where the set $\{i_1, i_2, \ldots\}$ is a permutation of $\{1, 2, \ldots\}$; and

$$Q(x_1, x_2, \ldots) = (0, x_2, x_3, \ldots).$$

(i) Show that A is always normal but A is self-adjoint if and only if $\alpha_i = \overline{\alpha_i}$ for each i, and A is unitary if and only if $|\alpha_i| = 1$ for all i.

(ii) Show that $S = T^*$ and $S^* = T$. Show that T is isometric but not unitary.

(iii) Show that U is unitary, but not a projection.

(iv) Show that Q is a projection but not isometric. (Hint: See Problem 7.3.7.)

× **7.4.2.** Prove Lemma 7.3 is false if H is a real Hilbert space. (Hint: Consider a rotation in R^2.) Show that the conclusion of Lemma 7.3 is true for all Hilbert spaces if T and S are self-adjoint.

7.4.3. (i) Prove that the product† (that is, composition) AB of two normal operators A and B is normal if and only if $\| A^*Bx \| = \| BA^*x \|$ for each $x \in H$.

† A normal, B normal and AB normal need not imply BA normal.

(ii) Prove that the product of two self-adjoint operators is self-adjoint if and only if the given operators commute.

7.4.4. Prove that every continuous linear operator $T: H \to H$, where H is a complex Hilbert space, can be expressed uniquely as $A + iB$, where A and B are self-adjoint, and that T is normal if and only if A and B commute.

× **7.4.5.** Let H be any Hilbert space and $(x_i)_{i \in I}$ be a basis for H. Let $T: H \to H$ be a bounded linear operator.

(i) Prove that T is isometric if and only if $\{Tx_i: i \in I\}$ is an orthonormal set.

(ii) Prove that T is unitary if and only if $\{Tx_i: i \in I\}$ is a basis for H.

× **7.4.6.** (i) Prove that if $T: H \to H$ is any bounded linear operator, then $\| T^*T \| = \| TT^* \| = \| T \|^2$.

(ii) Prove that if T is self-adjoint or normal, then for any positive integer n, $\| T^n \| = \| T \|^n$.

× **7.4.7.** If A and B are linear operators from H into H such that $(Ax \mid y) = (x \mid By)$ for all x and y in H, show that A is continuous and that $A^* = B$. [Hint: Show that A has a closed graph and use the Closed Graph Theorem (Theorem 6.9 in Chapter 6).]

7.4.8. If T_1 and T_2 are self-adjoint operators on H such that $T_1 \leq T_2$ and $T_2 \leq T_1$, show that $T_1 = T_2$ if H is a complex Hilbert space. Is this true if H is a real Hilbert space? (See Problem 7.4.2.)

× **7.4.9.** Suppose T is a linear operator on a separable Hilbert space H with orthonormal basis $(x_i)_N$ such that $Tx = \sum_{i=1}^{\infty} m_i(x \mid x_i)x_i$ when $x = \sum_{i=1}^{\infty}(x \mid x_i)x$ for a sequence $(m_i)_N$. Prove the following:

(i) T is bounded if and only if $(m_i)_N \in l_\infty$.

(ii) T is compact if and only if $(m_i)_N \in c_0$.

(iii) Suppose $m_i \to 1$, m_i are distinct, and $m_i \neq 1$ for all i. Then each m_i is an eigenvalue of T, 1 is in the continuous spectrum, and all other complex numbers are regular values of T.

7.4.10. If M is a closed subspace of an infinite-dimensional Hilbert space, prove that P_M, the projection on M, is compact if and only if M is finite dimensional.

7.4.11. If $T: H \to K$ is a bounded linear operator, show that T is compact if and only if T^* is compact.

7.4.12. Let T be the shift operator of Problem 7.4.1 and let $T_n = (T)^n$. Let V_i be the operator on the subspace N of l_2 of finitely nonzero sequences given by $V_i((x_1, x_2, \ldots)) = (1x_1, 2x_2, \ldots, ix_i, x_{i+1}, x_{i+2}, \ldots)$. Finally let U_i be the operator given by $U_i((x_1, x_2, \ldots)) = (0, 0, 0, \ldots, x_i, x_{i+1}, \ldots)$. Prove the following:

(i) The sequence (U_i) converges *strongly* to the zero operator (that is, $U_i x \to 0$ for each x in H), but (U_i) does not converge uniformly to 0 (that is, $\| U_i \| \not\to 0$).

(ii) (V_i) converges strongly to the operator V on N given by $V(x_1, x_2, \ldots) = (x_1, 2x_2, 3x_3, \ldots)$, but V is not bounded.

(iii) (T_n) converges neither strongly nor uniformly to the zero operator but (T_n) converges weakly to 0 [that is, $h^*(T_n f) \to 0$ for all $h^* \in l_2^*$ and f in l_2].

× **7.4.13.** Let T be the shift operator of Problem 7.4.1. Prove the following:

(i) T has no eigenvalue, but $0 \in \sigma(T)$. (Hint: If $Tx = \mu x$, show that $x = 0$.)

(ii) 0 is an eigenvalue of T^*.

(iii) μ is an eigenvalue for $T^* \Leftrightarrow |\mu| < 1$.

× **7.4.14.** (i) If T is a normal operator, prove that μ is an eigenvalue of T if and only if $\bar{\mu}$ is an eigenvalue of T^*.

(ii) If A is the operator of Problem 7.4.1, show $\{\alpha_i : i \in N\}$ are the only eigenvalues of A. Conclude that $\{\bar{\alpha}_i : i \in N\}$ are the only eigenvalues of A^*.

(iii) Prove $\sigma(A)$ is the closure of $\{\alpha_i : i \in N\}$.

7.4.15. Let T be the operator on the complex Hilbert space $L_2[0, 1]$ given by $Tf(x) = xf(x)$. Prove the following:

(i) T is self-adjoint.

(ii) $\| T \| = 1$.

(iii) T has no eigenvalue.

(iv) $\sigma(T) = [0, 1]$.

7.4.16. Let T be a linear operator on H. A complex number λ is called an *approximate* eigenvalue of T if there exists a sequence (h_n) of unit vectors in H (i.e., $\| h_n \| = 1$) such that $\lim_{n\to\infty}(T - \lambda)h_n = 0$.

(i) Prove that every approximate eigenvalue belongs to the spectrum $\sigma(T)$ of T.

(ii) Prove that if A is a bounded normal operator, then every point of the spectrum is an approximate eigenvalue.

× **7.4.17.** Let P be a nontrivial projection on the Hilbert space H. Prove that the spectrum of P consists of two eigenvalues, 0 and 1.

× **7.4.18.** Let $P \in L(H, H)$ and suppose $P^2 = P$. Prove $\operatorname{Ker} P \perp \operatorname{Range} P$ if and only if P is a projection (that is, $P^* = P$).

× **7.4.19.** If $P \in L(H, H)$, $P \neq 0$, and $P^2 = P$, prove $\| P \| \geq 1$. Prove moreover that $\| P \| = 1$ if and only if P is a projection.

7.4.20. Let $P_1, \ldots, P_n$ be projections onto $M_1, \ldots, M_n$, respectively. Prove $P \equiv P_1 + P_2 + \cdots + P_n$ is a projection if and only if $M_i \perp M_j$ when $i \neq j$. In this case P is the projection onto $M_1 \oplus M_2 \oplus \cdots \oplus M_n$.

× **7.4.21.** If $T \in L(H, H)$ and $T^2 = I$, prove that $T = T^*$ if and only if $\| T \| = 1$.

× **7.4.22.** Let H be a separable Hilbert space and let $S, T \in L(H, H)$. Let $(x_i)_N$ be an orthonormal basis for H.

(i) Show that with respect to $(x_i)_N$, T can be represented by a unique matrix $M = (m_{ij})$ with a countable number of rows and columns in the sense that if $x = \sum (x \mid x_i)x_i$ then $Tx = \sum_i(\sum_j m_{ij}(x \mid x_j))x_i$.

(ii) If $(\alpha_j)_N$ is a sequence of scalars such that $\sum_{j=1}^{\infty} | \alpha_j |^2 < \infty$, prove that

$$\sum_{i=1}^{\infty} \left| \sum_{j=1}^{\infty} m_{ij}\alpha_j \right|^2 \leq \| T \|^2 \sum_{j=1}^{\infty} | \alpha_j |^2.$$

(iii) Show that T^* is represented by the matrix $M^* = (m_{ij}^*)$ where $m_{ij}^* = \bar{m}_{ji}$.

(iv) If S is represented by $A = (n_{ij})$ with $n_{ij} = \bar{m}_{ij}$, prove that $T^* = S$.

(v) Prove T is unitary if and only if

$$\sum_{i=1}^{\infty} \bar{m}_{ij} m_{ik} = \sum_{i=1}^{\infty} m_{ji} \bar{m}_{ki} = \delta_{jk}.$$

7.4.23. Let T be a linear operator on a complex Hilbert space H. Prove that T is a bounded self-adjoint operator if and only if $(x \mid Tx)$ is real for all $x \in H$.

7.4.24. Suppose (T_n) is a sequence of bounded self-adjoint operators on H converging weakly [that is, $h^*(T_n f) \to h^*(Tf)$ for all $h^* \in H^*$ and $f \in H$] to a bounded linear operator T on H. Prove T is self-adjoint.

× **7.4.25.** Let T be a bounded normal (unitary) operator on H. Let $M_\lambda = \{x \mid Tx = \lambda x\}$ be the eigenspace of T corresponding to the eigenvalue λ. Prove

(i) if λ is eigenvalue of T then $\bar{\lambda}$ is an eigenvalue of $T^*(T^{-1})$;

(ii) M_λ is also the eigenspace of T^* corresponding to $\bar{\lambda}$; and

(iii) M_λ reduces T, i.e., $T(M_\lambda) \subset M_\lambda$ and $T(M_\lambda{}^\perp) \subset M_\lambda{}^\perp$.

7.4.26. Prove that if $\lambda \neq 0$ is an eigenvalue of a compact operator T on H, then the eigenspace M_λ is finite dimensional.

7.4.27. For $f \in L_2[0, 1]$, let $(Tf)(x) = \int_0^x f(t)\,dt$ for $x \in [0, 1]$. Show T is a compact operator on $L_2[0, 1]$ with no eigenvalues. Find $\sigma(T)$. [Hint: Compute $r(T)$, the spectral radius.]

7.4.28. Let T be a bounded normal operator on Hilbert space H.

(i) Prove $H = \operatorname{Ker} T \oplus \overline{R(T)}$.

(ii) Prove $\lambda \in \varrho(T)$ if $R(T - \lambda I) = H$.

(iii) Prove $\lambda \in C\sigma(T)$ if $R(T - \lambda I) \neq H$ but is dense in H.

(iv) Prove λ is an eigenvalue if $R(T - \lambda I)$ is not dense in H.

(v) Prove the residual spectrum is empty.

× **7.4.29.** If $A: H \to H$ is a compact self-adjoint operator, prove that a set of orthonormal eigenvectors (x_i) corresponding to the set of eigenvalues of A is a complete orthonormal set if 0 is not an eigenvalue of A. (Hint: Use Theorem 7.8.)

× **7.4.30.** Let T be a compact self-adjoint operator on H.

(i) Prove that T can be written as $\sum \lambda_k P_k$, where the λ_k are the nonzero eigenvalues of T, P_k is the projection onto the eigenspace M_{λ_k} corresponding to λ_k, and the convergence is uniform.

(ii) Show that if 0 is not an eigenvalue of T, T can be represented by a diagonal matrix with respect to a basis of orthonormal eigenvectors.

7.4.31. *Generalization of Theorem 7.8 to Normal Operators.* If $T: H \to H$ is a compact, normal operator, prove the following:

(i) H has an orthonormal basis (x_i) such that each x_i is an eigenvector of T.

(ii) T can be written as $Tx = \sum_{i=1}^{\infty} \lambda_i (x \mid x_i) x_i$, where (λ_i) is a sequence of complex numbers such that $\lambda_i \to 0$ as $i \to \infty$.

[Hint: To prove (i), use Zorn's Lemma to get a maximal orthonormal set of $\{x_i\}$ of eigenvectors of T. Let M be the closed linear span of $\{x_i\}$. Show M reduces T. If $M^\perp \neq \{0\}$, show $T_{M^\perp}$ is compact, normal, and nonzero with an eigenvector in M, contradicting the maximality of $\{x_i\}$. You might need the fact that $\gamma(T) = r \parallel T \parallel$ if T is normal and the Theorem 6.25.]

7.4.32. (*M. Rosenblum*). (i) If S and T are bounded linear operators on Hilbert spaces H and K, respectively, and if $\mathscr{F} \in L(L(K, H), L(K, H))$ is given by $\mathscr{F}(X) = SX - XT$ then prove $\sigma(\mathscr{F}) \subset \sigma(S) - \sigma(T)$. [Hint: If $\mathscr{S}$ and $\mathscr{E}$ are the operators on $L(K, H)$ given by $\mathscr{S}(X) = SX$ and $\mathscr{E}(X) = XT$, respectively, then show that if $\lambda \notin \sigma(S)$ then $\lambda \notin \sigma(\mathscr{S})$, and so $\sigma(\mathscr{S}) \subset \sigma(S)$. Likewise $\sigma(\mathscr{E}) \subset \sigma(T)$. Since $\mathscr{S}$ and $\mathscr{E}$ commute and $\mathscr{F} = \mathscr{S} - \mathscr{E}$, $\sigma(\mathscr{F}) \subset \sigma(\mathscr{S}) - \sigma(\mathscr{E})$ by 7.5.19(ii).]

(ii) If $\sigma(S) \cap \sigma(T) = \varnothing$ then prove that for each operator Y from K to H, there is a unique operator X from K to H such that $SX - XT = Y$.

× **7.4.33.** Let $(x_k)_{k=1}^{\infty}$ be a sequence of nonzero vectors in a Hilbert space H and $(\beta_k)_{k=1}^{\infty}$ a sequence of nonzero real scalars. Prove the following assertions:

(i) If for $x \in H, Tx = \sum_{k=1}^{n} \beta_k(x \mid x_k)x_k$, where n is a fixed positive integer, then T is a compact self-adjoint operator on H.

(ii) Suppose that $\{x_1, x_2, \ldots, x_n\}$ is a linearly independent set and let T be as in (i). Then 0 is an eigenvalue of T if and only if $\dim H > n$.

(iii) Suppose that $\{x_1, x_2, \ldots, x_n\}$ is an orthonormal set. Let T be as in (i). Then the set of all nonzero eigenvalues of T is the set $\{\beta_1, \beta_2, \ldots, \beta_n\}$.

(iv) Let $(x_k)_{k=1}^{\infty}$ be an orthonormal sequence and G be the closed subspace of H spanned by $(x_k)_{k=1}^{\infty}$. Then $(x_k)_{k=1}^{\infty}$ is an orthonormal basis of G.

(v) Let the x_k be as in (iv) and $\lim_{k\to\infty} |\beta_k| = 0$. Then the operator T defined by

$$Tx = \sum_{k=1}^{\infty} \beta_k(x \mid x_k)x_k, \qquad x \in H,$$

is a compact self-adjoint operator and the set of all nonzero eigenvalues of T is $\{\beta_1, \beta_2, \ldots\}$. [Hint: Suppose that $\beta \neq 0$, $x \neq 0$ and $Tx = \beta x$. Write $x = y + z$, $y \in G$ and $z \in G^{\perp}$, G as in (iv). Then $Tx = Ty = \beta y$.]

(vi) Let T be a nonzero compact self-adjoint operator on an infinite-dimensional Hilbert space. Then $M \geq 0$, where $M = \sup\{(Tx \mid x) : \|x\| = 1\}$. [Hint: Use Theorem 7.8 to see that T is as in (ii) when $M < 0$.]

7.4.34. Let $X = Y + Z$, where Y and Z are subspaces of a Banach space X and $Y \cap Z = \{0\}$. If $x = y + z$, then define $P(x) = y$. Prove that P is bounded if and only if both Y and Z are closed.

7.4.35. Let $T \in L(H, H)$. Prove that $\sigma(T) \subset$ the closure of N_T, where the numerical range $N_T = \{(Tx \mid x) : \|x\| = 1\}$. [Hint: If λ is not an approximate eigenvalue of T (see Problem 7.4.16) and $\lambda \in \sigma(T)$, then $R(T - \lambda I)$ is closed, but not all of H. Then there exists $x \in H$ such that $\|x\| = 1$ and $((T - \lambda I)x \mid x) = 0$.]

7.5. Spectral Decomposition of Self-Adjoint Operators

Our goal in this section is simply stated: to prove what is known as the Spectral Theorem for bounded self-adjoint operators. It can be said that all our preceding work in Hilbert space theory—although important in itself—has been preparatory to proving this outstanding theorem. As seen earlier, self-adjoint operators are generalizations of Hermitian matrices. The Spectral Theorem is a generalization of the diagonalization theorem for such matrices.

This theorem was originally proved by Hilbert between 1904 and 1910.† Proofs of this result were also given by F. Riesz in 1910 and 1913.‡ Since that time other proofs of spectral theorems have been given by many others for self-adjoint, unitary, and normal operators both in the bounded and unbounded cases. The interested reader may also consult for further study the works of Dunford and Schwartz [17], Halmos [23], Riesz and Sz-Nagy [48], Stone [59], Prugovečki [46], among others, as well as the Appendix.

The spectral theory of certain classes of operators has been studied extensively since it was initiated by Hilbert in the early 1900's. The theory has profound applications in the study of operators on Hilbert space and in areas of classical analysis such as differential equations. See for example the work of Dunford and Schwartz [17].

Although the Spectral Theorem can be generalized to broader classes of operators, as mentioned above, we consider here only self-adjoint bounded operators. In Chapter 8 other formulations of the Spectral Theorem for bounded and unbounded operators will be given—the proofs of which involve measure theoretic techniques. We will here prove the Spectral Theorem using an approach that makes use of what is known as the "functional" or "operational" calculus and involves no measure theory. After proving this theorem, some applications of this key result will be outlined in the Problems.

Recall from Theorem 7.8 that if $T: H \to H$ is a nonzero compact, self-adjoint operator, then for each x in H,

$$Tx = \sum_{k=1}^{\infty} \lambda_k (x \mid x_k) x_k, \tag{7.55}$$

where the λ_k are eigenvalues of T and the x_k form a corresponding set of eigenvectors. After Definition 7.13 we will show how the Spectral Theorem generalizes the formula (7.55) to arbitrary self-adjoint operators.

There is one stipulation we must make, however, before proceeding. *All Hilbert spaces considered in this section are complex Hilbert spaces.*

† D. Hilbert, Grundzüge einer allgemeinen Theorie der linearen Integralgleichungen, *Nachr. Akad. Wiss. Göttingen Math.-Phys. IV, Kl.* 157–227 (1906).

‡ F. Riesz, Über quadratische Formen von unendlich vielen Veränderlichen, *Nachr. Akad. Wiss. Göttingen Math.-Phys. Kl.* 190–195 (1910); *Les systèmes d'equations linéaires à une infinité d'inconnues*, Gauthier-Villars, Paris (1913).

The need for this stipulation arises from the fact that the complex numbers are algebraically closed while the real numbers are not—meaning every real polynomial has complex roots but not necessarily real roots. The necessity for using complex Hilbert spaces becomes already evident in the proof of Lemma 7.10.

In this section T will denote an arbitrary self-adjoint (bounded linear) operator on H, and m and M will denote real numbers such that $mI \leq T \leq MI$ (see Lemma 7.4).

To prove a form of the Spectral Theorem for bounded self-adjoint operators, we first prove a theorem that describes the so-called "Continuous Functional Calculus." It gives us much more information than actually needed to prove the Spectral Theorem for a bounded self-adjoint operator, but we prove it here since we will have need of it in our more general considerations of spectral theory in Chapter 8. We first must present two lemmas.

Lemma 7.10. Suppose $p(x) = \sum_{k=0}^{n} a_k x^k$ is a polynomial with real or complex coefficients. If T is a bounded self-adjoint operator, let $p(T) = \sum_{k=0}^{n} a_k T^k$. Then

$$\sigma(p(T)) = \{p(\lambda);\ \lambda \in \sigma(T)\} \equiv p(\sigma(T)). \qquad \blacksquare$$

For proof, see Theorem 6.24 in Chapter 6.

Lemma 7.11. Let T be a bounded self-adjoint operator. Then if p is as in Lemma 7.10,

$$\| p(T) \| = \sup_{\lambda \in \sigma(T)} | p(\lambda) |. \qquad \blacksquare$$

Proof.

$$\begin{aligned} \| p(T) \|^2 &= \| p(T)^* p(T) \| = \| (\bar{p}p)T \| \\ &= \sup_{\lambda \in \sigma(\bar{p}p(T))} | \lambda |, \quad \text{by Lemma 7.6 (ii)} \\ &= \sup_{\lambda \in \sigma(T)} | \bar{p}p(\lambda) |, \quad \text{by Lemma 7.10} \\ &= (\sup_{\lambda \in \sigma(T)} | p(\lambda) |)^2. \end{aligned}$$

[Here $\bar{p}(x) = \overline{p(x)}$, where $\overline{p(x)}$ is the complex conjugate of $p(x)$.] $\blacksquare$

If p is a polynomial and T is a bounded (self-adjoint) operator, $p(T)$ is defined as in Lemma 7.10.

Theorem 7.10. *Continuous Functional Calculus.* Let T be a self-adjoint operator on a Hilbert space H. There exists a unique map φ:

$C_1[\sigma(T)] \to L(H, H)$ with the following properties:

(i) $\varphi(\alpha f + \beta g) = \alpha\varphi(f) + \beta\varphi(g)$,
$\varphi(fg) = \varphi(f)\varphi(g)$,
$\varphi(1) = I_H$,
$\varphi(\bar{f}) = \varphi(f)^*$, for all f, g in $C_1[\sigma(T)]$ and scalars α and β.

(ii) $\| \varphi(f) \| \leq C \| f \|_\infty$, for all f, for some constant C.

(iii) If $f(x) = x$, $\varphi(f) = T$.

Moreover, φ satisfies the following additional properties.

(iv) If $Th = \lambda h$ for all h, $\varphi(f)h = f(\lambda)h$.

(v) $\sigma(\varphi(f)) = f(\sigma(T))$.

(vi) If $f \geq 0$, $\varphi(f) \geq 0$.

(vii) $\| \varphi(f) \| = \| f \|_\infty \equiv \sup\{| f(\lambda) |: \lambda \in \sigma(T)\}$.

[Part (v) is called the *Spectral Mapping Theorem*.] ∎

Proof. Define $\varphi(p)$ to be $p(T)$ for each polynomial p in $C_1(\sigma(T))$. By Lemma 7.11, $\| \varphi(p) \| = \| p \|_\infty$ so that φ has a unique continuous linear extension to the closure of the set of polynomials in $C_1(\sigma(T))$. By the Stone–Weierstrass Theorem (Theorem 1.26), this closure is all of $C_1(\sigma(T))$. Clearly, the extension of φ to $C_1(\sigma(T))$ satisfies (i), (ii), and (iii) and is the unique function satisfying these properties.

Since (iv) and (vii) are valid for polynomials, they are valid by the continuity of φ for all continuous functions. To prove (vi), observe that if $f \geq 0$, then $f = g^2$, where g is real-valued and $g \in C_1(\sigma(T))$. Hence, $\varphi(f) = \varphi(g^2) = [\varphi(g)]^2$, where $\varphi(g)$ is self-adjoint so that $\varphi(f) \geq 0$.

It remains to show (v). To this end suppose λ is a scalar and $\lambda \neq f(u)$ for any u in $\sigma(T)$. Let $g = (f - \lambda)^{-1}$. Then

$$\varphi(f-\lambda)\varphi(g) = \varphi((f-\lambda)g) = \varphi(1) = I_H = \varphi(g(f-\lambda)) = \varphi(g)\varphi(f-\lambda),$$

so that $\varphi(g) = [\varphi(f) - \lambda]^{-1}$. This means $\lambda \notin \sigma(\varphi(f))$. Hence $\sigma(\varphi(f)) \subset f(\sigma(T))$. Conversely, suppose $\lambda = f(\mu)$, $\mu \in \sigma(T)$ and $\lambda \notin \sigma(f(T))$. Then $f(T) - \lambda I$ is surjective. By Proposition 6.12, there exists $\delta > 0$ such that for $A \in L(H, H)$,

$$\| A - [f(T) - \lambda I] \| < \delta \Rightarrow A \text{ is surjective.}$$

Let $0 < \varepsilon < \delta$. Let p be a polynomial such that $\| p - f \|_\infty < \frac{1}{2}\varepsilon$. Then, $\| p(T) - p(\mu)I - [f(T) - \lambda I] \| < \varepsilon$. Consequently, $p(T) - p(\mu)I$ is surjective. Since $p(\mu) \in p(\sigma(T)) = \sigma(p(T))$, it follows that either $p(T) - p(\mu)I$ is not one-to-one or $[p(T) - p(\mu)I]^{-1}$, when it exists, is not bounded. In either case, by Proposition 6.5 there exists $h \in H$ such that $\| h \| = 1$ and $\| [p(T) - p(\mu)I]h \| < \frac{1}{2}\varepsilon$. But then $\| [f(T) - \lambda I]h \| < 2\varepsilon$. This contradicts our original assumption that $\lambda \notin \sigma(f(T))$. ∎

It should be emphasized that any real-valued continuous function is mapped by φ to a self-adjoint operator, as (i) shows; and, in particular, a nonnegative continuous function is mapped to a positive self-adjoint operator, as shown by (vi).

An immediate consequence of Theorem 7.10 is the following corollary.

Corollary 7.3. If $T \geq 0$, then there exists a positive operator S such that $S^2 = T$. (S is called the *square root* of T.)[†] ∎

Proof. If $T \geq 0$, then $\sigma(T) \subset [0, \infty)$ by Lemma 7.8. Let $f(x) = x^{1/2}$ on $[0, \infty)$. Then if $S = f(T)$, then $S^2 = T$. ∎

We wish now to "extend" (see Problem 7.5.2) the mapping φ to certain nonnegative discontinuous functions defined below. To do so we will need the following lemma.

Let P^+ denote the class of all (real-valued) nonnegative polynomials defined on $\sigma(T)$.

Lemma 7.12.

(i) If $p \in P^+$, $p(T)$ is a positive self-adjoint operator.

(ii) If (p_n) is a sequence in P^+ with $p_{n+1} \leq p_n$ for $n = 1, 2, \ldots$, then $p_n(T)$ converges strongly to a positive self-adjoint operator.

(iii) Let (p_n) and (q_n) be sequences in P^+ with $p_{n+1} \leq p_n$ and $q_{n+1} \leq q_n$ for $n = 1, 2, \ldots$, and let S_p and S_q be the strong limits of the sequences $(p_n(T))$ and $(q_n(T))$, respectively. If $\lim_{n\to\infty} p_n(t) \leq \lim_{n\to\infty} q_n(t)$ for all t in $\sigma(T)$, then $S_p \leq S_q$. ∎

[†] In Problem 8.1.7 it is established that there is exactly one positive operator S such that $S^2 = T$.

Proof.

(i) By (i) and (vi) of Theorem 7.10.

(ii) As in Lemma 7.9, the sequence $(p_n(T))$ converges strongly to a self-adjoint operator S_p. Since $p_n(T) \geq 0$,

$$(S_p x \mid x) = \lim_{n\to\infty} (p_n(T)x \mid x) \geq 0 \quad \text{or} \quad S_p \geq 0.$$

(iii) For each t in $\sigma(T)$, the sequences $(p_n(t))$ and $(q_n(t))$—as nonnegative, nonincreasing sequences—must converge. Let k be a fixed positive integer. For each t in $\sigma(T)$,

$$\lim_{n\to\infty} [p_n(t) - q_k(t)] \leq \lim_{n\to\infty} p_n(t) - \lim_{n\to\infty} q_n(t) \leq 0. \tag{7.56}$$

Let $m_n(t) = \max\{(p_n - q_k)(t), 0\}$ for each t and $n = 1, 2, \ldots$. From equation (7.56) it follows that $\lim_{n\to\infty} m_n(t) = 0$ for each t in $\sigma(T)$. Clearly, $m_{n+1} \leq m_n$. Hence, by Dini's Theorem, Problem 1.5.18, the sequence m_n converges uniformly to 0 in $\sigma(T)$. This means that for any $\varepsilon > 0$ there is a positive integer N so that whenever $n \geq N$,

$$p_n(t) - q_k(t) < \varepsilon,$$

for any t in $\sigma(T)$. Hence, by (vi) of Theorem 7.10,

$$p_n(T) \leq \varepsilon I + q_k(T),$$

for $n \geq N$. Hence $S_p \leq \varepsilon I + q_k(T)$ and since k is arbitrary $S_p \leq \varepsilon I + S_q$. The fact that ε is arbitrary means $S_p \leq S_q$. ∎

Now let L^+ represent the class of all real-valued functions f on $\sigma(T)$ for which there exists a sequence (p_n) of nonnegative polynomials defined on $\sigma(T)$ such that

$$\text{(i)} \quad 0 \leq p_{n+1}(t) \leq p_n(t), \qquad \text{for } n = 1, 2, \ldots,$$

and (7.57)

$$\text{(ii)} \quad \lim p_n(t) = f(t), \qquad \text{for each } t \text{ in } \sigma(T).$$

The preceding lemma enables us to define $f(T)$ for each f in L^+ as the strong limit of $(p_n(T))$, where (p_n) is a sequence as in condition (7.57). Indeed, if (q_n) is another sequence of nonnegative polynomials satisfying condition (7.57), then Lemma 7.12 assures us that the strong limit of

$(p_n(T))$ and $(q_n(T))$ are the same positive self-adjoint operator. Hence, $f(T)$ is well-defined as this positive self-adjoint operator.

Lemma 7.13. The mapping $f \to f(T)$ of L^+ into the class of positive self-adjoint operators satisfies the following properties:

(i) $(f+g)T = f(T) + g(T)$ for all f, g in L^+.

(ii) $(\alpha f)T = \alpha f(T)$ for all f in L^+ and $\alpha \geq 0$.

(iii) $(fg)T = f(T)g(T)$ for all f, g in L^+.

(iv) If $f \leq g$ then $f(T) \leq g(T)$ for all $f, g \in L^+$. ▮

Proof. The proofs of (i) and (ii) follow from the corresponding statements for p and q in P^+. Statement (iv) is an easy consequence of Lemma 7.12 (iii). We here prove (iii). Choose any nonincreasing sequences (p_n) and (q_n) in P^+ with $f(t) = \lim_n p_n(t)$ and $g(t) = \lim_n q_n(t)$. Then, $p_{n+1}q_{n+1} \leq p_n q_n$ and $\lim_n (p_n q_n)(t) = fg(t)$. For all x and y in H

$$(f(T)g(T)x \mid y) = (g(T)x \mid f(T)y) = \lim_{n\to\infty} (q_n(T)x \mid p_n(T)y)$$
$$= \lim_{n\to\infty} (p_n(T)q_n(T)x \mid y) = \lim_{n\to\infty} (p_n q_n(T)x \mid y) = ((fg)Tx \mid y).$$

Hence $f(T)g(T) = (fg)T$ by Remark 7.1. ▮

Let L be the set of all bounded real-valued functions on $\sigma(T)$ of the form $f - g$ with $f, g \in L^+$. Clearly L is a subspace of the real linear space of all bounded real-valued functions on $\sigma(T)$. If p is any real-valued polynomial on $\sigma(T)$, then $p \in L$ since $p + \alpha 1$ is in P^+ for some positive α and $p = (p + \alpha 1) - \alpha 1$.

Our development has led us to the point where we can extend the mapping $p \to p(T)$ of real-valued polynomials to L. Let $f \in L$ and choose g and h in L^+ so that $f = g - h$. As expected define $f(T)$ as $g(T) - h(T)$. Notice that if also $f = g' - h'$, then $g + h' = h + g'$ so that by Lemma 6.13 (i), $g(T) + h'(T) = h(T) + g'(T)$. Hence $g(T) - h(T) = g'(T) - h'(T)$ and $f(T)$ is well defined.

Proposition 7.22. The mapping $f \to f(T)$ on L is a linear mapping into the class of self-adjoint operators, and for all f and g in L,

(i) $f \leq g$ implies $f(T) \leq g(T)$,

(ii) $(fg)T = f(T)g(T)$. ▮

Proof. The proof follows readily from Lemma 7.13. ∎

There are special functions in L^+ that will be used to prove the Spectral Theorem. For each real number s define e_s on $\sigma(T)$ as follows: If $m \leq s \leq M$ put

$$e_s(t) = \begin{cases} 1, & \text{for } t \in [m, s] \cap \sigma(T) \\ 0, & \text{for } t \in (s, M] \cap \sigma(T). \end{cases} \tag{7.58}$$

If $s < m$, set $e_s = 0$; if $s \geq M$, set $e_s = 1$.

It is easy to see that e_s is in L^+ if $s < m$ or $s \geq M$. For $s \in [m, M]$, we appeal to the Weierstrass approximation theorem (Corollary 1.1, in Part A). For $s \in [m, M]$ let N be the least positive integer such that $s + 1/N \leq M$. For $n \geq N$ let

$$f_n(t) = \begin{cases} 1, & \text{for } t \in [m, s] \cap \sigma(T), \\ -nt + ns + 1, & \text{for } t \in (s, s + 1/n) \cap \sigma(T), \\ 0, & \text{for } t \in [s + 1/n, M] \cap \sigma(T). \end{cases} \tag{7.59}$$

For each such n, f_n is a real-valued continuous function on $\sigma(T)$ with $0 \leq f_{n+1} \leq f_n$ for $n = N, N + 1, \ldots$ $e_s(t)$ is the limit of $f_n(t)$ for each t in $[m, M]$. Utilizing the Weierstrass approximation theorem, we can find a sequence (p_n) in P with

$$e_s(t) \leq f_n(t) + 2^{-n-1} < p_n(t) < f_n(t) + 2^{-n}, \tag{7.60}$$

for each t in $\sigma(T)$ and $n = N, N + 1, \ldots$. From inequalities (7.60) it follows that $p_n \in P^+$ and $p_{n+1} \leq p_n$ for $n = N, N + 1, \ldots$. Also $e_s(t) = \lim_n p_n(t)$. Thus, $e_s \in L^+$.

Definition 7.12. For every real number s, define $E(s)$ to be the positive self-adjoint operator given by $E(s) = e_s(T)$, where e_s is the element of L^+ given by equation (7.58). ∎

Proposition 7.23. Let T be a self-adjoint operator on H and suppose $mI \leq T \leq MI$. For each real number s there is a projection $E(s)$ on H such that

(i) $E(s)T = TE(s)$,
(ii) $E(s) \leq E(s')$, for $s \leq s'$,
(iii) $E(s) = 0$, for $s < m$,
(iv) $E(s) = I$, for $s > M$,
(v) $\lim_{h \to 0^+} E(s + h) = E(s)$ (in the strong sense). ∎

Proof. For each s, $E(s)$ is the positive self-adjoint operator $e_s(T)$. Since $T = f(T)$, where $f(x) = x$ on $\sigma(T)$, (i) follows from Proposition 7.22 (ii). Since $e_s^2 = e_s$, $e_s \leq e_{s'}$, for $s \leq s'$, $e_s = 0$ for $s < m$, and $e_s = 1$ for $s \geq M$, it is clear that $E(s)$ is a projection and that statements (ii), (iii), and (iv) are true. To prove (v) we note that analogous to (7.60) we can construct a sequence (p_n) in P^+ with $p_{n+1} \leq p_n$ and $e_{s+1/n} \leq p_n$ such that $\lim_{n\to\infty} p_n(t) = e_s(t)$. This means $p_n(T) \geq E(s + 1/n) \geq E(s)$. Since $\lim_n p_n(T) = E(s)$, we have $E(s + 1/n) \to E(s)$ as $n \to \infty$. Using (ii) we see that $\lim_{h\to 0^+} E(s + h) = E(s)$. ∎

Any projection-valued function E satisfying the conditions of Proposition 7.23 is called a *resolution of the identity* associated with T.

Definition 7.13. Suppose E is a function on R that assigns to each real number s a self-adjoint operator $E(s)$ on H. Let a and b be real numbers with $a < b$ and let f be a real-valued function on $[a, b]$. *f is E-integrable* if and only if there is a self-adjoint operator S with the following property: For each $\varepsilon > 0$ there exists a $\delta > 0$ such that, for all partitions $\{a = s_0 < s_1 < \cdots < s_n = b\}$ of $[a, b]$ with $s_k - s_{k-1} < \delta$ for $k = 1, \ldots, n$ and for all numbers $t_1, t_2, \ldots, t_n$ with $s_{k-1} \leq t_k \leq s_k$ for $k = 1, \ldots, n$,

$$\left\| S - \sum_{k=1}^{n} f(t_k)[E(s_k) - E(s_{k-1})] \right\| < \varepsilon.$$

The operator S is called the *integral of f* with respect to E and is denoted by $\int_a^b f(t)\, dE(t)$. ∎

Example 7.6. *A Concrete Example of the Resolution of the Identity for a Compact Self-Adjoint Operator.* Let us pause a moment in our development and see that the function $f(t) = t$ is E-integrable with respect to some operator-valued function E on R and its integral $\int_a^b t\, dE(t)$ is equal to a given compact self-adjoint operator T. To this end let T be a fixed nonzero compact, self-adjoint operator on H, where $\dim H = \infty$. Then, by Problem 7.4.33(vi), $M \geq 0$. Recall from Theorem 7.8 that, for each x in H,

$$Tx = \sum_i \lambda_i (x \mid x_i) x_i, \tag{7.61}$$

where the λ_i are the eigenvalues of T and the x_i form an orthonormal set of eigenvectors with x_i corresponding to λ_i. For each real number s define $E(s)$ by

$$E(s)(x) = \begin{cases} \sum_{\lambda_k \leq s} (x \mid x_k)x_k\,, & \text{for } s < 0\,, \\ x - \sum_{\lambda_k > s} (x \mid x_k)x_k\,, & \text{for } s \geq 0. \end{cases} \tag{7.62}$$

It is understood that if there are no $\lambda_k \leq s$ for $s < 0$ or no $\lambda_k > s$ for $s \geq 0$, then the respective sums in equation (7.62) are zero. If B is the orthonormal set $\{x_i: i = 1, 2, \ldots\}$ and G is the closed linear span of B, then B is an orthonormal basis[†] for G and each element x in H can be written in the form

$$x = y_x + \sum_k (x \mid x_k)x_k\,, \tag{7.63}$$

where $y_x \in G^\perp$ by Theorem 7.2. Clearly for $s < 0$, $E(s)$ is the projection onto the closed linear span G_s of $\{x_k \mid \lambda_k \leq s\}$, and for $s \geq 0$, $E(s)$ is the projection on $G^\perp \oplus G_s$. It is left as an exercise (Problem 7.5.1) for the reader to show that E is a projection-valued function on R that satisfies the criteria of Proposition 7.23. In short, we may say that the values of E "increase" from the zero projection when $s < m$ to the identity projection when $s > M$ where $mI \leq T \leq MI$.

Now let a and b be real numbers with $a < m$ and $b \geq M$. If $P = \{a = s_0 < s_1 < \cdots < s_n = b\}$ is any partition of $[a, b]$, using equation (7.63) we easily see that

$$[E(s_i) - E(s_{i-1})](x) = \sum_{s_{i-1} < \lambda_k \leq s_i} (x \mid x_k)x_k\,, \qquad \text{if } s_i < 0, \tag{7.64}$$

$$[E(s_i) - E(s_{i-1})](x) = y_x + \sum_{s_{i-1} < \lambda_k \leq s_i} (x \mid x_k)x_k, \qquad \text{if } s_{i-1} < 0 \leq s_i, \tag{7.65}$$

$$[E(s_i) - E(s_{i-1})](x) = \sum_{s_{i-1} < \lambda_k \leq s_i} (x \mid x_k)x_k, \qquad \text{if } 0 \leq s_{i-1} < s_i. \tag{7.66}$$

We show now that the compact operator in equation (7.61) is the integral $\int_a^b t\, dE(t)$ as in Definition 7.13, where E is as in equation (7.62). Let $\varepsilon > 0$ be arbitrary. With $\delta = \varepsilon$, let $P = \{a = s_0 < s_1 < s_2 \cdots < s_n = b\}$ be any partition of $[a, b]$ with $s_i - s_{i-1} < \delta$ for $i = 1, 2, \ldots, n$. Let i_0

[†] If $y = \sum_{i=1}^m (y \mid x_i)x_i$ and $n \geq m$, then $\| x - \sum_{i=1}^n (x \mid x_i)x_i \| \leq 2 \| x - y \|$.

be the index with $s_{i_0-1} < 0 \leq s_{i_0}$. If then t_i is arbitrary in $[s_{i-1}, s_i]$, we have, using equation (7.61) and the orthonormality of the set $\{x_1, x_2, \ldots\}$, that

$$\left\| Tx - \sum_{i=1}^{n} t_i[E(s_i) - E(s_{i-1})](x) \right\|^2$$

$$= \left\| \sum_{\substack{i=1 \\ i \neq i_0}}^{n} \left[\sum_{s_{i-1} < \lambda_k \leq s_i} (\lambda_k - t_i)(x \mid x_k)x_k \right] + \sum_{s_{i_0-1} < \lambda_k \leq s_{i_0}} (\lambda_k - t_{i_0})(x \mid x_k)x_k + t_{i_0} \cdot y_x \right\|^2$$

$$= \sum_{\substack{i=1 \\ i \neq i_0}}^{n} \left[\sum_{s_{i-1} \leq \lambda_k \leq s_i} (\lambda_k - t_i)^2 \mid (x \mid x_k) \mid^2 \right] + \sum_{s_{i_0-1} < \lambda_k \leq s_{i_0}} (\lambda_k - t_{i_0})^2 \mid (x \mid x_k) \mid^2$$

$$+ \mid t_{i_0} \mid^2 \| y_x \|^2$$

$$\leq \varepsilon^2 \sum_{\substack{i=1 \\ i \neq i_0}}^{n} \sum_{s_{i-1} \leq \lambda_k \leq s_i} \mid (x \mid x_k) \mid^2 + \varepsilon^2 \sum_{s_{i_0-1} \leq \lambda_k \leq s_{i_0}} \mid (x \mid x_k) \mid^2 + \varepsilon^2 \| y_x \|^2$$

$$= \varepsilon^2 \left[\sum_k \mid (x \mid x_k) \mid^2 + \| y_x \|^2 \right] = \varepsilon^2 \| x \|^2.$$

Hence,

$$\left\| T - \sum_{i=1}^{n} t_i[E(s_i) - E(s_{i-1})] \right\| < \varepsilon,$$

provided $s_i - s_{i-1} < \delta$ and $t_i \in [s_{i-1}, s_i]$. Hence $T = \int_a^b t\, dE(t)$.

Using this example as a motivation, we now come to the goal—and the most outstanding theorem—of this section.

Theorem 7.11. *The Spectral Theorem* (*Resolution of the Identity Formulation*). Let T be a self-adjoint operator, and m and M be real numbers with $mI \leq T \leq MI$. Then there exists a unique resolution of the identity E on R associated with T such that if a and b are real numbers with $a < m$ and $b \geq M$, then the mapping $t \to t$ of $[a, b]$ into R is E-integrable and

$$T = \int_a^b t\, dE(t). \tag{7.67}$$

■

Proof. If e_s represents the function defined in equation (7.58), it is easy to verify that if $s < u$ then

$$e_u(t) - e_s(t) = \begin{cases} 1, & \text{for } t \in (s, u] \cap \sigma(T), \\ 0, & \text{for } t \in \sigma(T) - (s, u]. \end{cases}$$

Hence for all $t \in \sigma(T)$,

$$s[e_u(t) - e_s(t)] \leq t[e_u(t) - e_s(t)] \leq u[e_u(t) - e_s(t)].$$

Defining E by Definition 7.12, and using Proposition 7.22, we obtain

$$s[E(u) - E(s)] \leq T[E(u) - E(s)] \leq u[E(u) - E(s)]. \tag{7.68}$$

We know from the proof of Proposition 7.23 that E satisfies conditions (i)–(v) of Proposition 7.23. We must show that the mapping $t \to t$ of $[a, b]$ into R is E-integrable. To this end, let $\varepsilon > 0$ be arbitrary and let $\{a = s_0 < s_1 < \cdots < s_n = b\}$ be a partition of $[a, b]$ with $s_k - s_{k-1} < \varepsilon$ for $k = 1, 2, \ldots, n$. Then from (7.68) we get the inequality

$$\sum_{k=1}^{n} s_{k-1}[E(s_k) - E(s_{k-1})] \leq T \sum_{k=1}^{n} [E(s_k) - E(s_{k-1})] \leq \sum_{k=1}^{n} s_k[E(s_k) - E(s_{k-1})].$$

However, since $a < m$ and $b \geq M$, Proposition 6.23 tells us that

$$\sum_{k=1}^{n} [E(s_k) - E(s_{k-1})] = E(b) - E(a) = I. \tag{7.69}$$

Hence,

$$\sum_{k=1}^{n} s_{k-1}[E(s_k) - E(s_{k-1})] \leq T \leq \sum_{k=1}^{n} s_k[E(s_k) - E(s_{k-1})].$$

Now let t_k be a real number with $s_{k-1} \leq t_k \leq s_k$ for $k = 1, 2, \ldots, n$. Then

$$\begin{aligned} \sum_{k=1}^{n} (s_{k-1} - t_k)[E(s_k) - E(s_{k-1})] &\leq T - \sum_{k=1}^{n} t_k[E(s_k) - E(s_{k-1})] \\ &\leq \sum_{k=1}^{n} (s_k - t_k)[E(s_k) - E(s_{k-1})]. \end{aligned} \tag{7.70}$$

However, $s_{k-1} - t_k > -\varepsilon$, $s_k - t_k < \varepsilon$, and $E(s_k) - E(s_{k-1}) \geq 0$ so that using equation (7.69), the inequalities (7.70) reduce to

$$-\varepsilon I \leq T - \sum_{k=1}^{n} t_k[E(s_k) - E(s_{k-1})] \leq \varepsilon I,$$

so that from Lemma 7.4

$$\left\| T - \sum_{k=1}^{n} t_k[E(s_k) - E(s_{k-1})] \right\| \leq \varepsilon.$$

Hence $T = \int_a^b t \, dE(t)$.

It remains to show the uniqueness conclusion of the theorem. The proof of this is outlined in Problems 7.5.4 and 7.5.5. ∎

Two applications of the preceding results are outlined in Problems 7.5.6 and 7.5.7. For a further study of the spectral theory of self-adjoint operators in Hilbert space along with applications of this theory, the reader may wish to consult the exhaustive work of Dunford and Schwartz [12].

To conclude this section, we present an example to illustrate much of the preceding theory.

Example 7.7. *The Spectrum, the Eigenvalues, and the Resolution of Identity of a Bounded Self-Adjoint Operator.* Let $H = L_2([\alpha, \beta], \mu)$ where $-\infty < \alpha < \beta < \infty$ and μ is the Lebesgue measure of $[\alpha, \beta]$. Define T: $H \to H$ by $Tf = \lambda f(\lambda)$. T is called the multiplication operator. The equation

$$\int_\alpha^\beta | Tf |^2 \, d\mu = \int_\alpha^\beta | \lambda f(\lambda) |^2 \, d\mu(\lambda) \leq m^2 \int_\alpha^\beta | f(\lambda) |^2 \, d\mu(\lambda) = m^2 \| f \|^2 < \infty, \tag{7.71}$$

where $m = \max\{| \alpha |, | \beta |\}$, shows that $Tf \in L_2[\alpha, \beta]$ for every f and that T is a bounded operator. T is also self-adjoint, since

$$(Tf \mid g) = \int_\alpha^\beta \lambda f(\lambda)\overline{g(\lambda)} \, d\mu(\lambda) = \int_\alpha^\beta f(\lambda)\overline{\lambda g(\lambda)} \, d\mu(\lambda) = (f \mid Tg). \tag{7.72}$$

Let λ_0 be any scalar and let f be an element of $L_2[\alpha, \beta]$ such that $(T - \lambda_0 I)f = 0$. Then $f(\lambda) = 0$ almost everywhere or $f = 0$. Hence T has no eigenvalues.

Next suppose that $\lambda_0 \in [\alpha, \beta]$. For each positive integer n, let f_n be the characteristic function of $[\lambda_0 - 1/n, \lambda_0 + 1/n] \cap [\alpha, \beta]$. Noting that $\| f_n \| > 0$, let g_n be the function $f_n/\| f_n \|$. If $\lambda_0 \notin \sigma(T)$, then

$$\begin{aligned}
1 = \| g_n \|^2 &= \| (T - \lambda_0 I)^{-1}(T - \lambda_0 I)g_n \|^2 \leq \| (T - \lambda_0 I)^{-1} \|^2 \, \| (T - \lambda_0 I)g_n \|^2 \\
&= \| (T - \lambda_0 I)^{-1} \|^2 \int_\alpha^\beta | \lambda - \lambda_0 |^2 \, | g_n(\lambda) |^2 \, d\mu(\lambda) \\
&\leq \| (T - \lambda_0 I)^{-1} \|^2 \frac{1}{n^2} \int_\alpha^\beta | g_n(\lambda) |^2 \, d\mu(\lambda) \\
&= \| (T - \lambda_0 I)^{-1} \|^2 \frac{1}{n^2}.
\end{aligned}$$

This contradiction shows that $\lambda_0 \in \sigma(T)$ and that $[\alpha, \beta] \subset \sigma(T)$.

Finally, let ξ_0 be any scalar outside of $[\alpha, \beta]$. We want to show $\xi_0 \notin \sigma(T)$ and conclude $[\alpha, \beta] = \sigma(T)$. Let $\delta = \inf\{|\xi - \xi_0| : \xi \in [\alpha, \beta]\} > 0$. For any $g \in L_2[\alpha, \beta]$, define Sg to be $(\xi - \xi_0)^{-1}g(\xi)$. Since

$$\int_\alpha^\beta \left|\frac{g(\xi)}{(\xi - \xi_0)}\right|^2 d\mu(\xi) \leq \int_\alpha^\beta \frac{|g(\xi)|^2}{\delta^2} d\mu(\xi) = \frac{1}{\delta^2} \|g\|^2, \tag{7.73}$$

$Sg \in L_2[\alpha, \beta]$ for each g. Moreover,

$$\xi \frac{g(\xi)}{\xi - \xi_0} = g(\xi) + \xi_0 \frac{g(\xi)}{\xi - \xi_0},$$

which means that $(T - \xi_0 I)S = I$. In addition $S(T - \xi_0 I) = I$, as is easily verified. Moreover, the inequality (7.73) shows that $\|S\| \leq 1/\delta$. We conclude that $\xi_0 \notin \sigma(T)$.

To complete this example, let us calculate the unique resolution of the identity associated with T satisfying equation (7.67). Since $[\alpha, \beta] = \sigma(T) \subset [m_T, M_T]$ and m_T and M_T are in $\sigma(T)$, $\alpha = m_T$ and $\beta = M_T$. Define E: $R \to L(H, H)$ by

$$E(\lambda) = \begin{cases} 0, & \text{if } \lambda < \alpha, \\ I, & \text{if } \lambda \geq \beta, \end{cases}$$

and by

$$E(\lambda) f(\xi) = \begin{cases} f(\xi), & \text{if } \alpha \leq \xi \leq \lambda, \\ 0, & \text{if } \lambda < \xi < \beta. \end{cases}$$

It is easy to verify that E is a resolution of the identity associated with T. It remains to show that $T = \int_a^b t\, dE(t)$ if $a < \alpha$ and $b \geq \beta$.

Let $\varepsilon > 0$ be arbitrary and let $\{a = s_0 < s_1 < \cdots < s_n = b\}$ be a partition of $[a, b]$ with $s_i - s_{i-1} < \varepsilon$. For each i, let $s_{i-1} \leq t_i \leq s_i$. Now

$$\left\| Tf - \sum_{i=1}^n t_i[E(s_i) - E(s_{i-1})]f \right\|^2 = \left\| \lambda f(\lambda) - \sum_{i=1}^n t_i f(\lambda) \chi_{[\alpha,\beta]\cap[s_{i-1},s_i]}(\lambda) \right\|^2$$
$$= \sum_{i=1}^n \int_{[\alpha,\beta]\cap[s_{i-1},s_i]} |\lambda - t_i|^2 |f(\lambda)|^2 d\mu(\lambda) \leq \varepsilon^2 \int |f(\lambda)|^2 d\mu(\lambda) = \varepsilon^2 \|f\|^2.$$

Hence $T = \int_a^b \lambda\, dE(\lambda)$.

Problems

× **7.5.1.** Prove that the function of R defined by equation (7.62) is a resolution of the identity associated with the compact self-adjoint operator

T. [Hint: To prove (v) of Proposition 7.23 recall from Theorem 7.8 that $|\lambda_1| \geq |\lambda_2| > \cdots \to 0$.]

In the following problems T is a self-adjoint operator on H and $mI \leq T \leq MI$.

× **7.5.2.** (i) Prove that each real-valued continuous function f on $\sigma(T)$ is an element of L, that is, $f = g - h$, where g and h are in L^+. [Hint: If $f \geq 0$, then by the Weierstrass approximation theorem, there exist polynomials $p_n(t)$ such that $f(t) + 1/(n+1) \leq p_n(t) \leq f(t) + 1/n$ for all t in $\sigma(T)$.]

(ii) Prove that the mapping in Proposition 7.22 of L into the class of self-adjoint operators is an extension of the mapping φ in Theorem 7.10 on the class of continuous real-valued functions.

(iii) Let $\{E(t): t \in R\}$ be a family of projections in $L(H, H)$ such that $s \leq t \Rightarrow E(s) \leq E(t)$. Let $h_n \to 0^+$ and $k_n \to 0^+$. By Lemma 7.9, there are A_t and B_t in $L(H, H)$ such that $E(t + h_n) \to A_t$ and $E(t + k_n) \to B_t$ strongly. Show that $A_t = B_t$. Thus, the strong limit $\lim_{h \to 0^+} E(t + h) = E(t+)$ exists and equals A_t. Prove that $E(t-)$ also exists, and that $E(t+)$ and $E(t-)$ are both projections.

× **7.5.3.** Let E' be any projection-valued function on R satisfying the conditions (i), (ii), (iii), and (iv) of Proposition 7.23. If $a < m$ and $b \geq M$, suppose $T = \int_a^b t\, dE'(t)$. (This can be proved as in Theorem 7.11.) Prove the following:

(i) $T^m = \int_a^b t^m\, dE'(t)$ for every nonnegative integer m. {Hint: If $i < j$, $[E'(s_i) - E'(s_{i-1})][E'(s_j) - E'(s_{j-1})] = 0$. Hence

$$\left(\sum_{i=1}^{n} t_i[E'(s_i) - E'(s_{i-1})]\right)^m = \sum_{i=1}^{n} t_i^m[E'(s_i) - E'(s_{i-1})].$$

Show that

$$\left\| T^m - \sum_{i=1}^{n} t_i^m[E'(s_i) - E'(s_{i-1})] \right\| \leq K(m) \left\| T - \sum_{i=1}^{n} t_i[E'(s_i) - E'(s_{i-1})] \right\|,$$

where $K(m)$ is a constant dependent on m.}

(ii) $p_0(T) = \int_a^b p(t)\, dE'(t)$ for every real polynomial p on $[a, b]$, where $p_0 = p|_{\sigma(T)}$.

(iii) $f_0(T) = \int_a^b f(t)\, dE'(t)$ for every continuous real-valued function f on $[a, b]$, where $f_0 = f|_{\sigma(T)}$. [Note that $f_0(T)$ is defined by Proposition 7.22 since $f_0 \in L$ by Problem 7.5.2.] {Hint: There exists a real polynomial p so that $-\varepsilon/3 \leq p(t) - f(t) \leq \varepsilon/3$ for all t in $[a, b]$. By Lemma 7.4,

$\| p_0(T) - f_0(T) \| < \varepsilon/3$. Also

$$-(\varepsilon/3)I \leq \sum[p(t_i) - f(t_i)][E'(s_i) - E'(s_{i-1})] \leq (\varepsilon/3)I,$$

so that

$$\| \sum[p(t_i) - f(t_i)][E'(s_i) - E'(s_{i-1})] \| < \varepsilon/3.$$

Use part (ii).}

× **7.5.4.** This problem outlines a proof of the uniqueness conclusion of the Spectral Theorem.

(i) Suppose E' is a projection-valued function on R satisfying conditions (i)–(iv) of Proposition (7.23), and $T = \int_a^b t\, dE'(t)$ $(a < m < M \leq b)$. Suppose $m \leq s < M$ and f_n are the continuous functions on $[a, b]$ given by

$$f_n(x) = \begin{cases} 1, & \text{if } a \leq x \leq s, \\ n(s - x) + 1, & \text{if } s < x < s + 1/n, \\ 0, & \text{if } s + 1/n \leq x \leq b, \end{cases}$$

for $n = N + 1, N + 2, \ldots$, where N is the least positive integer so that $s + 1/N < M$. Note that $f_n|_{\sigma(T)}(T)$ is equal to $\int_a^b f_n(t)\, dE'(t)$ by (iii) of Problem 7.5.3. Prove that $\int_a^b f_n(t)\, dE'(t)$ converges strongly to $\lim_{h \to 0+} E'(s + h)$.

(ii) Establish the uniqueness conclusion of the Spectral Theorem.

7.5.5. (i) If E is any projection-valued function satisfying conditions (ii), (iii), and (iv) of Proposition 7.23, show that the function $t \to (E(t)x \mid y)$ is a bounded variation function on $[a, b]$, where x and y are fixed in H. [Hint: $(E(t)x \mid x)$ is a monotonic function; for any t, $(E(t)x \mid y)$ is a sesquilinear form on H. Use the polarization identity.]

(ii) If the assumptions are as in Problem 7.5.3 (using E instead of E'), prove that for all x and y in H

$$(f_0(T)x \mid y) = \int_a^b f(t)\, d(E(t)x \mid y)$$

in the Riemann–Stieltjes sense.

(iii) Give another proof of the uniqueness conclusion of Theorem 7.11 using part (ii).

7.5.6. Let E be the resolution of the identity satisfying Theorem 7.11.

(i) Prove that a real number λ is not in the spectrum $\sigma(T)$ of T if and only if there exists a positive number ε such that $E(s) = E(t)$ whenever $\lambda - \varepsilon \leq s < t \leq \lambda + \varepsilon$. {Hint: To prove the sufficiency, define a continuous function f on $[a, b]$ so that $f(t) = (t - \lambda)^{-1}$ for $t \notin [a, b] \cap [\lambda - \varepsilon,$

$\lambda + \varepsilon]$. Let $g(t) = t - \lambda$ for all t. Show $(T - \lambda)f(T) = g(T)f(T) = \int_a^b f(t)g(t)\, dE(t) = I$ whenever $\lambda \in [m, M]$. To prove the necessity, assume that for each $\varepsilon > 0$, s and t exist in $[\lambda - \varepsilon, \lambda + \varepsilon]$ such that $E(s) \neq E(t)$. Exhibit a y such that $E(s)y = 0$ and $E(t)y = y$ if $s < t$. Using Problem 7.5.5 (ii), show $\| (T - \lambda I)(y) \|^2 \leq \varepsilon^2 \| y \|^2$.}

(ii) Deduce from (i) that if $\lambda \notin \sigma(T)$, $(T - \lambda)^{-1} = \int_a^b f(t)\, dE(t)$, where $f(t)$ is a continuous function on $[a, b]$ equal to $(t - \lambda)^{-1}$ for $t \notin [a, b] \cap [\lambda - \varepsilon, \lambda + \varepsilon]$.

7.5.7. Prove that a real number λ is an eigenvalue of T if and only if $E(\lambda) \neq E(\lambda-)$, where E is the resolution of the identity satisfying Theorem 7.11. {Hint: To prove the sufficiency, by Problem 7.5.5 (ii) $\| (T - \lambda)x \|^2 = \int_a^b (t - \lambda)^2\, d \| E(t)x \|^2$ for any x in H. Apply this to a particular x, where $E(t)x = x$ if $t \geq \lambda$ and $E(t)x = 0$ if $t < \lambda$. To prove the necessity, if λ is an eigenvalue, suppose $Tx_0 - \lambda x_0 = 0$, where $x_0 \neq 0$. By Problem 7.5.5 (iii),

$$((T - \lambda)^2 x_0 \mid x_0) = \int_a^b (t - \lambda)^2 d(E(t)x_0 \mid x_0) = 0.$$

Since $\lambda \in (a, b]$, show $E(\lambda + \varepsilon)x_0 = x_0$ and $E(\lambda - \varepsilon)x_0 = 0$ by considering the integrals

$$\int_{\lambda+\varepsilon}^{M+\varepsilon} (t - \lambda)^2 d(E(t)x_0 \mid x_0) \quad \text{and} \quad \int_m^{\lambda-\varepsilon} (t - \lambda)^2 d(E(t)x_0 \mid x_0).\}$$

7.5.8. An operator $T \in L(H, H)$ is a *partial isometry* if there exists a closed subspace M in H such that $T|_M$ is isometric (that is $\| Tm \| = \| m \|$ for all $m \in M$) and $T|_{M^\perp} = 0$. Prove

(i) $\operatorname{Ker} T = M^\perp$.

(ii) If T is a partial isometry, so is T^*. (Hint: Show $T^*|_{T(H)}$ is isometric.)

(iii) T is a partial isometry if and only if T^*T is a projection. In this case T^*T is a projection on $(\operatorname{Ker} T)^\perp$.

(iv) T is a partial isometry if and only if $T = TT^*T$.

7.5.9. *Polar Decomposition.* (i) Prove that if $T \in L(H, H)$, then there exists a unique positive operator P and a unique partial isometry U such that $T = UP$ and $\operatorname{Ker} U = \operatorname{Ker} P = \operatorname{Ker} T$. This is called the polar decomposition of T. [Hint: Let $P = (T^*T)^{1/2}$ and $U(Pf) = Tf$ on the range of P.]

(ii) Show also that $T = QV$, where $Q = (TT^*)^{1/2}$ and V is a partial isometry.

(iii) Prove that if T is normal then $UP = PU$.

7.5.10. Let $T \in L(H, H)$. (i) Prove T is compact if and only if $(T^*T)^{1/2}$ is compact. [Hint: If $T = UP$ is the polar decomposition of T, $U^*T = (T^*T)^{1/2}$.]

(ii) Prove that if T is compact, then T is a Hilbert–Schmidt operator if and only if the eigenvalues of $(T^*T)^{1/2}$ are in l_2.

7.5.11. *The Square Root of a Positive Operator.* Let T be a bounded linear operator on a complex Hilbert space H such that $(Tx \mid x) \geq 0$ for every $x \in H$. Note that T is then self-adjoint (via a polarization identity argument). Let $P \in L(H, H)$, $P \geq 0$, such that $P^2 = T$. Let Q be the square root of T, that is, $Q = f(T)$, where $f(x) = x^{1/2}$. Since $PT = TP$, P commutes with Q. Since $(P - Q)P(P - Q)$ and $(P - Q)Q(P - Q)$ are both positive and their sum is zero, both are zero operators. Then $(P - Q)^3$, their difference, is also zero. Use Problem 7.4.6(ii) to show that $P = Q$. Prove also the following assertions:

(i) If $T \geq 0$ and T is compact, then $T^{1/2}$ is compact.

(ii) If $0 \leq T_1 \leq T_2$ and T_2 is compact, then T_1 is compact.

[Hint: By Theorem 7.8, $Tx = \sum_{k=1}^{\infty} \lambda_k (x \mid x_k) x_k$, $\lambda_k \geq 0$. Note that $T^{1/2}x = \sum_{k=1}^{\infty} \lambda_k^{1/2} (x \mid x_k) x_k$.]

8

Spectral Theory

8.1. Spectral Theory for Bounded Operators Revisited

In this section we utilize the knowledge of measure theory at our disposal to prove other versions of the Spectral Theorem for bounded self-adjoint operators—other than the resolution of the identity version given in Chapter 7. Although it is quite possible to generate a spectral measure from the resolution of the identity corresponding to a given self-adjoint operator, we prefer here to prove a more sophisticated spectral measure version of the Spectral Theorem making use of an elegant functional calculus version of the Spectral Theorem. Many of the results of this section will be utilized and duplicated in the next section, where we deal with unbounded operators.

Throughout this section, H denotes a complex Hilbert space.

Definition 8.1. If X is a set and $\mathscr{R}$ is a ring of subsets of X, then a *positive-operator-valued measure* E is a function $E: \mathscr{R} \to L(H, H)$ such that

(i) $E(M) \geq 0$ for all M in $\mathscr{R}$,
(ii) $E(\bigcup_{i=1}^{\infty} M_i) = \lim_n[\sum_{i=1}^{n} E(M_i)]$ (in the strong convergence sense) whenever (M_i) is a disjoint sequence of measurable sets whose union is also in $\mathscr{R}$.

If the values of E are projections, then E is called a *spectral measure*. If $X \in \mathscr{R}$ and $E(X) = I$, then spectral measure E is called *normalized*. ∎

Remark 8.1. If E is a positive-operator-valued measure, then

(i) $E(\varnothing) = 0$;

(ii) E is finitely additive;

(iii) if $M \subset N$, then $E(M) \leq E(N)$;

(iv) $E(M \cup N) + E(M \cap N) = E(M) + E(N)$;

(v) $E(M \cap N) = E(M)E(N)$ if and only if E is a spectral measure.

Theorem 8.1. Let $E: \mathscr{R} \to L(H, H)$ be a function whose values are positive operators. Then E is a positive-operator-valued measure if and only if, for each h in H, the formula $\mu_h(M) = (E(M)h \mid h)$ defines a measure on $\mathscr{R}$. ∎

Proof. The necessity is clear. For the converse, let μ_h be a measure on $\mathscr{R}$ for each h and let (M_i) be a disjoint sequence of sets in $\mathscr{R}$ with $\bigcup_{i=1}^{\infty} M_i$ in $\mathscr{R}$. Now

$$\left(E\left(\bigcup_{i=1}^{\infty} M_i\right)h \mid h\right) = \mu_h\left(\bigcup_{i=1}^{\infty} M_i\right) = \lim_{n\to\infty} \mu_h\left(\bigcup_{i=1}^{n} M_i\right)$$
$$= \lim_{n\to\infty} \sum_{i=1}^{n} \mu_h(M_i) = \lim_{n\to\infty} \left(\sum_{i=1}^{n} E(M_i)h \mid h\right). \tag{8.1}$$

If

$$A_n = E\left(\bigcup_{i=1}^{\infty} M_i\right) - E\left(\bigcup_{i=1}^{n} M_i\right),$$

then $\| A_n \|$ is bounded† by $\| E(\bigcup_{i=1}^{\infty} M_i) \|$, and we have by Lemma 7.5 in Chapter 7 that

$$\| A_n h \|^4 = (A_n h \mid A_n h)^2 \leq (A_n h \mid h)(A_n^2 h \mid A_n h) \leq (A_n h \mid h) \| A_n \|^3 \| h \|^2,$$

so that by equation (8.1), $A_n h \to 0$. Hence $E(\bigcup_{i=1}^{\infty} M_i)$ is the strong limit of $\sum_{i=1}^{n} E(M_i)$ as $n \to \infty$. ∎

Proposition 8.1. Let $\mathscr{R}$ be a ring of subsets of a set X, and suppose that for each vector h in H there is given a finite measure μ_h on $\mathscr{R}$. There exists a unique positive-operator-valued measure E on $\mathscr{R}$ such that $\mu_h(M) = (E(M)h \mid h)$ for all h in H and for all M in $\mathscr{R}$ if and only if for all vectors h and k and for each M in $\mathscr{R}$

(i) $[\mu_{h+k}(M)]^{1/2} \leq [\mu_h(M)]^{1/2} + [\mu_k(M)]^{1/2}$,

(ii) $\mu_{ch}(M) = | c |^2 \mu_h(M)$ for each scalar c,

† Since μ_h is a measure, A_n is a positive operator and $\| A_n \| = \sup_{\|h\| \leq 1} (A_n h \mid h)$.

(iii) $\mu_{h+k}(M) + \mu_{h-k}(M) = 2\mu_h(M) + 2\mu_k(M)$,

(iv) for each M in $\mathscr{R}$ there exists a constant k_M such that

$$\mu_h(M) \leq k_M \| h \|^2.$$ ∎

Proof. Let M be arbitrary in $\mathscr{R}$. Define the real-valued function $\| \ \|_M$ on H by

$$\| h \|_M = [\mu_h(M)]^{1/2}.$$

As in the proof of Proposition 7.2 the conditions (i)–(iii) are equivalent to saying that $\| \ \|_M$ is a pseudonorm ($\| h \|_M = 0$ does not necessarily mean that $h = 0$) satisfying the Parallelogram Law, which in turn is equivalent to saying that

$$B_M(h, k) = \tfrac{1}{4}(\| h + k \|_M^2 - \| h - k \|_M^2 + i \| h + ik \|_M^2 - i \| h - ik \|_M^2)$$

defines a Hermitian sesquilinear form on H such that $B_M(h, h) = \| h \|_M^2$. The boundedness of B_M is equivalent to (iv). Hence by Theorem 7.6, conditions (i)–(iv) are equivalent to the existence for each M of a bounded self-adjoint operator $E(M)$ with $(E(M)h \mid h) = B_M(h, h) = \mu_h(M)$. By Theorem 8.1, $E(M)$ defines a positive-operator-valued measure. ∎

The next result is easily verified.

Proposition 8.2. If E is a positive-operator-valued measure whose domain is a σ-ring, then

$$\sup\{\| E(M) \| \colon M \in \mathscr{R}\} < \infty.$$ ∎

From this point on through Theorem 8.2, E will denote a positive-operator-valued measure on a σ-ring $\mathscr{R}$. If μ_h is the measure

$$\mu_h(M) = (E(M)h \mid h),$$

then by Proposition 8.2 there exists a positive number K such that $\mu_h(M) \leq K \| h \|^2$ for all M and for all h.

If $f = g + ih$ is a complex-valued function on X, then f is *measurable* in case g and h are measurable. If μ is a measure on $\mathscr{R}$ and g and h are μ-integrable, then we say f is *μ-integrable* and define (as in Problem 3.2.24)

$$\int f \, d\mu = \int g \, d\mu + i \int h \, d\mu.$$

f is μ-integrable if and only if $|f|$ is μ-integrable, and in this case

$$\left| \int f \, d\mu \right| \leq \int |f| \, d\mu.$$

Definition 8.2. Suppose $\mathscr{F} \equiv \{\mu_h\}_{h\in H}$ is a family of measures on a σ-ring $\mathscr{R}$. [In particular $\mathscr{F}$ could be *generated by* E in the sense that $\mu_h(M) = (E(M)h \mid h)$ for each h and each M.] A measurable complex function on $(X, \mathscr{R})$ is *$\mathscr{F}$-integrable* (in particular *E-integrable* if $\mathscr{F}$ is generated by E) if it is μ_h-integrable for each h. If f is $\mathscr{F}$-integrable then for any ordered pair of vectors in H, we write

$$\int f\,d\mu_{h,k} = \tfrac{1}{4}\left(\int f\,d\mu_{h+k} - \int f\,d\mu_{h-k} + i\int f\,d\mu_{h+ik} - i\int f\,d\mu_{h-ik}\right). \tag{8.2}$$

∎

Lemma 8.1. For each pair of vectors h and k in H, the complex-valued function $L_{h,k}$ given by

$$L_{h,k}(f) = \int f\,d\mu_{h,k}$$

is a linear functional on the complex vector space of $\mathscr{F}$-integrable functions. Moreover if $\mathscr{F}$ is generated by E, then

$$\int \chi_M\,d\mu_{h,k} = (E(M)h \mid k), \tag{8.3}$$

for all M in $\mathscr{R}$. ∎

Proof. The fact that $L_{h,k}$ is linear follows from equation (8.2). To prove the latter part of the lemma, note that if $h \in H$, then

$$\int \chi_M\,d\mu_h = \mu_h(M) = (E(M)h \mid h).$$

Hence from equation (8.2) and the Polarization Identity (Proposition 7.1 in Chapter 7),

$$\begin{aligned}\int \chi_M\,d\mu_{h,k} &= \tfrac{1}{4}[(E(M)(h+k) \mid h+k) - (E(M)(h-k) \mid h-k) \\ &\quad + i(E(M)(h+ik) \mid h+ik) - i(E(M)(h-ik) \mid h-ik)] \\ &= (E(M)h \mid k).\end{aligned}$$

∎

Lemma 8.2. Let $\mathscr{F} = \{\mu_h\}_{h\in H}$ be any family of measures on a σ-ring $\mathscr{R}$. Let L be any linear functional on the vector space of $\mathscr{F}$-integrable functions. Then $L = L_{h,k}$ for some h and k if and only if

(i) whenever (f_n) is a sequence of $\mathscr{F}$-integrable functions such that $0 \le f_n \uparrow f$, where f is also $\mathscr{F}$-integrable, then $L(f_n) \to L(f)$, and
(ii) $L(\chi_M) = L_{h,k}(\chi_M)$ for all M in $\mathscr{R}$. ∎

Proof. To prove the necessity, if $0 \le f_n \uparrow f$ as in (i), then

$$\int f_n d\mu_h \to \int f\, d\mu_h$$

for each h by the Monotone Convergence Theorem. Hence

$$\int f_n\, d\mu_{h,k} \to \int f\, d\mu_{h,k}$$

by equation (8.2).

To prove the sufficiency, by (ii), $L = L_{h,k}$ for all simple functions. If $f \ge 0$, choose a sequence of simple functions (f_n) such that $0 \le f_n \uparrow f$. By (i)

$$Lf = \lim_{n\to\infty} L(f_n) = \lim_{n\to\infty} L_{h,k}(f_n) = L_{h,k}(f).$$

By linearity $Lf = L_{h,k}(f)$ for all $\mathscr{F}$-integrable functions f. ∎

Lemma 8.3. If for each vector h and for each M in $\mathscr{R}$,

$$\mu_h(M) = \tfrac{1}{4}[\mu_{h+h}(M) - \mu_{h-h}(M) + i\mu_{h+ih}(M) - i\mu_{h-ih}(M)],$$

then

$$\int f\, d\mu_{h,h} = \int f\, d\mu_h \tag{8.4}$$

for each $\mathscr{F}$-integrable function f. In particular, if $\mu_h(M) = (E(M)h \mid h)$ for each M and h, then equation (8.4) is true for each E-integrable function f. ∎

Proof. Define Lf to be $\int f\, d\mu_h$ for each f. L is linear and satisfies (i) of Lemma 8.2 by the Monotone Convergence Theorem. (ii) is satisfied since

$$L(\chi_M) = \mu_h(M) = \int \chi_M \mu_{h,h} = L_{h,h}(\chi_M).$$

Hence $L = L_{h,h}$. ∎

Proposition 8.3. Suppose $\mathscr{F} = (\mu_h)_{h\in H}$ is a family of measures on a σ-ring $\mathscr{R}$ such that, for each M in $\mathscr{R}$,

$$L_{\alpha h,k}(\chi_M) = \alpha L_{h,k}(\chi_M), \tag{8.5}$$

$$L_{h_1+h_2,k}(\chi_M) = L_{h_1,k}(\chi_M) + L_{h_2,k}(\chi_M), \tag{8.6}$$

$$L_{h,\alpha k}(\chi_M) = \bar{\alpha} L_{h,k}(\chi_M), \tag{8.7}$$

$$L_{h,k_1+k_2}(\chi_M) = L_{h,k_1}(\chi_M) + L_{h,k_2}(\chi_M), \tag{8.8}$$

for each scalar α and for all vectors in H. Then for each $\mathscr{F}$-integrable function f, the mapping

$$(h, k) \to \int f \, d\mu_{h,k} \equiv L_{h,k}(f)$$

is a sesquilinear form on H. Moreover if for each M in $\mathscr{R}$

$$L_{h,k}(\chi_M) = \overline{L_{k,h}(\chi_M)}, \tag{8.9}$$

then $L_{h,k}(f) = \overline{L_{k,h}(\bar{f})}$ for each $\mathscr{F}$-integrable function f. In addition, if equation (8.4) is true for some bounded measurable function f, and if $\mathscr{F}$ is a family of measures such that $\mu_h(X) \leq K \| h \|^2$ for some constant K independent of h, then the form above is also bounded. ∎

Proof. By use of Lemma 8.1, statements (8.5)–(8.9) are readily verified to hold for each $\mathscr{F}$-integrable function f. For instance, to verify equation (8.9) define L on the class on $\mathscr{F}$-integrable functions by $L(f) = \overline{L_{k,h}(\bar{f})}$. Then L is linear and satisfies the conditions (i) and (ii) of Lemma 8.1 for the pair (h, k) so that $L = L_{h,k}$. The last statement of the proposition follows from the inequality

$$\left| \int f \, d\mu_{h,h} \right| = \left| \int f \, d\mu_h \right| \leq \int |f| \, d\mu_h \leq \| f \|_\infty K \| h \|^2. \quad ∎$$

The last four lemmas have been preparatory to proving the next result. It will be most useful in the ensuing work of this section.

Theorem 8.2. Let E be a positive-operator-valued measure on $(X, \mathscr{R})$. For each bounded measurable complex-valued function f on X, there exists a unique bounded operator T_f on H such that

$$(T_f h \mid h) = \int f \, d\mu_h, \quad \text{for all } h \in H. \tag{8.10}$$

Moreover,

$$(T_f h \mid k) = \int f \, d\mu_{h,k}, \quad \text{for all } h \text{ and } k \text{ in } H.$$

[Here μ_h is the measure $\mu_h(M) = (E(M)h \mid h)$.] ∎

Proof. Since for each h, $\mu_h(M) = (E(M)h \mid h)$, each of the equations (8.5), (8.6), (8.7), and (8.8) is true. This can be verified by equation

(8.3). Moreover, by Lemma 8.2, equation (8.4) is true. Hence corresponding to the bounded sesquilinear form

$$(h, k) \to \int f \, d\mu_{h,k}$$

is a unique bounded operator T_f such that

$$(T_f h \mid k) = \int f \, d\mu_{h,k}, \quad \text{for all } h, k \text{ in } H.$$

The following identity shows that equation (8.10) is sufficient for uniqueness. If B is a bounded operator also satisfying equation (8.10), then for any h and k in H,

$$\begin{aligned} 4\int f \, d\mu_{h,k} &= \int f \, d\mu_{h+k} - \int f \, d\mu_{h-k} + i\int f \, d\mu_{h+ik} - i\int f \, d\mu_{h-ik} \\ &= (B(h+k) \mid h+k) - (B(h-k) \mid h-k) \\ &\quad + i(B(h+ik) \mid h+ik) - i(B(h-ik) \mid h-ik) \\ &= 4(Bh \mid k). \end{aligned}$$

∎

The operator corresponding to f in Theorem 8.2 is denoted by $\int f \, dE$. Thus for all h and k in H we have

$$\left(\left(\int f \, dE\right) h \mid k\right) = \int f \, d\mu_{h,k}.$$

Remark 8.2. Of interest are the following properties of the mapping $f \to \int f \, dE$.

(i) $\int (f+g) \, dE = \int f \, dE + \int g \, dE$.

(ii) $\int \alpha f \, dE = \alpha \int f \, dE$ for all scalars α.

(iii) $\int \bar{f} \, dE = (\int f \, dE)^*$.

(iv) $\int \chi_M \, dE = E(M)$ for each M in $\mathscr{R}$.

(v) $\int fg \, dE = \int f \, dE \int g \, dE$ if E is a spectral measure.

(vi) There exists a constant K such that $\| \int f \, dE \| \leq K \| f \|_\infty$ for all f.

Proof. Let us verify (iii), (vi), and (v), respectively. If we define L by $Lf = \overline{\int \bar{f} \, d\mu_{h,k}}$ for each E-integrable function, then the equation

$$L(\chi_M) = \overline{\int \chi_M \, d\mu_{h,k}} = \overline{(E(M)h \mid k)} = (E(M)k \mid h)$$

and Lemma 8.2 show that $L = L_{h,k}$. Hence we have

$$\left(\left(\int f\,dE\right)^* h \mid k\right) = \overline{\left(\left(\int f\,dE\right)k \mid h\right)} = \overline{\int f\,d\mu_{k,h}} = \int \bar{f}\,d\mu_{h,k} = \left(\left(\int \bar{f}\,dE\right)h \mid k\right);$$

this means $(\int f\,dE)^* = \int \bar{f}\,dE$.

If f is a real-valued function,

$$\left(\left(\int f\,dE\right)h \mid h\right) = \left| \int f\,d\mu_h \right| \leq \| f \| \, \mu_h(X).$$

If f is a complex function $k + ig$, then

$$\left\| \int f\,dE \right\| = \left\| \int (k + ig)\,dE \right\| = \left\| \int k\,dE + i \int g\,dE \right\| \leq 2 \, \| f \| \, \mu_h(X).$$

This equation verifies (vi).

Finally since $E(M \cap N) = E(M)E(N)$ when E is a spectral measure, (v) is valid if $f = \chi_M$ and $g = \chi_N$. If f and g are simple, (v) follows by linearity. If f and g are arbitrary bounded measurable functions, choose simple functions (f_n) and (g_n) such that $\| f_n - f \|_\infty \to 0$ and $\| g - g_n \|_\infty \to 0$. Then $\| f_n g_n - fg \|_\infty \to 0$ so that

$$\begin{aligned}
\left\| \int fg\,dE - \int f\,dE \int g\,dE \right\| &\leq \left\| \int fg\,dE - \int f_n g_n\,dE \right\| \\
&\quad + \left\| \int f_n g_n\,dE - \int f_n g\,dE \right\| \\
&\quad + \left\| \int f_n g\,dE - \int fg\,dE \right\| \\
&\leq 2K\{\| fg - f_n g_n \|_\infty + \| f_n g_n - f_n g \|_\infty \\
&\quad + \| f_n g - fg \|_\infty\} \to 0.
\end{aligned}$$

∎

Starting with a positive-operator-valued measure E on a σ-ring $\mathscr{R}$, we have constructed for each h in Hilbert space H a finite measure μ_h on $\mathscr{R}$ and then corresponded to E a mapping $f \to T_f$ on the class of all bounded complex-valued measurable functions on X such that equation (8.10) holds. Our next goal is to start with a fixed bounded self-adjoint operator T on H and associate with it a similar mapping from bounded measurable functions to operators in $L(H, H)$.

Associated with a spectral measure E defined on a measurable space $(X, \mathscr{R})$, where X is also a topological space, is the *spectrum* $\Sigma(E)$ of E. It is defined as the complement in X of the union of all those open sets

M in $\mathscr{R}$ for which $E(M) = 0$. Obviously $\Sigma(E)$ is a closed set in X. If $\Sigma(E)$ is compact, then E is called a *compact spectral measure*.

We need the information of the following lemma and proposition. As before, $B(R)$ denotes the σ-algebra of Borel sets on R.

Lemma 8.4. If E is a spectral measure on $B(R)$, then

(i) for any M in $B(R)$, $E(M)$ is the "smallest" projection,† denoted $\vee E(K)$, "greater than" all $E(K)$, where K is a compact subset of M [in symbols $E(M) = \vee E(K)$, where K is compact in M];

(ii) $E(R - \Sigma(E)) = 0$. ∎

Proof. (i) Since E is a spectral measure, $E(M) \geq E(K)$ for all compact subsets K of M; whereby $E(M) \geq \vee E(K)$. If $E(M) \neq \vee E(K)$, there exists a nonzero vector h in the range of $E(M)$ that is orthogonal to the range of each $E(K)$. Letting μ_h be the measure $\mu_h(M) = (E(M)h \mid h)$, which is regular by Theorem 5.2 in Chapter 5, we have $\mu_h(M) = \sup\{\mu_h(K): K \subset M,\ K \text{ compact}\}$. Since $\mu_h(K) = 0$ for each K, $\mu_h(M) = 0$ or $(E(M)h \mid h) = 0$. This means $h = E(M)h = 0$.

(ii) In view of (i), it suffices to show $E(K) = 0$ for K compact in $R - \Sigma(E)$. Since K can be covered by a finite collection of open sets, each of which has zero spectral measure, $E(K) = 0$. ∎

If E is a compact spectral measure on $B(R)$, let $f(\lambda) = \lambda\chi_{\Sigma(E)}$. Let $T = \int f\, dE$. The next proposition shows how $\sigma(T)$ is related to $\Sigma(E)$.

Proposition 8.4. If E is a compact normalized spectral measure on $B(R)$ and $T = \int \lambda\chi_{\Sigma(E)}(\lambda)\, dE(\lambda)$, then $\Sigma(E) = \sigma(T)$. ∎

Proof. Since E is normalized, $\Sigma(E) \neq \varnothing$. If $\lambda_0 \in \Sigma(E)$, then $E(M) \neq 0$ for every open set M containing λ_0. Suppose $\lambda_0 \in \Sigma(E)$ but $\lambda_0 \notin \sigma(T)$. If $M = \{\lambda: |\lambda - \lambda_0| < 1/(2\, \|(T - \lambda_0)^{-1}\|)\}$, then there is a unit vector h in the range of $E(M)$ as $E(M) \neq 0$. Thus

$$\|Th - \lambda_0 h\|^2 = ((T - \lambda_0)^*(T - \lambda_0)h \mid h) = \int \overline{(\lambda - \lambda_0)}(\lambda - \lambda_0)\chi_{\Sigma(E)}(\lambda)\, d\mu_h(\lambda)$$

$$= \int |\lambda - \lambda_0|^2\, d\mu_h(\lambda) \leq \left(\frac{1}{2\, \|(T - \lambda_0)^{-1}\|}\right)^2.$$

† See the remark following Proposition 7.21.

However,

$$\| h \| = \| (T - \lambda_0)^{-1}(T - \lambda_0)h \| \leq \| (T - \lambda_0)^{-1} \| \frac{1}{2 \| (T - \lambda_0)^{-1} \|} = \frac{1}{2}.$$

This contradiction shows $\Sigma(E) \subset \sigma(T)$.

Conversely, if $\lambda_0 \notin \Sigma(E)$, we show $\lambda_0 \notin \sigma(T)$. If $\lambda_0 \notin \Sigma(E)$, there is an open set M containing λ_0 such that $E(M) = 0$. If $\delta = \inf\{| \lambda - \lambda_0 |: \lambda \in M^c\}$, then for every h we have as before that

$$\| Th - \lambda_0 h \|^2 = \int | \lambda - \lambda_0 |^2 \, d\mu_h(\lambda) = \int_{M^c} | \lambda - \lambda_0 |^2 \, d\mu_h(\lambda) \geq \delta^2 \| h \|^2.$$

Hence from Proposition 6.5, $(T - \lambda_0)^{-1}$ exists and $\lambda_0 \notin \sigma(T)$. ▮

Starting with a compact normalized spectral measure E on $B(R)$, we have obtained a bounded self-adjoint operator $T = \int \lambda \chi_{\Sigma(E)}(\lambda) \, dE(\lambda)$ on H. Moreover $\sigma(T) = \Sigma(E)$. We now wish to reverse the process in starting with a given bounded self-adjoint operator and obtaining a spectral (compact, normalized) measure on $B(R)$. By means of the next theorem—sometimes called the functional calculus form of the Spectral Theorem—we will have a neat way of obtaining this spectral measure. It is also a beautiful extension of the Continuous Functional Calculus Theorem (Theorem 7.10) to bounded measurable functions.

Let T be a bounded self-adjoint operator and let $\sigma(T)$ be its spectrum—a compact subset of the real axis. If $f \in C_1(\sigma(T))$ and f is real-valued, let $f(T)$ be the bounded linear operator in $L(H, H)$, defined as the image of f under φ in Theorem 7.10. It is easy to verify that if $h \in H$, the function on the real space of real-valued functions in $C(\sigma(T))$ given by $L_h(f) = (f(T)h \mid h)$ is a positive linear functional. Hence by Theorem 5.7 there is a unique Borel measure μ_h on $B(\sigma(T))$ such that

$$L_h(f) = (f(T)h \mid h) = \int f \, d\mu_h, \tag{8.11}$$

for all $f \in C_1(\sigma(T))$ that are real-valued.

Let $\mathscr{F}$ now be the particular family of measures $\{\mu_h\}_{h \in H}$ satisfying equation (8.11). In particular if f is the function $\chi_{\sigma(T)}$ in $C_1(\sigma(T))$, then equation (8.11) gives for each μ_h

$$\mu_h(\sigma(T)) = \int \chi_{\sigma(T)} \, d\mu_h = (h \mid h) = \| h \|^2.$$

Each bounded measurable function on $\sigma(T)$ is thereby integrable for each μ_h, and for each such function g we can define $\int g \, d\mu_{h,k}$ as in equation

(8.2). Since for each continuous function f in $C_1(\sigma(T))$,

$$\begin{aligned}\int f\,d\mu_{h,h} &= \tfrac{1}{4}\left[\int f\,d\mu_{h+h} - \int f\,d\mu_{h-h} + i\int f\,d\mu_{h+ih} - i\int f\,d\mu_{h-ih}\right]\\ &= \tfrac{1}{4}[(f(T)(h+h) \mid h+h) - (f(T)(h-h) \mid h-h)\\ &\quad + i(f(T)(h+ih) \mid h+ih) - i(f(T)(h-ih) \mid h-ih)]\\ &= (f(T)h \mid h)\\ &= \int f\,d\mu_h;\end{aligned}$$

equation (8.4) is satisfied for continuous functions. Similarly a simple calculation shows that for each $f \in C_1(\sigma(T))$, equations (8.5)–(8.8) are valid (with f replacing χ_M). Since each bounded measurable function on $\sigma(T)$ can be approximated in L_1 (see Problem 5.3.12) by a continuous function on $\sigma(T)$, the equations (8.4)–(8.8) are seen to hold for all bounded measurable functions. This means (see Lemma 8.3 and Proposition 8.3) that for each bounded measurable function g on $\sigma(T)$ the mapping

$$(h, k) \to \int g\,d\mu_{h,k}$$

is a bounded sesquilinear form on H so that by Theorem 7.6, there exists a unique bounded operator, which we denote by $g(T)$, on H such that

$$(g(T)h \mid k) = \int g\,d\mu_{h,k}, \quad \text{for all } h, k \text{ in } H. \tag{8.12}$$

In particular for any h in H,

$$(g(T)h \mid h) = \int g\,d\mu_{h,h} = \int g\,d\mu_h, \tag{8.13}$$

and it is this equation for each h which determines $g(T)$ uniquely. This is shown by using the same equation employed in the proof of Theorem 8.2.

If f is in $C_1(\sigma(T))$, $f(T)$ is $\varphi(f)$, where φ is the function of Theorem 7.10. By the properties of φ, $\varphi(\bar{f}) = (\varphi(f))^*$ or in other notation $\bar{f}(T) = (f(T))^*$. This relation is also true for any bounded measurable function g on $\sigma(T)$. Indeed for $f \in C_1(\sigma(T))$,

$$\overline{\int \bar{f}\,d\mu_{h,k}} = \overline{(f(T)^*h \mid k)} = (f(T)k \mid h) = \int f\,d\mu_{k,h}; \tag{8.14}$$

and since the class $C_1(\sigma(T))$ is dense in L_1, equation (8.14) is also valid for g. This means

$$(k \mid \bar{g}(T)h) = \overline{(\bar{g}(T)h \mid k)} = (g(T)k \mid h),$$

so that $(g(T))^* = \bar{g}(T)$. In particular if g is real-valued, $g(T)$ is self-adjoint.

To prove our next theorem, we have need of the following lemmas.

Lemma 8.5. If g is a bounded nonnegative measurable function on $\sigma(T)$ and (g_n) is a sequence of measurable functions with $0 \leq g_n \leq g$ and $\int | g - g_n | \, d\mu_h \to 0$, then $g(T)h = \lim_{n\to\infty} g_n(T)h$. ∎

Proof. From equation (8.13) and the above hypothesis, $\lim(g_n(T)h \mid h) = (g(T)h \mid h)$. Also

$$\| g(T) - g_n(T) \| \leq \| g(T) \| + \| g_n(T) \| \leq 2 \| g(T) \|.^{\dagger}$$

Hence by Lemma 7.5 of Chapter 7,

$$\begin{aligned}
\| [g(T) - g_n(T)]h \|^4 &= ([g(T) - g_n(T)]h \mid [g(T) - g_n(T)]h)^2 \\
&\leq ([g(T) - g_n(T)]h \mid h)([g(T) - g_n(T)]^2 h \mid [g(T) - g_n(T)]h) \\
&\leq ([g(T) - g_n(T)]h \mid h) \| g(T) - g_n(T) \|^3 \| h \|^2 \to 0. \qquad (8.15)
\end{aligned}$$

∎

Lemma 8.6. If g is a nonnegative bounded measurable function and (f_n) is a sequence of continuous nonnegative functions such that $\| f_n \|_\infty \leq \| g \|_\infty$ and $\int | f_n - g | \, d\mu_h \to 0$ (such a sequence always exists), then $g(T)h = \lim f_n(T)h$. ∎

Proof. Let $g_n = f_n \wedge g$. $[g_n(x) = \min\{f_n(x), g(x)\}.]$ Since the g_n satisfy the conditions of Lemma 8.5, $g(T)h$ is the limit of $g_n(T)h$.

Also the sequence $(f_n - g_n)$ is a nonnegative sequence such that $\int (f_n - g_n) \, d\mu_h \to 0$. Moreover, the sequence $\| f_n(T) - g_n(T) \|$ is bounded since

$$\| f_n(T) - g_n(T) \| \leq \| f_n(T) \| + \| g_n(T) \| \leq 2 \| f_n(T) \|,$$

† $\| g_n(T) \| = \sup_{\|h\| \leq 1}(g_n(T)h \mid h) \leq \sup_{\|h\| \leq 1}(g(T)h \mid h) = \| g(T) \|$.

and $(\| f_n(T) \|)$ is a bounded sequence. [Since $f_n(T)$ is self-adjoint,

$$\| f_n(T) \| = \sup_{\|h\| \le 1} (f_n(T)h \mid h)$$

$$= \sup_{\|h\| \le 1} \int f_n \, d\mu_h \le \sup_{\|h\| \le 1} \int \| f_n \|_\infty \, d\mu_h \le \sup_{\|h\| \le 1} \int \| g \|_\infty \, d\mu_h$$

$$= \| g \|_\infty \sup_{\|h\| \le 1} \int d\mu_h = \| g \|_\infty .]$$

By an inequality similar to (8.15) again, $[f_n(T) - g_n(T)]h$ converges to 0. We can conclude that $f_n(T)h$ converges to $g(T)h$. ∎

We have collected sufficient information to prove the following theorem.

Theorem 8.3. *Functional Calculus Form of the Spectral Theorem.* Let T be a bounded self-adjoint operator on H. There is a unique mappig $\hat{\varphi}$ on the class of bounded measurable functions on $\sigma(T)$ into $L(H, H)$ such that

(i) $\hat{\varphi}(\alpha f + \beta g) = \alpha\hat{\varphi}(f) + \beta\hat{\varphi}(g)$,
$\hat{\varphi}(fg) = \hat{\varphi}(f)\hat{\varphi}(g)$,
$\hat{\varphi}(\bar{f}) = (\hat{\varphi}(f))^*$,
$\hat{\varphi}(1) = I$.

(ii) $\| \hat{\varphi}(f) \| \le \| f \|_\infty$.

(iii) If $f(\lambda) = \lambda$, then $\hat{\varphi}(f) = T$.

(iv) If $f_n(\lambda) \to f(\lambda)$ for each λ and $(\| f_n \|_\infty)$ is bounded, then $\hat{\varphi}(f_n) \to \hat{\varphi}(f)$ strongly.

In addition, $\hat{\varphi}$ satisfies

(v) If $Th = \lambda h$, then $\hat{\varphi}(f)h = f(\lambda)h$.

(vi) If $f \ge 0$, then $\hat{\varphi}(f) \ge 0$.

(vii) If $AT = TA$, then $A\hat{\varphi}(f) = \hat{\varphi}(f)A$. ∎

Proof. The mapping $\hat{\varphi}$ is given by $\hat{\varphi}(g) = g(T)$, where $g(T)$ is the unique bounded operator satisfying equation (8.12). $\hat{\varphi}$ is thus an extension of the mapping φ of Theorem 7.10 defined on $C_1(\sigma(T))$ and thereby satisfies (i), (ii), (iii), (v), and (vi) for continuous functions. The linearity of $\hat{\varphi}$ is readily verified using equation (8.13); the fact that $\hat{\varphi}(\bar{f}) = (\hat{\varphi}(f))^*$ and statement (vi) have been verified above.

To prove $\hat{\varphi}(fg) = \hat{\varphi}(f)\hat{\varphi}(g)$, we assume f and g are nonnegative and use linearity to treat the general case. Let (f_n) and (g_n) be sequences of continuous functions with $\| f_n \|_\infty \le \| f \|_\infty$, $\| g_n \|_\infty \le \| g \|_\infty$, $g_n \to g$ in

$L_1(\mu_h)$, and $f_n \to f$ in $L_1(\mu_{g(T)h} + \mu_h)$ for some h in H. By Lemma 8.6,

$$f(T)[g(T)h] = \lim f_n(T)[g(T)h]$$

and

$$g(T)h = \lim g_n(T)h.$$

Hence for any positive integer m,

$$(f_m g)(T)h = \lim_{n\to\infty} (f_m g_n)(T)h = \lim_{n\to\infty} [f_m(T)g_n(T)]h = f_m(T)g(T)h,$$

whereby

$$(fg)(T)h = \lim_{n\to\infty} (f_n g)(T)h = \lim_{n\to\infty} [f_n(T)g(T)]h = [f(T)g(T)]h.$$

The proofs of the other statements—(ii), (v), and (vii)—are similar.

The uniqueness of $\hat{\varphi}$ and statement (iv) remain to be verified. Clearly, if (f_n) converges pointwise to f and $(\| f_n \|_\infty)$ is bounded, then by the Dominated Convergence Theorem (Theorem 3.3),

$$\int f_n \, d\mu_h \to f \, d\mu_h,$$

for each h. Hence for each h, $f_n(T)h \to f(T)h$ as in equation (8.15), which means $f_n(T)$ converges strongly to $f(T)$.

Suppose ψ is any mapping satisfying (i)–(iv). By the uniqueness of φ of Theorem 7.10, $\hat{\varphi}$ and ψ agree on $C_1(\sigma(T))$. By linearity, to show ψ and φ are equal, it suffices to show they agree for real measurable functions. Let h be arbitrary but fixed in H and let VL denote the real vector lattice of real, bounded measurable functions on $\sigma(T)$. If we define I on VL by

$$I(f) = (\psi(f)h \mid h),$$

then I is a Daniell integral on VL [here we use (iv) to verify (D) of Definition 5.2]. By Proposition 5.4, there exists a unique (finite) measure ν_h on $\sigma(\text{VL}) = B(\sigma(T))$ such that

$$(\psi(f)h \mid h) = If = \int f \, d\nu_h, \qquad \text{for all } f \in \text{VL}.$$

Since μ_h is the unique measure on $B(\sigma(T))$ such that

$$(\hat{\varphi}(f)h \mid h) = \int f \, d\mu_h$$

for continuous real functions, and ψ and $\hat{\varphi}$ agree on continuous functions, it follows that $\nu_h = \mu_h$. As equation (8.13) determines $g(T)$ uniquely for nonnegative bounded measurable functions g, $\psi(g) = g(T) = \hat{\varphi}(g)$ for such functions. ∎

We now consider special bounded measurable functions on $\sigma(T)$ in order to generate a spectral measure on $B(R)$. For each Borel-measurable set M in R consider the characteristic function $\chi_{M\cap\sigma(T)}$. Define $P: B(R) \to L(H, H)$ by

$$P(M) = \hat{\varphi}(\chi_{M\cap\sigma(T)}).$$

In light of the properties of $\hat{\varphi}$, it is easy to convince oneself that P is a normalized spectral measure on $B(R)$.

By Theorem 8.2, corresponding to each bounded measurable function g on $\sigma(T)$ is a unique bounded operator $\int g\, dP$ such that

$$\left(\left(\int g\, dP\right)h \,\middle|\, h\right) = \int g\, d\mu_h, \quad \text{for all } h \text{ in } H, \tag{8.16}$$

where $\mu_h(M) = (P(M)h \mid h)$ for each h in H and for each M in $B(\sigma(T))$. By equation (8.13), $g(T)$ is the unique operator such that

$$(g(T)h \mid h) = \int g\, d\tilde{\mu}_h, \quad \text{for each } h \text{ in } H, \tag{8.17}$$

where we use $\tilde{\mu}_h$ momentarily to denote the unique Borel measure on $B(\sigma(T))$ such that

$$(f(T)h \mid h) = \int f\, d\tilde{\mu}_h,$$

for all continuous functions f on $\sigma(T)$. Since for every M in $B(\sigma(T))$

$$\tilde{\mu}_h(M) = \int \chi_M\, d\tilde{\mu}_h = (\hat{\varphi}(\chi_M)h \mid h) = (P(M)h \mid h) = \mu_h(M),$$

$\tilde{\mu}_h = \mu_h$. By the uniqueness of $g(T)$ in equation (8.17), we conclude that $g(T) = \int g\, dP$ for each bounded measurable function on $\sigma(T)$. In particular

$$T = \int \lambda\, dP(\lambda). \tag{8.18}$$

We are now ready to prove the following version of the Spectral Theorem.

Theorem 8.4. *Spectral Measure Formulation of the Spectral Theorem.* There is a one-to-one correspondence between bounded self-adjoint operators T on H and normalized compact spectral measures E on $B(R)$. It is the following correspondence: $T = \int \lambda \chi_{\Sigma(E)}(\lambda)\, dE(\lambda)$. ∎

Proof. If E is a normalized compact spectral measure, let us consider $T = \int \lambda \chi_{\Sigma(E)}(\lambda)\, dE(\lambda)$ as given in Theorem 8.2. It must be verified that this correspondence is a one-to-one correspondence.

To verify that the association of $E \to \int \lambda \chi_{\Sigma(E)}(\lambda)\, dE(\lambda)$ is onto, let S be any bounded self-adjoint operator. Using Theorem 8.3, let P be the normalized spectral measure on $B(R)$ given by $P(M) = \hat{\varphi}(\chi_{M \cap \sigma(S)})$. Clearly, P is compact and $\Sigma(P) = \sigma(S)$. Moreover, as we have established in equation (8.18)

$$S = \int \lambda \chi_{\sigma(S)}(\lambda)\, dP(\lambda),$$

so that P maps to S.

To verify the assertion $E \to \int \lambda \chi_{\Sigma(E)}(\lambda)\, dE(\lambda)$ is one-to-one, it suffices to show that when $T = \int \lambda \chi_{\Sigma(E)}(\lambda)\, dE(\lambda)$, then $E = P$, where P is the normalized compact spectral measure given by $P(M) = \hat{\varphi}(\chi_{M \cap \sigma(T)})$. We know from Proposition 8.4 that $\Sigma(P) = \sigma(T) = \Sigma(E)$. To show that $E = P$, it suffices to show that $E(M) = P(M)$ for $M \subset \sigma(T)$ or that $E(M) = \hat{\varphi}(\chi_M)$. Now by Theorem 8.2, $E(M)$ is the unique operator such that

$$(E(M)h \mid h) = \int \chi_M \, d\mu_h; \tag{8.19}$$

whereas, by Theorem 8.3, $\hat{\varphi}(\chi_M)$ is the unique operator such that

$$(\hat{\varphi}(\chi_M)h \mid h) = \int \chi_M \, d\tilde{\mu}_h, \tag{8.20}$$

where $\tilde{\mu}_h$ again denotes momentarily the unique Borel measure on $\sigma(T)$ such that

$$(f(T)h \mid h) = \int f \, d\tilde{\mu}_h,$$

for all $f \in C(\sigma(T))$. By virtue of Remark 8.2 and the fact that $T = \int \lambda \, dE$, the unique map of Theorem 7.10 in Chapter 7 is identical to the map of Theorem 8.2 restricted to $C_1(\sigma(T))$. Hence for every continuous function f in $C_1(\sigma(T))$,

$$(f(T)h \mid h) = \int f \, d\mu_h$$

and $\mu_h = \tilde{\mu}_h$. This means, by equations (8.19) and (8.20), that $E(M) = \hat{\varphi}(\chi_M)$. ∎

Our final consideration in this section is the multiplication operator form of the Spectral Theorem. We will prove it for self-adjoint and normal operators. The proof is relatively easy and depends primarily on the Continuous Functional Calculus Theorem of Chapter 7. We begin with a definition and a preliminary proposition.

Definition 8.3. A vector h in H is called a *cyclic vector* for T in $L(H, H)$ if the closed linear span of $\{T^n h: n = 0, 1, \dots\}$ is H. ∎

Proposition 8.5. Let T be a bounded self-adjoint operator with cyclic vector h_0. There then exists a unitary[†] operator $U: H \to L_2(\sigma(T), \mu_{h_0})$ such that

$$(UTU^{-1}f)(\lambda) = \lambda f(\lambda) \quad \text{a.e.}$$

[Here μ_{h_0} is the unique measure such that $(f(T)h_0 \mid h_0) = \int f\, d\mu_{h_0}$ for all f in $C(\sigma(T))$.] ∎

Proof. Define U on the dense subset

$$S = \{\varphi(f)h_0: f \in C_1(\sigma(T))\}$$

of H (φ is the unique map of Theorem 7.10 in Chapter 7) by $U(\varphi(f)h_0) = f$. To show U is well-defined on S suppose that $\varphi(f)h_0 = \varphi(g)h_0$. Then the equation

$$\begin{aligned} 0 = \| \varphi(f)h_0 - \varphi(g)h_0 \|^2 &= (\varphi(f-g)^*\varphi(f-g)h_0 \mid h_0) \\ &= (\varphi[\overline{(f-g)}(f-g)]h_0 \mid h_0) = \int |f-g|^2\, d\mu_{h_0} \end{aligned} \tag{8.21}$$

implies that $f = g$ a.e. with respect to μ_{h_0} or that $f = g$ in $L_2(\sigma(T), \mu_{h_0})$. An equation like (8.21) also shows that U is an isometric map from a dense subset of H onto a dense subset of $L_2(M, \mu_{h_0})$, namely, $C_1(\sigma(T))$. Hence U can be extended to an isometric map, also designated by U, from H onto $L_2(M, \mu_{h_0})$. If $f \in C_1(\sigma(T))$,

$$\begin{aligned} (UTU^{-1}f)(\lambda) &= [UT\varphi(f)h_0](\lambda) \\ &= [U\varphi(\lambda)\varphi(f)h_0](\lambda) \\ &= [U\varphi(\lambda f)h_0](\lambda) \\ &= \lambda f(\lambda). \end{aligned}$$

[†] By a unitary operator $U: H \to L_2$ we mean a linear isometry from H onto L_2.

By continuity, this equation is valid also for any f in $L_2(\sigma(T), \mu_{h_0})$ so that the proposition is proved. ∎

Inasmuch as not all bounded self-adjoint operators have cyclic vectors, the above proposition is not applicable in the general situation. Observe that if h is any vector in H, then the closed linear span M of $S = \{T^n h: n = 0, 1, 2, \ldots\}$ is *invariant* under T, that is, $TM \subset M$. Moreover h is a cyclic vector for M. Since $TM \subset M$ if and only if $T^*(M^\perp) \subset M^\perp$ by Proposition 7.14, it is also readily apparent that $T(M^\perp) \subset M^\perp$. To extend the above proposition to the arbitrary case, H must be decomposed into cyclic subspaces as in the next proposition.

If $(H_i)_{i\in I}$ is an arbitrary collection of Hilbert spaces each with inner product $(|)_i$, the *direct sum* $\bigoplus_{i\in I} H_i$ of $(H_i)_{i\in I}$ is the Hilbert space H of all families of element $(h_i)_{i\in I}$ with $h_i \in H_i$ such that the family $(\| h_i \|^2)_{i\in I}$ is summable. Addition, scalar multiplication, and the inner product in H are given by

$$x + y = (x_i + y_i)_{i\in I},$$

$$\alpha x = (\alpha x_i)_{i\in I},$$

$$(x \mid y) = \sum_{i\in I} (x_i \mid y_i)_i,$$

where $x = (x_i)_I$ and $y = (y_i)_I$.

Proposition 8.6. Let T be a bounded self-adjoint operator on H. Then $H = \bigoplus_I H_i$, where $(H_i)_I$ is a family of mutually orthogonal closed subspaces of H such that

(i) T leaves H_i invariant, that is, $TH_i \subset H_i$ for each i.

(ii) For each i there exists a cyclic vector h_i for H_i, implying $H_i = \overline{\{f(T)h_i: f \in C_1(\sigma(T))\}}$ for some vector h_i. ∎

Proof. If $h \neq 0$ is any vector in H, the closed linear span H_1 of $S = \{T^n h: n = 0, 1, \ldots\}$ is invariant with cyclic vector h. If $H = H_1$, the theorem is proved.

If $H_1 \neq H$, there exists a vector h_2 orthogonal to H_1 and by the same process a subspace H_2 of H exists such that H_2 is orthogonal to H_1 and satisfies (i) and (ii) with h_2 the cyclic vector. If $H = H_1 \oplus H_2$, the proposition is proved. If not, we proceed using Zorn's Lemma as follows.

Let $\mathscr{F}$ be the family of all systems $(H_i)_{i\in J}$ consisting of mutually orthogonal subspaces of H satisfying (i) and (ii). By Zorn's Lemma, the family $\mathscr{F}$, ordered by the inclusion relation $\subset$, has a maximal element $(H_i)_{i\in I}$. Now $H = \bigoplus_I H_i$, for if not there exists a vector h_0 orthogonal to $\bigoplus_I H_i$ and a cyclic subspace H_0 satisfying (i) and (ii), contradicting the maximality of $(H_i)_I$. ∎

Theorem 8.5. *Multiplication Operator Form of the Spectral Theorem.* Let T be a bounded self-adjoint operator on a Hilbert space H. Then there exists a family of finite measures $(\mu_i)_{i\in I}$ on $\sigma(T)$ and a unitary operator

$$U\colon H \to \bigoplus_{i\in I} L_2(\sigma(T), \mu_i),$$

such that

$$(UTU^{-1}f)_i(\lambda) = \lambda f_i(\lambda),$$

where $f = (f_i)_I$ is in $\bigoplus_{i\in I} L_2(\sigma(T), \mu_i)$. ∎

Proof. By Proposition 8.6, $H = \bigoplus_{i\in I} H_i$ where each H_i is invariant and cyclic with respect to T. By Proposition 8.5, for each i there exists a unitary operator $U_i\colon H_i \to L_2(\sigma(T), \mu_i)$ such that

$$(U_i T|_{H_i} U_i^{-1} f_i)(\lambda) = \lambda f_i(\lambda),$$

for $f_i \in L_2(\sigma(T), \mu_i)$. Define $U\colon \bigoplus_{i\in I} H_i \to \bigoplus_{i\in I} L_2(\sigma(T), \mu_i)$ by $U(h) = (U_i h_i)_{i\in I}$. It is easy to see that U satisfies the criteria of the theorem. ∎

If H is a separable Hilbert space, it is clear that the index set I in Proposition 8.6 and hence in Theorem 8.5 is the finite set $\{1, 2, \ldots, n\}$ for some $n \in N$ or the set N itself. Using this observation, we have the following important corollary of Theorem 8.5.

Corollary 8.1. Let T be a bounded self-adjoint operator on a Hilbert space H. Then there exists a measure space $(X, \mathscr{A}, \mu)$, a bounded function F on X, and a unitary map $U\colon H \to L_2(X, \mu)$ so that

$$(UTU^{-1}f)(x) = F(x) f(x) \text{ a.e.}$$

If H is separable, then μ can be chosen to be finite. Otherwise μ is semi-finite. ∎

Proof. Let X be the disjoint union of card(I) copies of $\sigma(T)$ and define μ on X by requiring that its restriction to the ith copy of $\sigma(T)$ be μ_i. If H is separable, in Proposition 8.6 we can choose the cyclic vectors h_i so that $\| h_i \| = 2^{-i}$. In this case clearly $\mu(X) = \sum_{i=1}^{\infty} \mu_i(\sigma(T)) < \infty$ since $\mu_i(\sigma(T)) \leq 2^{-i}$ (see Proposition 8.5). Letting U be the composition of the unitary operator from Theorem 8.5 and the natural unitary operator from $\bigoplus_{i\in I} L_2(\sigma(T), \mu_i)$ onto $L_2(X, \mu)$, and letting F be the function whose restriction to the ith copy of $\sigma(T)$ is the identity function, the corollary follows. ∎

To extend the multiplication operator form of the Spectral Theorem to normal operators, we extend the continuous functional calculus to continuous functions of two variables in the following fashion (see also [25]).

Suppose T_1 and T_2 are two commuting self-adjoint operators. (Actually the argument below applies to any finite collection of commuting self-adjoint operators.) Let $S = \sigma(T_1)\times\sigma(T_2) \subset R\times R$. Let $\hat{\varphi}_1$ and $\hat{\varphi}_2$ be the unique mappings corresponding to T_1 and T_2, respectively, as in Theorem 8.3 and let P_1 and P_2 be the respective compact spectral measures. Thus for $i = 1$ or 2, $P_i(M) = \hat{\varphi}_i(\chi_M)$ for all measurable M contained in $\sigma(T_i)$. Since by Lemma 8.6, $\hat{\varphi}_i(\chi_M)$ is the strong limit of a sequence of polynomials in T_i, it is clear that for each measurable set A in $\sigma(T_1)$ and for each measurable set B in $\sigma(T_2)$, $P_1(A)$ and $P_2(B)$ commute.

Let f be any finite linear combination of functions of the form $\chi = \chi_{A\times B}$, where A and B are measurable in $\sigma(T_1)$ and $\sigma(T_2)$, respectively. Define $\chi(T_1, T_2)$ as $P_1(A)P_2(B)$ and define $f(T_1, T_2)$ by linearity. It is straightforward to check that f is well defined and also to verify that f can be written as $\sum \alpha_i\chi_{A_i\times B_i}$, where $(A_i\times B_i)\cap(A_j\times B_j) = \varnothing$ for $i\neq j$. Since f can be so written and

$$\|\chi(T_1, T_2)\| \leq \sup_{\lambda\in S} |\chi(\lambda)|,$$

it is also true that for f

$$\|f(T_1, T_2)\| \leq \sup_{\lambda\in S} |f(\lambda)|.$$

The continuous mapping $f\to f(T_1, T_2)$ can thus be extended to uniform limits of functions such as f. By virtue of the fact that each function of the form $p(x, y) = x^i y^j$ on S can be approximated uniformly by functions like f and that polynomials in the variables x and y approximate uniformly continuous functions on S by Corollary 1.1 the extension to uniform limits includes in particular continuous functions on S.

We thereby have a mapping ψ from $C_1(\sigma(T_1)\times\sigma(T_2)) \to L(H, H)$. This mapping is easily seen to be linear, to preserve multiplication, and to satisfy $\psi(\bar{f}) = (\psi(f))^*$. Moreover the mapping $f(x, y) = x + iy$ is mapped by ψ to $T_1 + iT_2$, as is evident from (iv) of Theorem 8.3 if we consider sequences of simple functions converging uniformly to $f_1(x, y) = x$ and $f_2(x, y) = y$, respectively.

Analogous to Definition 8.3 we can consider for $h_0 \in H$ the closed linear span of the set $\{T_1{}^i T_2{}^j h_0: i = 0, 1, 2, \ldots, j = 0, 1, 2, \ldots\}$. If this closed span is H, we can define as in Proposition 8.5 a unitary operator U: $H \to L_2(\sigma(T_1) \times \sigma(T_2), \mu)$ where μ is the unique measure such that

$$(\psi(f)h_0 \mid h_0) = \int f\, d\mu,$$

for all f in $C_1(\sigma(T_1) \times \sigma(T_2))$, and U is defined by the rule $U(\psi(f)h_0) = f$ for continuous functions.

Now if T is any normal operator, then T can be written as $T_1 + iT_2$, where T_1 and T_2 are the commuting self-adjoint operators given by

$$T_1 = \tfrac{1}{2}(T + T^*) \quad \text{and} \quad T_2 = \tfrac{1}{2}i(T - T^*).$$

Since $\psi(x + iy) = T_1 + iT_2 = T$, the unitary operator U satisfies for f in $C_1(\sigma(T_1) \times \sigma(T_2))$ the equation

$$\begin{aligned}(UTU^{-1}f)(x, y) &= [UT\psi(f)h_0](x, y)\\ &= [U\psi(x + iy)\psi(f)h_0](x, y)\\ &= [U\psi[(x + iy) \cdot f(x, y)]h_0](x, y)\\ &= (x + iy) f(x, y).\end{aligned}$$

By continuity, this equation is also valid for any $f \in L_2(\sigma(T_1) \times \sigma(T_2), \mu)$.

Thus for any normal T, if the closed linear span of $\{T_1{}^i T_2{}^j h_0: i = 0, 1, 2, \ldots, j = 0, 1, 2, \ldots\}$ is H for some h_0 in H, then $T = T_1 + iT_2$ is unitarily equivalent to a multiplication operator just as for self-adjoint operators. By a similar application of Zorn's Lemma to that used in Proposition 8.6 we arrive at the following generalization of Theorem 8.5.

Theorem 8.6. Let $T = T_1 + iT_2$ be a bounded normal operator on H. Then there exists a family of finite measures $(\mu_i)_{i \in I}$ on $\sigma(T_1) \times \sigma(T_2)$ and a unitary operator

$$U: H \to \bigoplus_{i \in I} L_2(\sigma(T_1) \times \sigma(T_2), \mu_i),$$

such that

$$(UTU^{-1}f)_i(x, y) = (x + iy) f_i(x, y) \quad \text{a.e.},$$

where $f = (f_i)_I$ is in $\bigoplus_{i \in I} L_2(\sigma(T), \mu_i)$. ∎

Immediately we obtain the following corollary of Theorem 8.6, which is analogous to Corollary 8.1.

Corollary 8.2. Let T be a bounded normal operator on a Hilbert space H. Then there exists a measure space $(X, \mathscr{A}, \mu)$, a bounded complex function G on X, and a unitary map U: $H \to L_2(X, \mu)$ so that

$$(UTU^{-1}f)(\lambda) = G(\lambda)f(\lambda) \text{ a.e.}$$

If H is separable, μ can be taken to be a finite measure. Otherwise μ is semi-finite. ∎

Using this representation of a bounded normal operator we can prove a spectral measure version of the spectral theorem for bounded normal operators similar to Theorem 8.4 for self-adjoint operators. The proof we give here, however, uses a different approach than that used in Theorem 8.4.

Theorem 8.7. *Spectral Measure Version of the Spectral Theorem for Normal Operators.* If T is a bounded normal operator on Hilbert space H, then there exists a unique normalized compact spectral measure P on $B(R^2)$, the (weakly) Borel subsets of $R^2 (= R \times R)$, such that $T = \int \lambda \chi_{\sigma(T)} \, dP$. ∎

Proof. By Corollary 8.2, we can assume that T is a linear operator on $L_2(X, \mu)$ given by $Tf(x) = G(x)f(x)$ for some bounded complex function G on X. For each $M \in B(R^2)$ define $P(M) \in L(L_2, L_2)$ by $P(M)f = \chi_{G^{-1}(M)}f = (\chi_M \circ G)f$. Then P is a normalized compact spectral measure on $B(R^2)$. [Note: $\Sigma(P) \subset \sigma(T)$ since $\sigma(T)$ is the essential range (Example 6.17) of G and so $P(0)$ is 0 for $0 \subset \sigma(T)^c$.] We must show that for each $f \in L_2(X, \mu)$, $(Tf \mid f) = \int \lambda \chi_{\sigma(T)}(\lambda) \, d\mu_f$, where $\mu_f(M) = (P(M)f \mid f)$ for $M \in B(R^2)$. Theorem 8.2 then gives us $T = \int \lambda \chi_{\sigma(T)}(\lambda) \, dP$.

Fix f in $L_2(X, \mu)$ and let $M \in B(R^2)$. Then

$$\int \chi_M \, d\mu_f = (P(M)f \mid f) = \int (\chi_M \circ G) f\bar{f} \, d\mu,$$

whereby, for all simple functions $h = \sum \alpha_i \chi_{M_i}$, $M_i \in B(R^2)$,

$$\int h \, d\mu_f = \int (h \circ G) f\bar{f} \, d\mu.$$

Therefore we have

$$\int \lambda\chi_{\sigma(T)}\, d\mu_f = \int (\lambda\chi_{\sigma(T)} \circ G) f\bar{f}\, d\mu = \int G f\bar{f}\, d\mu$$
$$= (Tf \mid f).$$

To show uniqueness, suppose also E is a normalized compact spectral measure on $B(R^2)$ such that $T = \int \lambda\chi_{\sigma(T)}(\lambda)\, dE$. Using Remark 8.2 (iii),

$$\int \bar{\lambda}\chi_{\sigma(T)}(\lambda)\, dE = T^* = \int \bar{\lambda}\chi_{\sigma(T)}(\lambda)\, dP.$$

Hence for all polynomials $p(\lambda, \bar{\lambda})$ on $\sigma(T)$

$$\int p\chi_{\sigma(T)}\, dE = \int p\chi_{\sigma(T)}\, dP,$$

so that, for any $f \in L_2$,

$$\int p\chi_{\sigma(T)}\, d\tilde{\mu}_f = \int p\chi_{\sigma(T)}\, d\mu_f$$

where $\tilde{\mu}_f(M) = (E(M)f \mid f)$ and $\mu_f(M) = (P(M)f \mid f))$ if $M \in B(R^2)$. Using Theorem 1.27, if $g \in C_1(\sigma(T))$,

$$\int g\chi_{\sigma(T)}\, d\tilde{\mu}_f = \int g\chi_{\sigma(T)}\, d\mu_f .$$

Since $\tilde{\mu}_f$ and μ_f are finite regular (weakly) Borel measures on $B(R^2)$, $\tilde{\mu}_f = \mu_f$ (See Theorem 5.10 or Theorem 5.11) for each f. Therefore $P = E$. ∎

Problems

8.1.1. Let T be a bounded linear operator on H. Prove

(i) $\sigma(T) \subset \overline{\{(Tx \mid x: \| x \| = 1\}}$, the closure of the numerical range of T;

(ii) if $\omega(T) \equiv \sup\{| (Tx \mid x) | : \| x \| = 1\}$, then $r(T) \leq \omega(T) \leq \| T \|$, where $r(T)$ is the spectral radius of T; and

(iii) if T is normal, $r(T) = \omega(T) = \| T \|$.

[Hint: Prove $r(T) = \| T \|$ by showing $\lim \| T^{2n} \|^{1/2n} = \| T \|$.]

8.1.2. *A spectral function* on R is a function $E\colon R \to L(H, H)$ whose values $E(\lambda)$ are projections and that satisfies

(1) $E(\lambda) \leq E(\lambda')$ if $\lambda \leq \lambda'$,

(2) $E(\lambda) = (s) \lim\limits_{\substack{\lambda' \to \lambda \\ \lambda' \geq \lambda}} E(\lambda')$ (in the strong sense),

(3) $E_{-\infty} \equiv (s) \lim\limits_{\lambda \to -\infty} E(\lambda) = 0$ and $E_{+\infty} \equiv (s) \lim\limits_{\lambda \to \infty} E(\lambda) = I$.

A *normalized spectral measure* on R is a function P from the Borel sets $B(R)$ of R into $L(H, H)$ whose values are projections such that

(a) $P(R) = I$,

(b) $P(\varnothing) = 0$,

(c) $P(M \cup N) = P(M) + P(N)$ whenever $M \cap N = \varnothing$,

(d) $P\left(\bigcup_{i=1}^{\infty} M_i\right) = (s) \lim_n \sum_{i=1}^{n} P(M_i)$ whenever $M_i \cap M_j = \varnothing$ for $i \neq j$.

Prove that to every spectral measure P on $B(R)$, there corresponds a unique spectral function E such that $E(\lambda) = P\big((-\infty, \lambda]\big)$, and conversely.

8.1.3. Let $(x_k)_N$ be an orthonormal sequence in a Hilbert space H. Let $(\lambda_k)_N$ be a bounded nondecreasing sequence of real numbers. Define for each real number λ the operator $E(\lambda)$ on H by

$$E(\lambda)h = 0, \qquad \text{if } \lambda < \lambda_1,$$

$$E(\lambda)h = \sum_{k=1}^{n} (h \mid x_k)x_k, \qquad \text{if } \lambda_n \leq \lambda < \lambda_{n+1}$$

$$E(\lambda)h = h, \qquad \text{if } \lambda \geq \sup_k \lambda_k$$

(i) Show that $E(\lambda)$ is a spectral function on R.

(ii) Let P be the normalized spectral measure on $B(R)$ corresponding to E as in Problem 8.1.2. Show that the measures μ_h corresponding to P (and hence E) as in Theorem 8.1 are Lebesgue–Stieltjes measures restricted to $B(R)$ and determine a generating function for each μ_h.

(iii) Calculate the spectrum $\Sigma(P)$ of P.

(iv) Let $T = \int \lambda \chi_{\Sigma(P)}(\lambda)\, dP(\lambda)$. Calculate $(Th \mid h)$ and $(Th \mid k)$ for h, k in H.

8.1.4. Let $T \in L(H, H)$. Prove T is compact if and only if the range $R(T)$ contains no infinite-dimensional closed subspace. [Hint: For sufficiency, let $T = UP$ be the polar decomposition of T. Assuming $R(T)$ does not contain any infinite-dimensional closed subspaces, prove P is compact by using Corollary 8.1 and proving $M_F\colon L_2(X, \mu) \to L_2(X, \mu)$ given by $M_F f = Ff$ is compact. To show M_F is compact, let $\varepsilon > 0$, let $X_\varepsilon =$

$\{x \in X: F(x) \geq \varepsilon\}$ and let M_ε be the closed subspace $\{f \in L_2(X, \mu): f = 0$ a.e. on $X_\varepsilon^c\}$. Show M_ε is finite dimensional and prove $\| M_F - P_\varepsilon M_F \| < \varepsilon$, where P_ε is the projection onto M_ε.]

8.1.5. Let $(X, \mathscr{A}, \mu)$ be a semifinite measure space. Let $F \in L_\infty(\mu)$. Define $T_F: L_2(\mu) \to L_2(\mu)$ by $T_F(g) = Fg$. Prove the following:

(i) The mapping $F \mapsto T_F$ is a one-to-one linear map from L_∞ into $L(L_2, L_2)$;

(ii) T_F is normal;

(iii) T_F is self-adjoint if and only if F is real-valued a.e.;

(iv) T_F is unitary if and only if $| F | = 1$ a.e.;

(v) $\| T_F \| = \| F \|_\infty$;

(vi) If G is a measurable function, finite a.e., such that $Gg \in L_2(\mu)$ for all $g \in L_2(\mu)$, then $G \in L_\infty(\mu)$.

8.1.6. Let $(X, \mathscr{A}, \mu)$ be a finite measure space. Let $T \in L(L_2, L_2)$. If for $G \in L_\infty(\mu)$, $T_G \in L(L_2, L_2)$ is the operator $T_G f = Gf$, prove that if $TT_G = T_G T$ for all $G \in L_\infty(\mu)$, then $T = T_F$ for some $F \in L_\infty(\mu)$. Deduce that $\{T_G: G \in L_\infty(\mu)\}$ is a maximal commutative family in $L(L_2, L_2)$. [Hint: Let $F = T(\chi_X)$.]

8.1.7. Using Corollary 8.1 show that each positive operator T has a unique positive square root $\sqrt{T}$. [Hint: We may assume T operates on $L_2(X, \mu)$ by the rule $T(f) = Ff$, where $F \geq 0$. For existence, let $A(f) = \sqrt{F} f$. For uniqueness, suppose also $T = B^2$. Show the following: A and B commute with T; A is the uniform limit of a sequence of polynomials in T since the square root function on $\sigma(T)$ is a uniform limit of a sequence of polynomials; A and B commute; Ker T and (Ker $T)^\perp$ are invariant under T, A, and B; Ker T = Ker A = Ker B = Ker$(A + B)$; $(A + B)$(Ker $T)^\perp$ is dense in (Ker $T)^\perp$; and $(A - B)(A + B) = 0$.]

8.1.8. Prove a bounded normal operator on H is (i) self-adjoint, (ii) unitary, (iii) positive, (iv) a projection if and only if its spectrum is respectively (i′) real, (ii′) on the unit circle, (iii′) on the nonnegative real axis, or (iv′) the set $\{0, 1\}$.

8.2. Unbounded Operators and Spectral Theorems for Unbounded Self-Adjoint Operators

Most of the operators that occur in applications of the theory of Hilbert space to differential equations and quantum mechanics are unbounded. For this reason, we consider in this section basic definitions and theorems necessary to deal with unbounded operators and also we prove versions of the spectral theorem for unbounded operators. Let us first define precisely the type of operator we consider in this section.

Definition 8.4. A linear operator T *in* (in contrast to "on") a Hilbert space H is a linear transformation on a linear subspace D_T of H into H. D_T is called the *domain* of T and $R_T \equiv \{Tf: f \in D_T\}$ is the *range* of T. ▮

If S and T are operators in H, we say T extends S; written $S \subset T$, if $D_S \subset D_T$ and $Sh = Th$ for all h in D_S. T is also called an *extension* of S. T is *bounded* if T is bounded as a linear transformation from D_T into H.

Example 8.1. Let $H = L_2(-\infty, \infty)$ and let T be defined in H on

$$D_T = \{f \in L_2(-\infty, \infty): \lambda f(\lambda) \in L_2(-\infty, \infty)\}$$

by $Tf(\lambda) = \lambda f(\lambda)$. Since D_T contains all functions in $L_2(-\infty, \infty)$ which vanish outside a finite interval, D_T is a dense subset of H and contains moreover the characteristic functions $\chi_{[n,n+1)}$. Since $\| \chi_{[n,n+1)} \| = 1$ and

$$\| T\chi_{[n,n+1)} \|^2 = \int_n^{n+1} x^2\, dx \geq n^2, \quad \text{for } n \geq 0,$$

T is clearly unbounded.

Analogous to bounded linear operators on H, we can also consider in many cases the adjoint of a linear operator in H whether it be bounded or unbounded by means of the following definition.

Definition 8.5. Let T: $D_T \to H$ be an operator in H. Let D_{T^*} be the possibly empty set given by

$D_{T^*} = \{h \in H$: corresponding to h is a unique element h^* in H such that $(Tk \mid h) = (k \mid h^*)$ for all k in $D_T\}$.

If $D_{T^*} \neq \varnothing$, the mapping T^*: $D_{T^*} \to H$ given by $T^*h = h^*$ is called the *adjoint* of T. ▮

The questions that naturally arise are: When does T^* exist ($D_{T^*} \neq \varnothing$?) as an operator in H and when is $D_{T^*} = H$? Obviously, if T is a bounded linear operator on H, then $D_{T^*} = H$ and T^* is the "ordinary" adjoint defined earlier. The following proposition tells us in general when T^* has any meaning at all.

Proposition 8.7. $D_{T^*} \neq \varnothing$ and T^* is a linear mapping in H with domain D_{T^*} if and only if D_T is dense in H. ∎

Proof. Suppose $\bar{D}_T = H$ and h and h^* are two vectors satisfying

$$(Tk \mid h) = (k \mid h^*), \quad \text{for all } k \text{ in } D_T. \tag{8.22}$$

Then h^* is uniquely determined by h and equation (8.22), for if $h_1{}^*$ also satisfies equation (8.22), then

$$(k \mid h^* - h_1{}^*) = 0, \quad \text{for all } k \text{ in } D_T$$

and as $\bar{D}_T = H$, $h^* = h_1{}^*$. This means $D_{T^*} \neq \varnothing$ since $0 \in D_{T^*}$. To show T^* is linear, let h_1 and h_2 be in D_{T^*}. Then

$$\begin{aligned}(k \mid \alpha_1 T^*h_1 + \alpha_2 T^*h_2) &= \bar{\alpha}_1(k \mid T^*h_1) + \bar{\alpha}_2(k \mid T^*h_2)\\ &= \bar{\alpha}_1(Tk \mid h_1) + \bar{\alpha}_2(Tk \mid h_2)\\ &= (Tk \mid \alpha_1 h_1 + \alpha_2 h_2),\end{aligned}$$

so that not only is $\alpha_1 h_1 + \alpha_2 h_2$ in D_{T^*} but $T^*(\alpha_1 h_1 + \alpha_2 h_2) = \alpha_1 T^*h_1 + \alpha_2 T^*h_2$.

Conversely, suppose D_T is not dense in H. Let h be in $H - \bar{D}_T$. By Theorem 7.2, $h = h_1 + h_2$, where $h_1 \perp \bar{D}_T$ and $h_1 \neq 0$. Hence

$$(Tk \mid 0) = 0 = (k \mid h_1), \quad \text{for all } k \text{ in } D_T,$$

and corresponding to 0 are two vectors 0 and h_1 that satisfy equation (8.22). Moreover this means that for every vector x in H for which there exists a vector x^* satisfying equation (8.22), there are two vectors x^* and $x^* + h_1$, that satisfy equation (8.22). This means $D_{T^*} = \varnothing$ and T^* cannot exist. ∎

The notion of the adjoint of an operator in H gives rise to the idea of self-adjointness of operators in H. This idea generalizes our previous concept of self-adjointness for bounded operators on H.

Definition 8.6. An operator in H with $\bar{D}_T = H$ is *symmetric* if $T^* \supseteq T$. A symmetric operator is *self-adjoint* if $T^* = T$. ∎

The next result shows in particular that a self-adjoint operator with $D_T = H$ is always bounded.

Proposition 8.8. The adjoint T^* of a linear operator on H ($D_T = H$) is a bounded operator in H. ▮

Proof. Assume T^* is not bounded. There then exists a sequence (h_n) in D_{T^*} such that $\| h_n \| = 1$ and $\| T^*h_n \| \to \infty$ as $n \to \infty$. Define the functionals φ_n on H by

$$\varphi_n(h) = (Th \mid h_n).$$

The φ_n are clearly bounded linear functionals on H. Moreover for each h, the sequence $(\varphi_n(h))$ is also bounded as shown by

$$| \varphi_n(h) | = | (Th \mid h_n) | \leq \| Th \| \; \| h_n \| = \| Th \|.$$

By the Principle of Uniform Boundedness (Theorem 6.11) we have for some constant C,

$$| \varphi_n(h) | \leq C \| h \|, \qquad n = 1, 2, \ldots. \tag{8.23}$$

Letting $h = T^*h_n$ in inequality (8.23) we have

$$| \varphi_n(T^*h_n) | = (T(T^*h_n) \mid h_n) = (T^*h_n \mid T^*h_n) = \| T^*h_n \|^2 \leq C \| T^*h_n \|,$$

which implies $\| T^*(h_n) \| \leq C$. This is a contradiction to the selection of (h_n). ▮

Corollary 8.3. Any symmetric operator defined on H is bounded. ▮

Below are listed a few more important but trivially demonstrable facts.

Remarks

8.3. If S and T are linear operators in H, $S \subset T$ and $\bar{D}_S = H$, then $T^* \subset S^*$.

8.4. If T is a linear operator in H with $\bar{D}_T = \bar{D}_{T^*} = H$, then $T \subset T^{**}$.

8.5. Suppose T is a one-to-one operator in H so that T^{-1} is defined on $D_{T^{-1}} \equiv T(D_T)$. If $\bar{D}_T = \bar{D}_{T^{-1}} = H$, then T^* is one-to-one and $(T^*)^{-1} = (T^{-1})^*$.

Recall from Chapter 6 that an operator T in H is *closed* if whenever $h_1, h_2, \ldots$ in D_T converges to h in H and $Th_1, Th_2, \ldots$ converges to k in H, then $h \in D_T$ and $Th = k$. Equivalently T is closed if its graph $G_T = \{(h, Th)\colon h \in D_T\}$ is closed in the direct sum $H \oplus H$. [$H \oplus H = \{(h, k)\colon h \in H, k \in H\}$ with addition and scalar multiplication defined componentwise and the inner product given by $((h, k) \mid (h_1, k_1)) = (h \mid h_1) + (k \mid k_1)$.]

Proposition 8.9. The adjoint T^* of T in H is closed. ∎

Proof. Let $g_1, g_2 \ldots$ be a sequence in D_{T^*} converging to g in H and suppose $T^*g_1, T^*g_2, \ldots$ converge to h in H. Then for any k in D_T,

$$(Tk \mid g) = \lim_n (Tk \mid g_n) = \lim_n (k \mid T^*g_n) = (k \mid h),$$

implying that $g \in D_{T^*}$ and $h = T^*g$. ∎

Definition 8.7. An operator T in H is called *closable* if there is a closed operator S in H which extends T. The *closure* $\bar{T}$ of a closable operator T is the smallest closed operator extending T, that is, any closed operator extending T also extends $\bar{T}$. ∎

Obviously the closure exists for every closable operator.

Proposition 8.10. Suppose T is an operator in H with $\bar{D}_T = H$. Then

(i) if T is closed, $\bar{D}_{T^*} = H$ and $T = T^{**}$,
(ii) $\bar{D}_{T^*} = H$ if and only if T is closable, in which case $\bar{T} = T^{**}$, and
(iii) if T is closable, $(\bar{T})^* = T^*$. ∎

Proof. Observe from the definition of T^* that the set $A = \{(T^*g, -g)\colon g \in D_{T^*}\}$ is the set of all pairs $(g^*, -g)$ with $g \in D_{T^*}$ such that

$$(Tk \mid g) = (k \mid g^*), \qquad \text{for all } k \text{ in } D_T.$$

In other words A is the set of points $(g^*, -g)$ such that

$$((k, Tk) \mid (g^*, -g))_{H\oplus H} = 0, \qquad \text{for all } k \text{ in } D_T,$$

where $(\,\mid\,)_{H\oplus H}$ is the inner product in $H \oplus H$. This means A is the orthogonal complement $G_T{}^\perp$ of the linear space G_T. If T is closed then $A^\perp = (G_T{}^\perp)^\perp = G_T$.

Secondly, observe that if $\bar{D}_{T^*} = H$ so that T^{**} exists, then the graph of T^{**} consists of all the points (f, f^*) of $H \oplus H$ such that

$$(T^*h \mid f) = (h \mid f^*), \quad \text{for all } h \in D_{T^*},$$

or for which

$$((T^*h, \ -h) \mid (f, f^*))_{H\oplus H} = 0.$$

This implies $G_{T^{**}} = (G_T{}^\perp)^\perp$, and in particular $G_{T^{**}} = G_T$ if T is closed.

The proofs of (i), (ii), and (iii) follow from these observations. For (i), assume $\bar{D}_{T^*} \subsetneqq H$ and let h be a nonzero element of H such that $(h \mid g) = 0$ for all $g \in D_{T^*}$. Then

$$((0, h) \mid (T^*g, \ -g))_{H_1\oplus H_2} = 0, \quad \text{for all } g \text{ in } D_{T^*}.$$

Since $A^\perp = G_T$, $(0, h) \in G_T$ so that $h = 0$. Hence $\bar{D}_{T^*} = H$. Observing that always $A \subset A^{**}$, the relation $G_{T^{**}} = G_T$ above shows that $A = A^{**}$.

In (ii), if T is closable then by (i), $\bar{D}_{(\bar{T})^*} = H$ and $\bar{T} = T^{**}$. However, $D_{(\bar{T})^*} \subset D_{T^*}$ so $\bar{D}_{T^*} = H$. Conversely, if $\bar{D}_{T^*} = H$ then T^{**} exists and $T \subset T^{**}$. If S is any closed extension of T, then G_S is closed and contains G_T. From the observations above, $G_{T^{**}} = (G_T{}^\perp)^\perp$. Hence $G_{T^{**}}$ is the closure of G_T and $G_S \supset G_{T^{**}}$. Hence $\bar{T} = T^{**}$.

To prove (iii) notice that, if T is closable,

$$T^* = (\overline{T^*}) = T^{***} = (\bar{T})^*. \qquad \blacksquare$$

A symmetric operator is always closable since $D_{T^*} \supset D_T$ and D_T is dense in H. If T is symmetric, T^* is a closed extension of T, so the closure T^{**} of T satisfies $T \subset T^{**} \subset T^*$. If T is self-adjoint, $T = T^{**} = T^*$. If T is closed and symmetric, $T = T^{**} \subset T^*$. Clearly, a closed symmetric operator is self-adjoint if and only if T^* is symmetric.

The following is an illustrative example of three closed operators in H.

Example 8.2. As in Theorem 4.3, a complex-valued function f is absolutely continuous on $[\alpha, \beta]$, where $-\infty < \alpha < \beta < \infty$, if there exists an integrable function g on $[\alpha, \beta]$ such that

$$f(x) = \int_\alpha^x g(t)\, dt + f(\alpha).$$

Such a function f is continuous on $[\alpha, \beta]$ and differentiable a.e. with

$f'(x) = g(x)$ a.e. on $[\alpha, \beta]$. Yet its derivative need not belong to $L_2(\alpha, \beta)$, as is shown if one considers the function $f(x) = x^{1/2}$ in $L_2[0, 1]$.

For the purpose of this example we consider three separate Hilbert spaces as follows:

$$H_1 = L_2[\alpha, \beta], \qquad \text{where } -\infty < \alpha < \beta < \infty,$$
$$H_2 = L_2[\alpha, \infty), \qquad \text{where } -\infty < \alpha < \infty,$$
$$H_3 = L_2(-\infty, \infty).$$

Also we consider three operators T_1 in H_1, T_2 in H_2, and T_3 in H_3 defined, respectively, on the following three domains:

$D_1 = \{g \in H_1\colon g = f$ a.e. where f is absolutely continuous on $[\alpha, \beta]$, $f(\alpha) = 0 = f(\beta)$, and $f' \in L_2[\alpha, \beta]\}$,

$D_2 = \{g \in H_2\colon g = f$ a.e. where f is absolutely continuous on $[\alpha, \beta]$ for each $\beta > \alpha$, $f(\alpha) = 0$, and $f' \in L_2[\alpha, \beta]\}$,

$D_3 = \{g \in H_3\colon g = f$ a.e. where f is absolutely continuous on $[\alpha, \beta]$ for each $-\infty < \alpha < \beta < \infty$, and $f' \in H_3\}$.

By definition, $T_1 g = if'$, $T_2 g = if'$, and $T_3 g = if'$ (g, f as above).

For each $i = 1, 2, 3$, $\bar{D}_i = H_i$. To show this for D_1 we recall that the linear subspace spanned by the set $\{x^n\colon n = 0, 1, 2, \ldots\}$ is dense in $L_2[\alpha, \beta]$ since the class of all complex polynomials is dense in $L_2[\alpha, \beta]$. However, each x^n is in $\bar{D}_1$ since each x^n can be approximated in $L_2[\alpha, \beta]$ by a function f in D_1 as illustrated in Figure 8.1. This means $\bar{D}_1 = H$. To

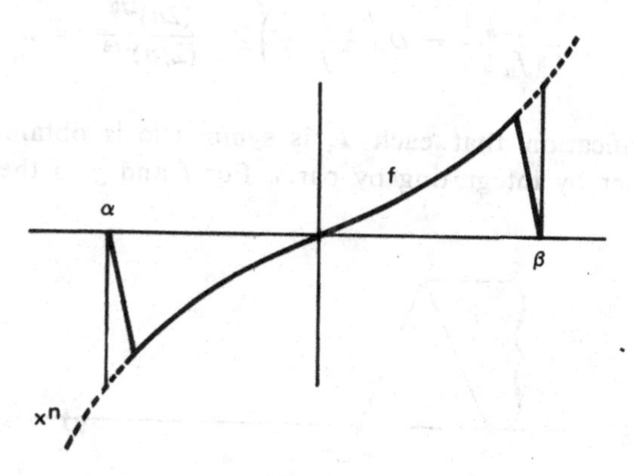

Fig. 8.1

prove $\bar{D}_2 = H$ and $\bar{D}_3 = H$, it is sufficient to observe analogously that the linear subspace spanned by the set $\{x^n e^{-x^2/2}: n = 0, 1, 2 \ldots\}$ is dense in $L_2(-\infty, \infty)$ {and hence their restrictions to $[\alpha, \infty)$ are dense in $L_2(\alpha, \infty)$} and then to approximate analogously each $x^n e^{-x^2/2}$ by a function f in D_1 or D_2.

The important observations to make here about T_1, T_2, and T_3 are the following: Each is an unbounded symmetric operator in H_i, respectively, and in particular T_3 is self-adjoint, whereas T_1 and T_2 are not self-adjoint. Moreover in each case $T_i = T_i^{**}$.

The verification that each T_i is unbounded is accomplished by considering functions in D_i of the following form. For $\alpha < \beta$ and $n \geq 2/(\beta - \alpha)$ define f_n by (see Figure 8.2)

$$f_n(x) = \begin{cases} n(x - \alpha), & \text{if } x \in [\alpha, \alpha + 1/n], \\ 2 - n(x - \alpha), & \text{if } x \in [\alpha + 1/n, \alpha + 2/n], \\ 0, & \text{if } x \in [\alpha + 2/n, \infty). \end{cases}$$

Clearly $f_n'(x) = n$, $-n$, or 0 on the respective intervals and

$$\| f_n \|^2 = \int_\alpha^{\alpha+2/n} | f_n(x) |^2 \leq 2/n,$$

while

$$\| if_n' \|^2 = \int_\alpha^{\alpha+2/n} n^2 \, dx = 2n.$$

These equations imply that

$$\frac{\| D_i f_n \|}{\| f_n \|} = D_i\left(\frac{f_n}{\| f_n \|}\right) \geq \frac{(2n)^{1/2}}{(2/n)^{1/2}} = n.$$

The verification that each T_i is symmetric is obtained in the following manner by integrating by parts. For f and g in the domain of T_1

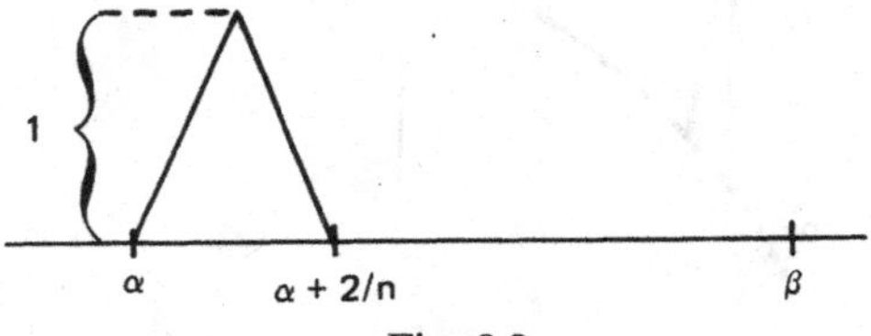

Fig. 8.2

and for $-\infty < \alpha < \beta < \infty$,

$$(if' \mid g) - (f \mid ig') = i \int_\alpha^\beta f'(\xi)\overline{g(\xi)}\, d\xi + i \int_\alpha^\beta f(\xi)\overline{g'(\xi)}\, d\xi$$

$$= if(\xi)\overline{g(\xi)} \Big|_\alpha^\beta = 0, \tag{8.24}$$

whereby $(T_1 f \mid g) = (f \mid T_1 g)$. The verifications for T_2 and T_3 are similar. {Observe that if $f \in D_2$ then $\lim_{x\to\infty} f(x) = 0$ and likewise if $f \in D_3$ then $\lim_{x\to\pm\infty} f(x) = 0$. Indeed, for H_2 the equation

$$(f \mid f') + (f' \mid f) = \lim_{x\to\infty} \int_\alpha^x [f(\xi)\overline{f'(\xi)} + f'(\xi)\overline{f(\xi)}]\, d\xi$$

$$= \lim_{x\to\infty} [|f(x)|^2 - |f(\alpha)|^2]$$

implies that $\lim_{x\to\infty} |f(x)|^2$ exists and since $f \in L_2(\alpha, \beta)$, $f(x) \to 0$ as $x \to \infty$.}

Let us next calculate the adjoint T_1^* of T_1. Let D_1^* be the set

$D_1^* = \{g \in H_1\colon g = f$ a.e. where f is absolutely continuous on $[\alpha, \beta]$, $f' \in H_1\}$.

Since equation 8.24 still is valid for g in D_1^*, the domain of T_1^* contains D_1^* and $T_1^* g = if'$ for g in D_1^*, where g and f are as above. We wish to show that the domain of T_1^* is D_1^*, which will clearly show that $T_1 \subset T_1^*$ and $T_1 \neq T_1^*$. To this end, let f be in the domain of T_1^*. Let h be the absolutely continuous function given by

$$h(x) = \int_\alpha^x T_1^* f(\xi)\, d\xi + C,$$

where C is a constant chosen so that

$$\int_\alpha^\beta [f(\xi) + ih(\xi)\, d(\xi)] = 0.$$

For every g in D_1, an integration by parts gives that

$$\int_\alpha^\beta ig'(\xi)\overline{f(\xi)}\, d\xi = (T_1 g \mid f) = (g \mid T_1^* f) = \int_\alpha^\beta g(\xi)\overline{T_1^* f(\xi)}\, d\xi$$

$$= g(\xi)\overline{h(\xi)} \Big|_\alpha^\beta - \int_\alpha^\beta g'(\xi)\overline{h(\xi)}\, d\xi = i \int_\alpha^\beta ig'(\xi)\overline{h(\xi)}\, d\xi.$$

Hence

$$\int_\alpha^\beta g'(\xi)\overline{[f(\xi) + ih(\xi)]}\, d\xi = 0.$$

In particular, letting g be the function in D_1 given by

$$g(x) = \int_\alpha^x [f(\xi) + ih(\xi)]\, d\xi ,$$

we obtain that

$$\int_\alpha^\beta | f(\xi) + ih(\xi) |^2 = 0$$

or that a.e. we have

$$f(x) = -ih(x) = -i \int_\alpha^x T_1{}^*f(\xi)\, d\xi - iC$$

and h is absolutely continuous with $h'(x) = T_1{}^*f(x)$. Hence f is in $D_1{}^*$.

In an almost identical fashion the verification that $T_2{}^*g = if'$ on the domain

$$D_2{}^* = \{ f \in H_2\colon f \text{ is absolutely continuous in each } [\alpha, \beta] \text{ with } \beta > \alpha \text{ and } f' \in H_2\}$$

and that $T_3{}^*g = if'$ on the domain

$$D_3{}^* = D_3$$

can be carried out. Since $D_1 \subsetneqq D_1{}^*$, $D_2 \subsetneqq D_2{}^*$, and $D_3 = D_3{}^*$, clearly T_3 is self-adjoint and $T_1{}^{**}$, $T_2{}^{**}$, and $T_3{}^{**}$ are all defined.

It remains to show that $T_j = T_j{}^{**}$ for $j = 1$ and 2. In either case since $T_j \subset T_j{}^*$, we can say that $T_j \subset T_j{}^{**} \subset T_j{}^*$. It suffices to show therefore that $D_{T_j^{**}} \subset D_j$. Let $f \in D_{T_j^{**}}$. Then for all g in $D_j{}^*$ we have

$$(T_j{}^{**}f \mid g) = (f \mid T_j{}^*g),$$

and moreover since $T_j{}^{**}f = if'$ (because $T_j{}^{**} \subset T_j{}^*$) we have

$$0 = (if' \mid g) - (f \mid ig').$$

If $j = 1$, this means

$$0 = i \int_\alpha^\beta f'(\xi)\overline{g(\xi)}\, d(\xi) + i \int_\alpha^\beta f(\xi)\overline{g'(\xi)}\, d\xi$$

$$= if(\xi)\overline{g(\xi)} \Big|_\alpha^\beta = i[f(\beta)\overline{g(\beta)} - f(\alpha)\overline{g(\alpha)}].$$

By first letting $g(x) = (x - \alpha)/(\beta - \alpha)$ in $D_1{}^*$ and then letting $g = (\beta - x)/(\beta - \alpha)$ in $D_1{}^*$, we obtain $f(\alpha) = 0 = f(\beta)$ implying that $f \in D_1$. If $j = 2$, let $g(x) = e^{-(x-\alpha)}$ to obtain $f(\alpha) = 0$ so that f is again in D_2.

Interestingly enough, T_1 has uncountably many different self-adjoint extensions. Let $\gamma \in C$ with $|\gamma| = 1$ and define T_γ in H_1 on

$$D_{T_\gamma} = \{g \in L_2(\alpha, \beta): g = f \text{ a.e. where } f \text{ is absolutely continuous on } [\alpha, \beta], f' \in H_1, \text{ and } f(\beta) = \gamma f(\alpha)\}$$

by $T_\gamma g = if'$. Each T_γ is self-adjoint and extends T_1. For each γ, we have $T_1 \subset T_\gamma \subset T_1^*$.

The next example presents a type of self-adjoint operator which will be shown in Theorem 8.7 to be the "prototype" of all self-adjoint operators.

Example 8.3. Let $(X, \mathscr{A}, \mu)$ be a measure space with μ a finite measure. Suppose that f is an extended real-valued measurable function on X which is finite a.e. Then the operator T_f in $L_2(X, \mu)$ defined by $T_f(g) = fg$ on

$$D_f = \{g \in L_2(X, \mu): fg \in L_2(X, \mu)\}$$

is self-adjoint. Indeed, by considering the functions f_n defined as

$$f_n = \begin{cases} 1, & \text{on } N_n = \{x: |f(x)| < n\}, \\ 0, & \text{otherwise,} \end{cases}$$

one can easily verify that $gf_n \in D_f$ for each $n = 1, 2, \ldots$ and for each g in $L_2(X, \mu)$. Moreover, since the functions gf_n are dense in $L_2(X, \mu)$, $\overline{D_f} = H$. T_f is thus clearly symmetric. In addition if h is an element in $D_{T_f^*}$, then by the Monotone Convergence Theorem,

$$\begin{aligned} \| T_f^* h \| &= \lim_{n\to\infty} \| f_n T_f^* h \| \\ &= \lim_{n\to\infty} [\sup_{\|k\|=1} (k \mid f_n T_f^*(h))] \\ &= \lim_{n\to\infty} [\sup_{\|k\|=1} (T_f(kf_n) \mid h)] \\ &= \lim_{n\to\infty} [\sup_{\|k\|=1} (k \mid f_n fh)] = \lim_{n\to\infty} \| f_n fh \| = \| fh \|. \end{aligned}$$

Hence $fh \in L_2(X, \mu)$ and $h \in D_{T_f}$. Therefore, $D_{T_f} = D_{T_f^*}$ and T_f is self-adjoint.

As before for bounded linear operators on H, a scalar λ is an *eigenvalue* of an operator T in H if there exists a nonzero vector h in D_T such that $Th = \lambda h$. If T is a symmetric operator, the equation

$$\lambda(h \mid h) = (\lambda h \mid h) = (Th \mid h) = (h \mid Th) = (h \mid \lambda h) = \bar{\lambda}(h \mid h)$$

shows that eigenvalue λ is always real.

Again, the *resolvent set* $\varrho(T)$ of an operator in H with $\overline{D_T} = H$ is the set of all scalars λ for which $R_{T-\lambda I}$ is dense in H and for which $(T - \lambda I)$ has a bounded inverse defined on $R_{T-\lambda I}$. The *spectrum* $\sigma(T)$ of T is the complement of $\varrho(T)$ and is clearly decomposed into three disjoint sets: the set of eigenvalues of T (sometimes called the *point spectrum* $P\sigma(T)$), the set of scalars λ for which $R_{T-\lambda I}$ is dense in H but for which $(T - \lambda I)^{-1}$ exists and is not bounded (called the *continuous spectrum* $C\sigma(T)$), and the set of scalars λ for which $(T - \lambda I)^{-1}$ exists but its domain $R_{T-\lambda I}$ is not dense (called the *residual spectrum* $R\sigma(T)$).

Proposition 8.11. If T is a closed linear operator in H with $\overline{D_T} = H$ and $\lambda \in \varrho(T)$, then $(T - \lambda I)^{-1}$ is a bounded linear operator *on* (all of) H. ∎

Proof. $R_{T-\lambda I}$ is dense in H and since $(T - \lambda I)^{-1}$ is bounded on $R_{T-\lambda I}$, there exists a positive constant C such that

$$\| h \| \leq C \| (T - \lambda I)h \|, \qquad \text{for all } h \text{ in } D_T. \tag{8.25}$$

If $k = \lim_n (T - \lambda I)h_n$, then by inequality (8.25) the $\lim_{n\to\infty} h_n$ exists, say h. Since T is closed, $(T - \lambda I)h = k$ and $k \in R_{T-\lambda I}$. Hence $R_{T-\lambda I} = H$. ∎

In light of Proposition 8.11, we can see that the resolvent of a closed operator in H is the set of scalars λ for which $T - \lambda I$ is a bijection from D_T to H and for which $(T - \lambda I)^{-1}$ is bounded on H.

The next result establishes that the spectrum of a self-adjoint operator in H is contained in the real numbers.

Proposition 8.12. Let T be a self-adjoint operator in H. Then $\varrho(T)$ contains all complex numbers with nonzero imaginary part. Moreover if $\operatorname{Im} \lambda \neq 0$ then

$$\| (T - \lambda I)^{-1} \| \leq \frac{1}{|\operatorname{Im} \lambda|} \tag{8.26}$$

and

$$\operatorname{Im}\big((T - \lambda I)h \mid h\big) = \operatorname{Im}(-\lambda) \| h \|^2, \qquad \text{for all } h \in D_T. \tag{8.27}$$

∎

Proof. Since T is self-adjoint, $(Th \mid h)$ is real for all h in D_T. Clearly equation (8.27) follows and by the Cauchy–Schwarz inequality,

$$|\operatorname{Im}(\lambda)| \; \| h \|^2 \leq |\big((T - \lambda I)h \mid h\big)| \leq \| (T - \lambda I)h \| \; \| h \|,$$

which implies that

$$| \operatorname{Im}(\lambda) | \; \| h \| \leq \| (T - \lambda I)h \|. \tag{8.28}$$

If $\operatorname{Im}(\lambda) \neq 0$, this inequality implies that $(T - \lambda I)^{-1}$ exists as a linear operator on $R_{T-\lambda I}$ since $T - \lambda I$ is one-to-one by inequality (8.28). Now $R_{T-\lambda I}$ is also dense in H. Indeed, if not, there exists a nonzero vector k in D_T (which is dense in H) such that $((T - \lambda I)h \mid k) = 0$ for all h in D_T. However, then $(h \mid (T^* - \bar{\lambda} I)k) = (h \mid (T - \bar{\lambda} I)k) = 0$ for all h in D_T; and since $\bar{D}_T = H$, $(T - \bar{\lambda} I)k = 0$. This means $Tk = \bar{\lambda} k$ and $(Tk \mid k) = \bar{\lambda}(k \mid k)$, which contradicts the fact that $(Tk \mid k)$ is real. Therefore $R_{T-\lambda I}$ is dense in H. Moreover from inequality (8.28), $(T - \lambda I)^{-1}$ is bounded on $R_{T-\lambda I}$ with bound satisfying inequality (8.26). All this means $\lambda \in \varrho(T)$ when $\operatorname{Im} \lambda \neq 0$. ∎

The next proposition also shows that the resolvent set is open for a self-adjoint operator.

Proposition 8.13. If T is a closed linear operator in H with $\bar{D}_T = H$, then the resolvent set is open and if λ and μ are in $\varrho(T)$,

$$R_\mu - R_\lambda = (\lambda - \mu) R_\lambda R_\mu,$$

where, as in Proposition 6.27 of Chapter 6, $R_\lambda = (T - \lambda I)^{-1}$. Moreover R_λ as a function on $\varrho(T)$ to $L(H, H)$ has derivatives of all orders. ∎

Proof. Observing that from Proposition 8.11, R_λ is defined on H for $\lambda \in \varrho(T)$, the proof is identical to that of Proposition 6.25 of Chapter 6. ∎

Remarks. Here are some other easy to prove facts concerning the eigenvalues and the spectrum of an operator in H.

8.6. If $\bar{D}_T = \bar{D}_{T^*} = H$ and $T = T^{**}$, then $\lambda \in \sigma(T)$ if and only if $\bar{\lambda} \in \sigma(T^*)$.

8.7. If $\bar{D}_T = \bar{D}_{T^*} = H$ and $T = T^{**}$, then

$$\{h \in D_T \colon Th = \lambda h\} = [(T^* - \bar{\lambda} I) D_{T^*}]^\perp.$$

In particular if T is self-adjoint, λ is an eigenvalue of T if and only if $(T - \lambda I) D_T$ is not dense in H.

8.8. If T is a self-adjoint operator, $\lambda \in \varrho(T)$ if and only if $(T - \lambda I) D_T = H$.

Proof. (8.6) If $\lambda \in \varrho(T)$, then by Proposition 8.11 $D_{(T-\lambda I)^{-1}} = H$. Since $(T-\lambda I)^* = T^* - \bar{\lambda} I$, by Remark 8.5 we have $(T^* - \bar{\lambda} I)^{-1} = [(T-\lambda I)^{-1}]^*$. Since $[(T-\lambda I)^{-1}]^*$ is bounded, $\bar{\lambda} \in \varrho(T^*)$. Conversely, by reversing the argument, $\bar{\lambda} \in \varrho(T^*)$ implies $\lambda \in \varrho(T)$.

$$\text{(8.7)} \quad \begin{aligned} Th - \lambda h = 0 &\Leftrightarrow (Th - \lambda h \mid k) = 0, && \text{for all } k \text{ in } D_{T^*}, \\ &\Leftrightarrow (h \mid (T^* - \bar{\lambda})k) = 0, && \text{for all } k \text{ in } D_{T^*}, \\ &\Leftrightarrow h \in [(T^* - \bar{\lambda} I) D_{T^*}]^{\perp}. \end{aligned}$$

(8.8) If $\lambda \in \varrho(T)$, we know from Proposition 8.11 that $(T-\lambda I)D_T = H$. Conversely, suppose $(T - \lambda I)D_T = H$. If $\lambda \notin R$, $\lambda \in \varrho(T)$ by Proposition 8.12. If $\lambda \in R$, λ is not an eigenvalue by (8.7) and $T - \lambda I$ is self-adjoint. By Remark 8.5, $(T - \lambda I)^{-1}$ is also self-adjoint and by Proposition 8.8, $(T - \lambda I)^{-1}$ is bounded. Hence $\lambda \in \varrho(T)$. ∎

Example 8.4. To illustrate some of the preceding theory let us consider three versions of the differentiation operator id/dt. Let T_1, T_2, T_3 be the operator id/dt on the respective domains†:

$D_{T_1} = \{f \in L_2[0, 2\pi]$: f is absolutely continuous on $[0, 2\pi]$ and $f(0) = 0\}$

$D_{T_2} = \{f \in L_2[0, 2\pi]$: f is absolutely continuous on $[0, 2\pi]$ and $f(0) = f(2\pi)\}$

$D_{T_3} = \{f \in L_2[0, 2\pi]$: f is absolutely continuous on $[0, 2\pi]\}$.

From Example 8.2 and the fact that D_1 of that example is contained in D_{T_1}, D_{T_2}, and D_{T_3}, each of these domains is dense in H. The spectrum of T_1 is empty, the spectrum of T_2 is the set of integers (which is also the set of eigenvalues of T_2), and the spectrum of T_3 is the whole complex plane. That the spectrum of T_1 is empty follows from the fact that for each λ the operator S_λ given by

$$(S_\lambda g)(t) = \int_0^t e^{-i\lambda(t-s)} g(s)\, ds, \qquad g \in L_2[0, 2\pi]$$

is the inverse of $T_1 - \lambda I$. To calculate the spectrum of T_2, observe first that $\sigma(T_2)$ is a set of real numbers since T_2 is self-adjoint as in Example 8.2. For each integer k, the function $f(t) = e^{-ikt}$ is a solution of $T_2 f = kf$ so that each integer is an eigenvalue. It is easy to verify that for each non-

† Though not explicitly stated, these domains are here assumed to have the property $g = f$ a.e. and $f \in D_{T_j} \Rightarrow g \in D_{T_j}$ and $T_j g = if'$.

integer real number λ the equation

$$if'(t) - \lambda f(t) = g(t)$$

is solvable for each g in $L_2[0, 2\pi]$ so that by Remark 8.8 each such λ is in $\varrho(T)$. Finally to calculate the spectrum $\sigma(T_3)$ observe that for each λ in C, the function $f(\lambda) = e^{-i\lambda t}$ is a solution of $if'(t) - \lambda f(t) = 0$ so that each λ in C is in fact an eigenvalue of T_3.

Let us now turn our attention to several versions of the Spectral Theorem for unbounded operators. In this case a multiplication operator form of the theorem leads nicely into a functional calculus form and then into a spectral measure version of the Spectral Theorem. The interested reader may also consult references [25] and [31].

The next theorem shows that all self-adjoint operators in H are unitarily equivalent in the sense of (ii) below to a self-adjoint operator of the type given in Example 8.3.

Theorem 8.8. *Multiplication Operator Form of the Spectral Theorem.* Let T be a self-adjoint operator in a Hilbert space H. There exists then a measure space $(X, \mathscr{A}, \mu)$, a unitary† operator U: $H \to L_2(X, \mu)$, and a measurable function F on X which is real a.e. such that

(i) $h \in D_T$ if and only if $F(\cdot)Uh(\cdot)$ is in $L_2(X, \mu)$,

(ii) if $f \in U(D_T)$, then $(UTU^{-1}f)(\cdot) = F(\cdot)f(\cdot)$. ∎

Proof. To achieve the proof we utilize the multiplication operator form of the Spectral Theorem for bounded normal operators (Corollary 8.2) by applying it to the operator $(T + i)^{-1}$. Let us first establish that this is a bounded normal operator.

By Propositions 8.11 and 8.12, $(T \pm i)^{-1}$ exist as bounded linear operators on H. In particular $R_{T\pm i} = H$ and $T \pm i$ are one-to-one operators. For any h and k in D_T, since T is self-adjoint,

$$((T - i)h \mid (T + i)^{-1}(T + i)k) = ((T - i)^{-1}(T - i)h \mid (T + i)k).$$

This implies that $((T + i)^{-1})^* = (T - i)^{-1}$. Since $(T + i)^{-1}$ and $(T - i)^{-1}$ commute by Proposition 8.13, we have

$$(T + i)^{-1}((T + i)^{-1})^* = (T + i)^{-1}(T - i)^{-1} = ((T + i)^{-1})^*(T + i)^{-1},$$

and $(T + i)^{-1}$ is seen to be a normal operator.

† By a unitary operator U: $H \to L_2$ we mean a linear isometry from H onto L_2. (See also Proposition 7.17.)

By Corollary 8.2, there is a measure space $(X, \mathscr{A}, \mu)$, a unitary operator $U: H \to L_2(X, \mu)$, and a bounded, measurable complex function G on X so that

$$(U(T + i)^{-1}U^{-1}f)(x) = G(x)f(x) \text{ a.e.} \tag{8.29}$$

for all f in $L_2(X, \mu)$.

Since Ker $(T + i)^{-1} = \{0\}$, $G(x) \neq 0$ a.e. Therefore if we define $F(x)$ as $G(x)^{-1} - i$ for each x in X, $|F(x)|$ is finite a.e. Now if $f \in U(D_T)$, then there exists a function g in $L_2(X, \mu)$ such that $f(\cdot) = G(\cdot)g(\cdot)$ in L_2. That this is so follows from the inclusions

$$U(D_T) \subset U(T + i)^{-1}(H) \subset U(T + i)^{-1}U^{-1}(L_2(X, \mu)). \tag{8.30}$$

Observing that $U(T + i)^{-1}U^{-1}$ is an injection, for any g in the range of $U(T + i)^{-1}U^{-1}$ we have from equation (8.29)

$$[U(T + i)^{-1}U^{-1}]^{-1}g(x) = [1/G(x)] \cdot g(x) \in L_2(X, \mu).$$

In particular for f in the set $U(D_T)$,

$$[U(T + i)^{-1}U^{-1}]^{-1}f(x) = [1/G(x)] \cdot f(x) \in L_2(X, \mu)$$

or

$$U(T + i)U^{-1}f(x) = [1/G(x)] \cdot f(x) \in L_2(X, \mu)$$

or

$$UTU^{-1}f(x) = [1/G(x)]f(x) - if(x) = F(x)f(x) \in L_2(X, \mu).$$

This proves (ii) and the necessity of (i) provided F is real-valued, which we show below. For the converse of (i), if $F(x)Uh(x)$ is in $L_2(X, \mu)$, then there is a k in H so that $Uk = [F(x) + i]Uh(x)$. Thus $G(x)Uk(x) = G(x)[F(x) + i]Uh(x) = Uh(x)$, so $h = (T + i)^{-1}k$, whereby $h \in D_T$.

To finish the proof it must be established that F is real valued a.e. Observe that the operator in $L_2(X, \mu)$ defined by multiplication by F is self-adjoint since by (ii) it is "unitarily equivalent" to T. Hence for all χ_M, M a measurable subset of X, $(\chi_M \mid F\chi_M)$ is real. However, if Im $F > 0$ on a set of positive measure, then there exists a bounded set B in the plane so that $M = F^{-1}(B)$ has nonzero measure. Clearly $F\chi_M$ is in $L_2(X, \mu)$ since B is bounded and $\text{Im}(\chi_M \mid F\chi_M) > 0$. This contradiction shows that Im $F = 0$ a.e. ∎

Using the foregoing theorem we can prove the following result.

Theorem 8.9. *Functional Calculus Form of the Spectral Theorem.* Let T be a self-adjoint operator in H. There is a unique map $\hat{\varphi}$ from the class of bounded Borel measurable functions on R into $L(H, H)$ so that

(i) $\hat{\varphi}(\alpha f + \beta g) = \alpha\hat{\varphi}(f) + \beta\hat{\varphi}(g)$,
$\hat{\varphi}(fg) = \hat{\varphi}(f)\hat{\varphi}(g)$,
$\hat{\varphi}(\bar{f}) = (\hat{\varphi}(f))^*$, $\hat{\varphi}(1) = I$.

(ii) $\| \hat{\varphi}(f) \| \leq K \| f \|_\infty$ for some $K > 0$.

(iii) If (f_n) is a sequence of bounded Borel functions converging pointwise to the identity function on R and $| f_n(x) | \leq | x |$ for all x and n, then for any $h \in D_T$, $\lim_n \hat{\varphi}(f_n)h = Th$.

(iv) If (f_n) converges pointwise to g and $(\| f_n \|_\infty)_N$ is bounded, then $\hat{\varphi}(f_n) \to \hat{\varphi}(g)$ strongly.

In addition:

(v) If $Th = \lambda h$, $\hat{\varphi}(g)h = g(\lambda)h$.

(vi) If $h \geq 0$, then $\hat{\varphi}(h) \geq 0$. ∎

Proof. Define $\hat{\varphi}$ by

$$\hat{\varphi}(g) = U^{-1}\tau_{g(F)}U,$$

where F and U are as in the previous theorem and $\tau_{g(F)}: L_2(X, \mu) \to L_2(X, \mu)$ is given by $\tau_{g(F)}\psi = g\big(F(\cdot)\big)\psi(\cdot)$. Using the previous theorem, the verification that $\hat{\varphi}$ satisfies the conditions of the theorem is routine but arduous. We illustrate by verifying $\hat{\varphi}(\bar{f}) = (\hat{\varphi}(f))^*$ and condition (iii).

First we show $\hat{\varphi}(\bar{f}) = (\hat{\varphi}(f))^*$. For any h and k in H we have

$$\begin{aligned}(\hat{\varphi}(f)h \mid k) &= (U^{-1}f[F(\cdot)]U(h)(\cdot) \mid k)\\ &= (U^{-1}f[F(\cdot)]U(h)(\cdot) \mid U^{-1}[U(k)(\cdot)])\\ &= (f[F(\cdot)]U(h)(\cdot) \mid U(k)(\cdot))\\ &= \int f[F(\cdot)]U(h)(\cdot)\overline{U(k)(\cdot)}\, d\mu(\cdot).\end{aligned}$$

Similarly,

$$\begin{aligned}(h \mid \hat{\varphi}(\bar{f})k) &= (U^{-1}[U(h)(\cdot)] \mid U^{-1}\bar{f}[F(\cdot)]Uk(\cdot))\\ &= \int U(h)(\cdot)\overline{\bar{f}[F(\cdot)]U(k)(\cdot)}\, d\mu(\cdot)\\ &= \int f[F(\cdot)U(h)(\cdot)\overline{U(k)(\cdot)}\, d\mu(\cdot),\end{aligned}$$

whereby $(\hat{\varphi}(f))^* = \hat{\varphi}(\bar{f})$.

Next we verify condition (iii). First observe that since $|f_n(x)| \leq |x|$ and $\lim_{n\to\infty} f_n(x) = x$, for any h in D_T,

$$|f_n(F(\cdot))U(h)(\cdot) - F(\cdot)U(h)(\cdot)|^2 \to 0 \tag{8.31}$$

and

$$|f_n(F(\cdot))U(h)(\cdot) - F(\cdot)U(h)(\cdot)|^2 \leq 4\,|F(\cdot)U(h)(\cdot)|^2, \tag{8.32}$$

where the right side of equation (8.32) is integrable by (i) of the previous theorem. Hence for any h in D_T

$$\int |f_n(F(\cdot))U(h)(\cdot) - F(\cdot)U(h)(\cdot)|^2 \, d\mu \to 0. \tag{8.33}$$

by the Dominated Convergence Theorem. However, for any h in D_T,

$$\begin{aligned}
&\| U^{-1}[f_n(F(\cdot))]U(h)(\cdot) - Th \|_H \\
&= \| U^{-1}[f_n(F(\cdot))]U(h)(\cdot) - U^{-1}UTU^{-1}U(h) \|_H \\
&\overset{(a)}{=} \| U^{-1}[f_n(F(\cdot))]U(h)(\cdot) - U^{-1}F(\cdot)U(h)(\cdot) \|_H \\
&\overset{(b)}{=} \| f_n(F(\cdot))U(h)(\cdot) - F(\cdot)U(h)(\cdot) \|_{L_2(X,\mu)},
\end{aligned}$$

where we have used (ii) of the previous theorem at (a) and the fact that U is unitary at (b). Since the final expression equals the expression in (8.33), $\hat{\varphi}(f_n)h \to Th$.

The uniqueness of $\hat{\varphi}$ must yet be established. The proof is not trivial and we need to do some preliminary work first. This we proceed to do.

Observe that corresponding to each mapping ψ from the class of bounded Borel functions on R into $L(H, H)$ that satisfies (i)–(iv) of Theorem 8.9 there is a normalized spectral measure E on $B(R)$ given by $E(M) = \psi(\chi_M)$. In particular, E satisfies the following properties:

(i) $E(M)$ is a projection for each M in $B(R)$,

(ii) $E(\varnothing) = 0$ and $E(R) = I$,

(iii) If $M = \bigcup_{i=1}^{\infty} M_i$ where $M_i \cap M_j = \varnothing$ for $i \neq j$, then

$$E(M) = \lim_{n\to\infty} \sum_{i=1}^{n} E(M_i) \text{ (strong sense)},$$

(iv) $E(M \cap N) = E(M)E(N)$.

The spectral measure which corresponds to $\hat{\varphi}$ is denoted by P.

By Theorem 8.2, if E is *any* normalized spectral measure on $B(R)$

(in particular E could correspond to ψ or be P), then for each bounded complex-valued measurable function f on R, there exists a unique bounded operator T_f such that

$$(T_f h \mid h) = \int_{-\infty}^{\infty} f(\lambda)\, d\mu_h(\lambda)\,, \quad \text{for all } h \text{ in } H, \tag{8.34}$$

where μ_h is the finite measure $\mu_h(M) = (E(M)h \mid h)$. T_f is denoted by $\int f\, dE$. Observe that if E is P, then for any bounded Borel function f on R, $\int f\, dP = \hat{\varphi}(f)$. This is readily verified for characteristic functions χ_M for M in $B(R)$ since

$$(\hat{\varphi}(\chi_M)h \mid h) = (P(M)h \mid h) = \mu_h(M) = \int \chi_M \, d\mu_h.$$

From this it follows for simple functions, and from (iv) of the theorem, it follows for nonnegative bounded functions. By additivity it is true for all bounded measurable functions.

If E is any normalized spectral measure on $B(R)$, we can make equation (8.34) be valid for arbitrary measurable functions by the following procedure. If g is any measurable function ($|\, g\,|$ finite), then the set

$$D_g = \left\{ h \,\middle|\, \int_{-\infty}^{\infty} |\, g\,|^2 \, d\mu_h < \infty \right\}$$

is dense in H. This follows from the fact that, for each h in H,

$$h = \lim_{n\to\infty} \sum_{i=1}^{n} E(S_i)h\,,$$

where $S_i = \{\lambda \in R \colon i - 1 \le |\, g(\lambda)\,|^2 < i\}$ for $i = 1, 2, 3, \ldots$ and the fact that for each n, $E(\bigcup_{i=1}^{n} S_i)h$ is in D_g. Therefore if g is a nonnegative measurable function, defining $T_g h$ for each h in D_g by

$$T_g h = \lim_{n\to\infty} T_{g\wedge n\cdot\chi_{[-n,n]}}\, h,$$

where $T_{g\wedge n\cdot\chi_{[-n,n]}}$ is given by equation (8.34), we define a linear operator T_g on a dense linear subspace of H. By the Monotone Convergence Theorem T_g satisfies for each h in D_g the equation

$$(T_g h \mid h) = \int g \, d\mu_h. \tag{8.35}$$

Now if g is measurable and $|\, g\,|$ finite we write g as $g_1 - g_2 + i(g_3 - g_4)$, where each g_i is nonnegative, measurable, and integrable with respect to μ_h for each h in D_g. On D_g we define T_g as the linear operator

$$T_g = T_{g_1} - T_{g_2} + i(T_{g_3} - T_{g_4}).$$

By virtue of the fact that each T_{g_i} satisfies equation (8.35), T_g is easily seen to be well-defined and to satisfy equation (8.35).

Equation (8.35) for each h in D_g actually determines T_g uniquely. To see this, we define for each h and k in D_g the sum $\int g\, d\mu_{h,k}$ of the integrals as in Definition 8.2. By a straightforward calculation, it is easy to check that

$$4\int g\, d\mu_{h,k} \equiv \int g\, d\mu_{h+k} - \int g\, d\mu_{h-k} + i\int g\, d\mu_{h+ik} - i\int g\, d\mu_{h-ik} = 4(Sh \mid k) \tag{8.36}$$

for any operator S satisfying equation (8.35) for all h in D_g. Hence for any h and k in D_g, $(T_g h \mid k) = (Sh \mid k)$ and since $\overline{D_g} = H$, $T_g h = Sh$ for any h in H.

Let us summarize this discussion in a lemma.

Lemma 8.7. Let E be a normalized spectral measure on $B(R)$. Corresponding to each complex-valued measurable function g is a unique linear operator T_g in H with domain $D_g = \{h \mid \int_{-\infty}^{\infty} |g|^2\, d\mu_h < \infty\}$ such that

$$(T_g h \mid h) = \int g\, d\mu_h. \quad \blacksquare$$

As before, we write $T_g = \int g\, dE$ if T_g is the unique linear operator on D_g satisfying equation (8.35).

Observe that if g is a real-valued function, then T_g is a symmetric operator on D_g. Indeed, writing g as $g_1 - g_2$, where g_1 and g_2 are nonnegative functions, for all h and k in D_g,

$$\begin{aligned}
(T_g h \mid k) &= (T_{g_1} h \mid k) - (T_{g_2} h \mid k) \\
&= \lim_{n\to\infty} (T_{g_1 \wedge n \cdot \chi_{[-n,n]}} h \mid k) - \lim_{n\to\infty} (T_{g_2 \wedge n \cdot \chi_{[-n,n]}} h \mid k) \\
&= \lim_{n\to\infty} (h \mid T_{g_1 \wedge n \cdot \chi_{[-n,n]}} k) - \lim_{n\to\infty} (h \mid T_{g_2 \wedge n \cdot \chi_{[-n,n]}} k) \\
&= (h \mid T_{g_1} k) - (h \mid T_{g_2} k) \\
&= (h \mid T_g k).
\end{aligned}$$

If T is a given self-adjoint operator in H and E is a normalized spectral measure corresponding to T via $E(M) = \psi(\chi_M)$, where ψ is a map from the class of bounded Borel measurable functions R into $L(H, H)$ as in Theorem 8.9, the question arises whether $T = \int \lambda \, dE$. An affirmative answer is easily obtained from (iii) of Theorem 8.8. To show this we must first establish that D_T the domain of T is actually $\{h \mid \int_{-\infty}^{\infty} |\lambda|^2 \, d\mu_h < \infty\}$, which we denote by D_λ.

First we show $D_T \subset D_\lambda$. Recall that, if f_n is the function $\lambda\chi_{[-n,n]}$, then

$$\psi(f_n) = \int f_n \, dE,$$

$$[\psi(f_n)]^2 = \psi(f_n^2) = \int f_n^2 \, dE.$$

Hence if $h \in D_T$, then

$$\begin{aligned}(Th \mid Th) &= \lim_n \, (\psi(f_n)h \mid \psi(f_n)h) \\ &= \lim_n \, (\psi(f_n)^*\psi(f_n)h \mid h) \\ &= \lim((\psi(f_n))^2 h \mid h) \\ &= \lim_n \int \lambda^2 \chi_{[-n,n]} \, d\mu_h \\ &= \int \lambda^2 \, d\mu_h,\end{aligned}$$

which implies $\int \lambda^2 \, d\mu_h < \infty$ and $h \in D_\lambda$.

Now if $T_1 = \int \lambda \, dE$, then T_1 is symmetric with domain the set D_λ. Also for any h in D_T we have by (iii) of Theorem 8.9,

$$\begin{aligned}(Th \mid h) &= \lim_n \, (\psi(f_n)h \mid h) = \lim_n \left(\left(\int f_n \, dE\right)h \mid h\right) \\ &= \lim_n \int \lambda\chi_{[-n,n]} \, d\mu_h = \int \lambda \, d\mu_h = (T_1 h \mid h).\end{aligned}$$

Hence $T \subset T_1$. This means $T_1^* \subset T^* = T$ and since $T_1 \subset T_1^*$, $T_1 \subset T$. Consequently $T = T_1$ and $D_T = D_\lambda$. More importantly, $T = \int \lambda \, dE$.

We have proved the following

Lemma 8.8. If T is a self-adjoint operator in H and E is a normalized spectral measure corresponding to T as in Theorem 8.9 via some map ψ, then $T = \int \lambda \, dE$ and for every $h \in D_T$

$$\| Th \|^2 = \int_{-\infty}^{\infty} \lambda^2 \, d\mu_h ,$$

with $D_T = \{h \mid \int_{-\infty}^{\infty} |\lambda|^2 \, d\mu_h < \infty\}$. ∎

In view of the rather lengthy discussion above, it should be clear that to prove $\hat{\varphi}$ is unique it suffices to show that when E is any normalized spectral measure such that

$$T = \int \lambda \, dP = \int \lambda \, dE , \tag{8.37}$$

then $P = E$. Indeed, if P corresponds to $\hat{\varphi}$ and E corresponds to ψ, then equation (8.37) is true and if $E = P$ then $\hat{\varphi}$ and ψ agree on characteristic functions, simple functions, and then all bounded measurable functions. Proof of the uniqueness of $\hat{\varphi}$ in this manner will also enable us to quickly prove the final formulation of the spectral theorem—Theorem 8.10 below.

We seek to show then that if E is any normalized spectral measure such that

$$T = \int \lambda \, dP = \int \lambda \, dE ,$$

$E = P$. First we need some lemmas.

Lemma 8.9. If E is a normalized spectral measure and A is a bounded self-adjoint operator such that

$$A = \int \lambda \, dE,$$

then E is compact. ∎

Proof. Suppose $\sigma(A) \subset (m, M)$. Let h be any vector in H and let $k = E(-\infty, m)h$. Then

$$E(-\infty, \lambda)k = \begin{cases} E(-\infty, \lambda)k , & \text{if } \lambda < m , \\ k , & \text{if } \lambda \geq m. \end{cases}$$

Hence

$$\begin{aligned} (Ak \mid k) &= \int_{-\infty}^{\infty} \lambda \, d\mu_k , \qquad \text{where } \mu_k(M) = (E(M)k \mid k), \\ &= \int_{-\infty}^{m} \lambda \, d\mu_k \leq m \, \| E(-\infty, m)k \|^2 = m \, \| k \|^2. \end{aligned}$$

However if $k \neq 0$,

$$(Ak \mid k) \geq \inf_{\|h\|=1} (Ah \mid h)(k \mid k) > m(k \mid k),$$

so that $k = 0$. Hence $E(K) = 0$ for all $K \subset (-\infty, m)$. Similarly $E(K) = 0$ if $K \subset (M, \infty)$. Hence E is compact. ∎

Lemma 8.10. Let T be an operator in H and Q be a projection. If $QT \subset TQ$ (meaning $D_T = D_{QT} \subset D_{TQ} \equiv \{h \in H\colon\ Qh \in D_T\}$ and $QTh = TQh$ for each h in D_T), then $QT = TQ$ (meaning $D_{QT} = D_{TQ}$). ∎

Proof. Since $QT \subset TQ$, then $QTQ \subset TQ^2 = TQ$. However, since $D_{QTQ} = D_{TQ}$, $QTQ = TQ$. Inasmuch as $(I - Q)T \subset T(I - Q)$ also, it follows in the same way that $(I - Q)T(I - Q) = T(I - Q)$. Moreover

$$T = QT + (I - Q)T \subset TQ + T(I - Q) = T,$$

since if $h \in D_{TQ}$ and $h \in D_{T(I-Q)}$, then $h \in D_T$ as $h = Qh + (I - Q)h$. Since the extremes of this inequality are equal,

$$QT + (I - Q)T = TQ + T(I - Q).$$

Applying Q to both sides, one obtains

$$QT + Q(I - Q)T = QTQ + QT(I - Q)$$

or

$$QT = TQ + Q(I - Q)T(I - Q) = TQ. \quad ∎$$

Lemma 8.11. For any normalized spectral measure E such that

$$T = \int \lambda \, dE, \tag{8.38}$$

$ET = TE$, that is, $E(M)T = TE(M)$ for all M in $B(R)$. ∎

Proof. First observe that, for any M, $E(M)$ maps D_T into D_T. For, if $k \in D_T$, then

$$\int_{-\infty}^{\infty} |\lambda|^2 \, d\mu_{E(M)k} \leq \int |\lambda|^2 \, d\mu_k < \infty,$$

where $\mu_{E(M)k}(N) = (E(N)E(M)k \mid k) \leq (E(N)k \mid k) = \mu_k(N)$ for all N in

$B(R)$. Hence $E(M)k \in D_T$ by definition of D_T. Secondly, for any k in D_T, we have

$$\begin{aligned}(Tk \mid k) &= \int \lambda \, d\mu_k \\ &= \sum_{n\in Z} \int \lambda \, \chi_{[n-1,n)} \, d\mu_k \\ &= \lim_{n\to\infty} (T_0k + T_1k + T_{-1}k + \cdots + T_nk + T_{-n}k \mid k), \end{aligned} \tag{8.39}$$

where $T_n = \int \lambda\chi_{[n-1,n]} \, dE$, a bounded self-adjoint operator on H. Now

$$\| T_nk \|^2 = (T_n^2k \mid k) = \int | \lambda |^2 \chi_{[n-1,n)} \, d\mu_k ,$$

so

$$\sum_{n\in Z} \| T_nk \|^2 = \int | \lambda |^2 \, d\mu_k < \infty$$

since $k \in D_T$. Hence $\sum T_nk$ converges in H and from equation (8.39), $Tk = \sum T_nk$. Now $E(M)T_n = T_nE(M)$ since

$$E(M)T_n = \int \chi_E \cdot \lambda \cdot \chi_{[n-1,n)} \, dE = T_nE(M).$$

Since $E(M)$ is continuous, for $k \in D_T$,

$$TE(M)k = \sum_{n\in Z} T_nE(M)k = \sum_{n\in Z} E(M)T_nk = E(M)(\sum_n T_nk) = E(M)Tk.$$

∎

Now apply Lemma 8.10.

Lemma 8.12. Let P be the spectral measure corresponding to the self-adjoint operator T in H via $\hat{\varphi}$. Let A be any operator in $L(H, H)$ such that $AT = TA$. Then for any integer n, $AH_n \subset H_n$, where $H_n = P_n(H)$ and $P_n = \int \chi_{[n-1,n]} \, dP$. ∎

Proof. By definition, for any M in $B(R)$ and $f \in L_2(X, \mu)$,

$$[UP(M)U^{-1}]f(\cdot) = (UU^{-1}\tau_{\chi_M(F)}UU^{-1})f(\cdot) = \chi_M(F(\cdot))f(\cdot),$$

where F, U, and μ are given in Theorem 8.8. Let R_M be the range of $UP(M)U^{-1}$ in $L_2(X, \mu)$. First let $M = [-1, 1] \subset R$. Then

$$F^{-1}(M) = \{x \in X\colon -1 \le F(x) \le 1\}.$$

Clearly $R_M = \{f \in L_2(X, \mu): f(x) = 0 \text{ a.e. if } x \notin F^{-1}(M)\}$. It is easy to see that also

$$R_M = \{f \in L_2(\mu): \| [F(\cdot)]^n f(\cdot) \|_2 \text{ is bounded for } 1 \le n < \infty\}.$$

[If $f(x) = 0$ for $x \notin F^{-1}(M)$, then

$$\int | [F(x)]^n f(x) |^2 \, d\mu \le \int | f(x) |^2 \, d\mu.$$

Conversely, suppose $g \in L_2$, $\| F^n g \|_2$ is bounded, and $g(x) \neq 0$ for $x \in B$ for some set B of positive measure on which $| F(x) | > 1$. Then

$$\int | F(x)^n g(x) |^2 \, d\mu \to \infty \text{ as } n \to \infty.]$$

Now suppose $f \in R_M$. Then $\| F(\cdot)^n f(\cdot) \|_2$ is bounded and for $f \in U(D_T)$,

$$\begin{aligned} \| F(\cdot)^n (UAU^{-1})[f(\cdot)] \|_2 &= \| (UT^n U^{-1})(UAU^{-1}) f(\cdot) \|_2 \\ &= \| UAU^{-1} UT^n U^{-1} f(\cdot) \|_2 \le \| A \| \, \| F(\cdot)^n f(\cdot) \|_2, \end{aligned}$$

which is bounded. Hence $UAU^{-1}(R_M) \subset R_M$ when $M = [-1, 1]$

If $M = [a, b] = \{\lambda \in R: | \lambda - \lambda_0 | \le r\}$, then

$$F^{-1}(M) = \left\{ x \in X: \left| \frac{F - \lambda_0}{r} (x) \right| \le 1 \right\}.$$

Since A commutes with T, UAU^{-1} will also commute with multiplication by $(F - \lambda_0)/r$; again in this case R_M will be invariant under UAU^{-1}. If now $M = [n - 1, n)$, then $M = \bigcup_{k=1}^{\infty} M_k$ when (M_k) is an increasing sequence of closed intervals. Since $R_M = \{f \in L_2(X, \mu): f(x) = 0 \text{ a.e. if } x \notin F^{-1}(M)\}$ clearly $R_{M_1} \subset R_{M_2} \subset \cdots \subset R_M$ and by the Dominated Convergence Theorem, $\overline{[\bigcup_{k=1}^{\infty} R_{M_k}]} = R_M$. For any k and f in R_{M_k}, $UAU^{-1} f \in R_{M_k} \subset R_M$. Since R_M is closed, $UAU^{-1}(R_M) \subset R_M$ and R_M is invariant under UAU^{-1}. This means

$$UAU^{-1}[UP_n U^{-1}(L_2)] \subset UP_n U^{-1}(L_2)$$

or

$$AP_n(H) \subset P_n(H).$$ ∎

For each integer n, let us continue to let $P_n = \int \chi_{[n-1,n]} \, dP$, that is

$P_n = P([n-1, n))$, and let $H_n = P_n(H)$. Since

$$I = P(R) = \lim_{n\to\infty} [P_0 + P_1 + P_{-1} + \cdots + P_n + P_{-n}]$$

in the strong sense, for each h in H, $h = \sum_{n\in Z} h_n$, where $h_n = P_n(h)$. This means $H = \oplus H_n$, the direct sum of the orthogonal family of subspaces H_n.

As shown above, if E is a normalized spectral measure satisfying equation (8.38), then $TE = ET$. From Lemma 8.12, $E(M)(H_n) \subset H_n$ for all $M \in B(R)$ and each n. Hence each H_n is invariant under $P(M)$ and $E(M)$, $M \in B(R)$. Thereby we can define $\hat{E}_n$ and $\hat{P}_n$ by

$$\hat{E}_n(M) = E(M)|_{H_n} \quad \text{and} \quad \hat{P}_n(M) = P(M)|_{H_n}$$

{Note that for each M in $B(R)$, $\hat{P}_n(M) = P([n-1, n) \cap M)$.} Since

$$T = \int \lambda \, dE,$$

we have for $h \in H_n \cap D_T$

$$(T|_{H_n} h \mid h) = \int \lambda \, dv_h(\lambda),$$

where $v_h(M) = (E(M)h \mid h) = (\hat{E}_n(M)h \mid h)$, so that $T|_{H_n} = \int \lambda \, d\hat{E}_n$. Similarly $T|_{H_n} = \int \lambda \, d\hat{P}_n$.

Since $\hat{P}_n$ is a compact normalized spectral measure, the equality $T|_{H_n} = \int \lambda \, d\hat{P}_n$ also tells us that $T|_{H_n}$ is a bounded self-adjoint operator defined on H_n. This follows from the one-to-one correspondence established between compact spectral measures and bounded self-adjoint operators established in Theorem 8.4. Since

$$T|_{H_n} = \int \lambda \, d\hat{P}_n = \int \lambda \, d\hat{E}_n,$$

$\hat{P}_n$ and $\hat{E}_n$ are both compact normalized spectral measures (Lemma 8.9) corresponding to $T|_{H_n}$. As there can be only one such measure,

$$E|_{H_n} = \hat{E}_n = \hat{P}_n = P|_{H_n}.$$

Since $H = \oplus H_n$ and for each M, $E(M)$ and $P(M)$ are bounded operators, if $h = \sum_{i \in Z} h_i$,

$$E(M)h = \lim_{n\to\infty} E(M)\left(\sum_{i=-n}^{n} h_i\right)$$
$$= \lim_{n\to\infty} \sum_{i=-n}^{n} E(M)h_i = \lim_{n\to\infty} \sum_{i=-n}^{n} \hat{E}_i(M)h_i$$
$$= \lim_{n\to\infty} \sum_{i=-n}^{n} \hat{P}_i(M)h_i = P(M)h.$$

Hence $E = P$.

We have at last completed the proof of Theorem 8.9. ∎

Our next result summarizes our preceding discussion and proof of the uniqueness of $\hat{\varphi}$ in a nutshell.

Theorem 8.10. *Spectral Measure Version of the Spectral Theorem.* There is a one-to-one correspondence between self-adjoint operators T in H and normalized spectral measures P on $B(R)$. The correspondence is given by $T = \int \lambda \, dP$. Moreover for each real-valued measurable function f [with respect to $B(R)$] there is a unique self-adjoint operator $f(T)$ given by

$$f(T) = \int f \, dP,$$

where $T = \int \lambda \, dP$. If f is bounded, $f(T) = \hat{\varphi}(f)$, where $\hat{\varphi}$ is given by Theorem 8.9. ∎

The remaining portion of this chapter is devoted to giving some applications of the Spectral Theorem.

Knowledge of the spectral measure corresponding to an operator T can be most valuable in obtaining complete knowledge of the operator T in regard to determining its domain, the value of inner product $(Th \mid k)$ for h and k in H, the spectrum and eigenspaces of T, and operator functions $f(T)$ of T for f measurable. In Theorem 8.11 below we show how the spectrum of T is related to the spectral measure corresponding to T.

A few crucial observations should be made regarding the Spectral Theorem. First, for any measurable (real or complex valued) function f on R, the unique operator $f(T)$ satisfying equation (8.35) is defined. Its domain is the set $\{h \mid \int_{-\infty}^{\infty} |f|^2 \, d\mu_h < \infty\}$, dense in H. Secondly, for any h in $D_{f(T)}$,

$$\| f(T)h \|^2 = \int_{-\infty}^{\infty} | f(\lambda) |^2 \, d\mu_h(\lambda). \tag{8.40}$$

Equation (8.40) is easily verified for the case when f is a nonnegative measurable function. From the discussion preceding equation (8.35), we have

$$\begin{aligned}\|f(T)h\|^2 &= \lim_{n\to\infty} \|f\wedge n\cdot\chi_{[-n,n]}(T)h\|^2 \\ &= \lim_{n\to\infty} ([f\wedge n\cdot\chi_{[-n,n]}(T)]^2h \mid h) \\ &= \lim_{n\to\infty}\int_{-\infty}^{\infty} [f\wedge n\cdot\chi_{[-n,n]}(\lambda)]^2\, d\mu_h(\lambda) \\ &= \int_{-\infty}^{\infty} f^2\, d\mu_h.\end{aligned}$$

In case f is any measurable function, $f=f_1-f_2+i(g_3-g_4)$ and $|f|^2 = f_1^2+f_2^2+g_3^2+g_4^2$. In this situation equation (8.40) can easily be seen to hold.

Recall from Problem 8.1.2 that there is a one-to-one correspondence between spectral functions on R and normalized spectral measures on $B(R)$. It is analogous to the correspondence between Borel measures on R and distribution functions. Given a spectral measure P on $B(R)$, the spectral function E on R is given by $E(x)=P(-\infty, x]$. Theorem 8.10 thus implies a one-to-one correspondence between self-adjoint operators T in H and spectral functions E on R. If f is a measurable function, then for any h in H

$$\int f(\lambda)\, d\mu_h(\lambda) = \int f(\lambda)\, d(E(\lambda)h \mid h),$$

which accords with Definition 3.10 in Chapter 3.

The applications given below are more easily stated in terms of spectral functions than spectral measures.

Using equation (8.40), the following theorem relating the spectrum of a self-adjoint operator T to properties of its spectral function is obtained.

Theorem 8.11. Ler T be a self-adjoint operator in H and let E be the spectral function on R corresponding to T. Then we have:

(i) the spectrum $\sigma(T)$ is a subset of the real numbers;

(ii) λ_0 is an eigenvalue of T if and only if

$$E(\lambda_0) \neq E(\lambda_0-) \equiv \lim_{\substack{\lambda<\lambda_0\\ \lambda\to\lambda_0}} E(\lambda);$$

moreover the eigenspace of H corresponding to λ_0 is $R_{E(\lambda_0)-E(\lambda_0-)}$;

(iii) λ_0 is in the continuous spectrum if and only if $E(\lambda_0) = E(\lambda_0-)$, but $E(\lambda_1) < E(\lambda_2)$ whenever $\lambda_1 < \lambda_0 < \lambda_2$; and

(iv) the residual spectrum of T is empty. ∎

Proof. (i) has been proved in Proposition 8.12. Clearly

$$T - \lambda_0 I = \int (\lambda - \lambda_0)\, dE(\lambda)$$

and by (8.40) for h in D_T

$$\| (T - \lambda_0 I)h \|^2 = \int | \lambda - \lambda_0 |^2 \, d(E(\lambda)h \mid h). \tag{8.41}$$

Hence $Th = \lambda_0 h$ if and only if $E(\lambda_0+) = E(\lambda_0) = E(\lambda)$ for all $\lambda \geq \lambda_0$ and $E(\lambda_0-) = E(\lambda)$ for all $\lambda < \lambda_0$. [Recall $E(\lambda)h$ is right continuous in λ for each h, $E(\lambda)h \to 0$ as $\lambda \to -\infty$, and $E(\lambda)h \to h$ as $\lambda \to \infty$.] In other words, $Th = \lambda_0 h$ if and only if $h = [E(\lambda_0) - E(\lambda_0-)]h$. This verifies statement (ii).

We next verify (iv). If the residual spectrum is not empty, then there is a real number λ_0 for which $(T - \lambda_0 I)^{-1}$ exists but its domain $R_{T-\lambda_0 I}$ is not dense. This means that there is a nonzero h_0 in H that is orthogonal to $R_{T-\lambda_0 I}$; that is, $((T - \lambda_0 I)h \mid h_0) = 0$ for all h in D_T. Hence

$$(Th \mid h_0) = (\lambda_0 h \mid h_0) = (h \mid \lambda_0 h_0),$$

so that $\lambda_0 h_0$ is in D_{T^*} and $T^* h_0 = \lambda_0 h_0$. Since T is self-adjoint, $Th_0 = \lambda_0 h_0$ and λ_0 is an eigenvalue of T. Since the point spectrum and residual spectrum are disjoint, this is a contradiction.

It remains to show (iii). From (ii) and (iv), if $E(\lambda_0) = E(\lambda_0-)$ then λ_0 is either in the resolvent of T or in the continuous spectrum. Now λ_0 is in the resolvent if and only if there exists a positive constant k such that

$$\| (T - \lambda_0 I)h \| \geq k \| h \|, \quad \text{for all } h \text{ in } D_T.$$

In other words from equation (8.41) it is necessary and sufficient that

$$\int_{-\infty}^{\infty} (\lambda - \lambda_0)^2 \, d(E(\lambda)h \mid h) \geq k^2 \| h \|^2. \tag{8.42}$$

Now if there exist λ_1 and λ_2 with $\lambda_1 < \lambda_0 < \lambda_2$ such that $\lambda_0 - \lambda_1 = \lambda_2 - \lambda_0 < k$ and $E(\lambda_1) \neq E(\lambda_2)$, then

$$\int_{-\infty}^{\infty} (\lambda - \lambda_0)^2 \, d(E(\lambda)h \mid h) < k^2 \int_{-\infty}^{\infty} d(E(\lambda)h \mid h) = k^2 \| h \|^2,$$

with $h = [E(\lambda_2) - E(\lambda_1)]x$ for x in H. Since this contradicts inequality (8.42) if $E(\lambda_1) \neq E(\lambda_2)$, λ_0 is in the continuous spectrum of T. Conversely, if λ_0 is in the continuous spectrum, $E(\lambda_0-) = E(\lambda_0)$ by (ii). Moreover if there exists λ_1 and λ_2 with $\lambda_1 < \lambda_0 < \lambda_2$ and $E(\lambda_1) = E(\lambda_2)$ {implying $E(\lambda)$ is constant on $[\lambda_1, \lambda_2]$}, then the function $f(\lambda) = 1/(\lambda - \lambda_0)$ is bounded almost everywhere and $f(T) = (T - \lambda_0 I)^{-1}$ by Theorem 8.9 is the bounded inverse defined on H of $T - \lambda_0 I$. This means λ_0 is in the resolvent of T, a contradiction. Hence for all λ_1 and λ_2 with $\lambda_1 < \lambda < \lambda_2$, $E(\lambda_1) \neq E(\lambda_2)$. ▌

The following example illustrates the use of the preceding theorem.

Example 8.5. Let T be the multiplication operator in $L_2(-\infty, \infty)$ considered in Example 8.1. It is routine to show T is symmetric. To show T is self-adjoint it must be shown that $D_{T^*} \subset D_T$. Suppose $g \in D_{T^*}$. Then for every $f \in D_T$

$$\int_{-\infty}^{\infty} xf(x)\overline{g(x)}\, dx = (Tf \mid g) = (f \mid T^*g) = \int_{-\infty}^{\infty} f(x)\overline{T^*g(x)}\, dx\,,$$

whence

$$\int_{-\infty}^{\infty} f(x)[\overline{g(x)}x - \overline{T^*g(x)}]\, dx = 0.$$

Let $[a, b]$ be a finite interval and define h by $h(x) = [xg(x) - T^*g(x)]\chi_{[a,b]}$. Then $\int_{-\infty}^{\infty}[h(x)]^2\, dx = 0$ so that $h(x) = 0$ almost everywhere. Since $[a, b]$ is an arbitrary interval $xg(x) = T^*g(x)$ almost everywhere or $xg(x) = T^*g(x) \in L_2(-\infty, \infty)$. Hence $g \in D_T$.

Let E be the function on R into $L(L_2(-\infty, \infty), L_2(-\infty, \infty))$ given by

$$E(\lambda)g = g\,\chi_{(-\infty,\lambda]}.$$

E is easily checked to be a projection-valued, nondecreasing, right-continuous (in the strong sense) function with $E(\lambda) \to 0$ as $\lambda \to -\infty$ and $E(\lambda) \to I$ as $\lambda \to \infty$. Moreover, for any f in D_T and g in $L_2(-\infty, \infty)$,

$$\begin{aligned}(Tf \mid g) &= \int_{-\infty}^{\infty} xf(x)\overline{g(x)}\, dx \\ &= \int_{-\infty}^{\infty} x d\int_{-\infty}^{x} f(t)\overline{g(t)}\, dt \\ &= \int_{-\infty}^{\infty} x d\int_{-\infty}^{\infty} \overline{g(t)}E(x)[f(t)]\, dt \\ &= \int_{-\infty}^{\infty} x\, d(E(x)f \mid g).\end{aligned}$$

This means E is the unique spectral function such that $T = \int \lambda \, dE$.

What is the spectrum of T? Note that for all λ, $E(\lambda) = E(\lambda-)$ and $E(\lambda_1) < E(\lambda_2)$ if $\lambda_1 < \lambda_2$. Hence the continuous spectrum makes up the entire set of real numbers.

It is known that the position operator T of the preceding example is unitarily equivalent to the so-called momentum operator, the operator T_3 of Example 8.2. This means there exists a unitary operator F on $L_2(-\infty, \infty)$ such that

$$F(D_{T_3}) = D_T \quad \text{and} \quad T_3 = F^{-1}TF.$$

The operator F is known as the Fourier–Plancherel operator. The proof of this unitary equivalence is not trivial and not given here.† However, the spectrum of T_3 can be easily analyzed to be exactly that of T by means of the following theorem.

Theorem 8.12. If S and T are unitarily equivalent operators in a Hilbert space H, then the point spectrum, continuous spectrum, and residual spectrum of S are the same as that of T. ∎

The proof is trivial and is omitted.

Example 8.6. *Application of the Spectral Theorem in Solving the Schrödinger Equation.* An equation that occurs in quantum mechanics is the time-dependent Schrödinger equation given by

$$i \frac{du}{dt} = Au(t),$$

where $u(t)$ is an element of a Hilbert space H, A is a self-adjoint operator in H, and t is a time variable with $u(t) \in D_A$. An initial condition is $u(0) = u_0 \in D_A$. The derivative of u is given as the

$$\lim_{\Delta t \to 0} \frac{u(t + \Delta t) - u(t)}{\Delta t}$$

in the strong topology of H.

The Spectral Theorem enables us to solve the Schrödinger equation. Let e^{-itA} be the bounded operator on H given by

$$e^{-itA} = \int_{-\infty}^{\infty} e^{-it\lambda} \, dP(\lambda),$$

† The proof is given in [16], page 135.

where $A = \int \lambda\, dP$. We wish to show that

$$\frac{d}{dt}(e^{-itA}h) = -iA(e^{-itA}h), \tag{8.43}$$

for every h in D_A.

To prove this, compute the following limit:

$$\begin{aligned}
&\lim_{\Delta t\to 0} \left\| \left[\frac{e^{-i(t+\Delta t)A} - e^{-itA}}{\Delta t} + ie^{-itA}A\right]h \right\|^2 \\
&= \lim_{\Delta t\to 0} \int_{-\infty}^{\infty} \left| \frac{e^{-i(t+\Delta t)\lambda} - e^{-it\lambda}}{\Delta t} + ie^{-it\lambda}\lambda \right|^2 d(E(\lambda)h \mid h) \\
&= \lim_{\Delta t\to 0} \int_{-\infty}^{\infty} \left| \frac{e^{-i(\Delta t)\lambda} - 1}{\Delta t} + i\lambda \right|^2 d(E(\lambda)h \mid h).
\end{aligned}$$

Letting $M = \max$ of $|\,[e^{-i(\Delta t)} - 1]/\Delta t + i\,|^2$ for $\Delta t \in R$, the integrand above is bounded by $M\lambda^2$, which is integrable since $h \in D_A$. Using the Lebesgue Dominated Convergence Theorem, the limit can be taken inside to the integrand. Since the limit of the integrand is zero, the above limit is zero.

Hence, for every h in D_A,

$$\frac{d}{dt}(e^{-itA}h) = -ie^{-itA}Ah. \tag{8.44}$$

Equation (8.43) follows from (A.44) since, for h in D_A,

$$e^{-itA}Ah = Ae^{-itA}h. \tag{8.45}$$

This follows from the fact that if h is in D_A, then $e^{-itA}h$ is in D_A since by equation (8.40)

$$\begin{aligned}
\| E(M)e^{-itA}h \|^2 &= \int \chi_M \mid e^{-it\lambda} \mid^2 dE_h(\lambda) \\
&= \int \chi_M \, dE_h(\lambda) = \| E(M)h \|^2.
\end{aligned}$$

Equation (8.45) then follows by an argument similar to that after Lemma 8.4.

The solution $u(t) = e^{-itA}u_0$ of the Schrödinger equation is unique. To prove this, suppose $v(t)$ in D_A is a solution. Then for any k in H

$$\frac{d}{ds}(e^{-i(t-s)A}v(s) \mid k) = \lim_{\Delta s \to 0} \frac{(e^{-i[t-(s+\Delta s)]A}v(s+\Delta s) \mid k) - (e^{-i(t-s)A}v(s) \mid k)}{\Delta s}$$

$$= \lim_{\Delta s \to 0}\left(\frac{e^{-i[t-(s+\Delta s)]A} - e^{-i(t-s)A}}{\Delta s} v(s+\Delta s) \,\middle|\, k\right)$$

$$+ \lim_{\Delta s \to 0}\left(e^{-i(t-s)A}\frac{v(s+\Delta s) - v(s)}{\Delta s} \,\middle|\, k\right)$$

$$= \left(-\frac{d}{dt}e^{-i(t-s)A}v(s) \,\middle|\, k\right) + \left(e^{-i(t-s)A}\frac{dv}{ds} \,\middle|\, k\right)$$

$$= (ie^{-i(t-s)A}Av(s) \mid k) + (e^{-i(t-s)A}[-iAv(s)] \mid k) = 0.$$

Hence, for all k in H,

$$0 = \int_0^t \frac{d}{ds}(e^{-i(t-s)A}v(s) \mid k)\,ds = (e^{-i0A}v(t) \mid k) - (e^{-itA}v(0) \mid k),$$

and since $v(0) = u_0$ and $e^{-i0A} = I$, we have

$$v(t) = e^{-itA}u_0.$$

Uniqueness is thus proved.

Problems

8.2.1. Let H be an infinite-dimensional Hilbert space with orthonormal basis $(x_i)_N$. Define $T: D_T \subset H \to H$ by

$$T\left(\sum_{k=1}^{\infty}(x \mid x_k)x_k\right) = \sum_{k=1}^{\infty}(x \mid x_k)kx_k,$$

where

$$D_T = \left\{x \in H: \sum_{k=1}^{\infty} |(x \mid x_k)k|^2 < \infty\right\}.$$

(i) Show that T is a closed linear operator but T is not bounded on D_T.

(ii) Prove that T is self-adjoint.

8.2.2. Prove that a closed and unbounded linear operator T in a Hilbert space H cannot have $D_T = H$.

8.2.3. Prove that every self-adjoint operator T is a maximal symmetric operator; that is, there is no symmetric operator extending T with a larger domain.

8.2.4. Let T and S be operators in H with domains D_T and D_S, respectively. Suppose U is a unitary operator on H such that $U(D_S) = D_T$ and $S = U^{-1}TU$. (S is said to be *unitarily equivalent* to T.) Prove the following:

(i) T is bounded if and only if S is bounded.
(ii) T is symmetric if and only if S is symmetric.
(iii) T is self-adjoint if and only if S is self-adjoint.

8.2.5. Let $M = (m_{ij})$ be a matrix with rows and columns in $l_2(N)$. Let T be the operator in $l_2(N)$ represented by M in the sense that $Tf = \sum_{i=1}^{\infty}(\sum_{j=1}^{\infty} m_{ij}(f \mid e_j))e_i$ for all f in the domain of $f \equiv \{f \in l_2(N) \mid Tf \in l_2(N)\}$ [Here $e_i(j) = \delta_{ij}$.] Show that

(i) T is a closed operator with dense domain;
(ii) if $N = (n_{ij})$, where $n_{ij} = \bar{m}_{ji}$ and S is the operator represented by N with domain the set $\{f \in l_2(N) \mid Sf \in l_2\}$, then S extends T;
(iii) S is not necessarily the adjoint of T.

8.2.6. Let T be a symmetric operator in a Hilbert space H. Prove the following statements are equivalent:

(i) T is self-adjoint.
(ii) T is closed and the null space of $T \pm i$ is $\{0\}$.
(iii) $R(T \pm i) = H$. [Hint: Ker $(T^* \pm iI) =$ [Range $(T \pm iI)]^{\perp}$.]

8.2.7. An operator in H is said to be *essentially self-adjoint* if its closure is self-adjoint. If T is a symmetric operator in H, prove the following are equivalent:

(i) T is essentially self-adjoint;
(ii) T^* is self-adjoint;
(iii) $\bar{T} = T^*$;
(iv) $\text{Ker}(T^* \pm i) = \{0\}$;
(v) $\overline{R(T \pm i)} = H$.

8.2.8. Let M be a closed subspace of H and let P be the projection operator in H onto M. Prove M and M *completely reduce* T [that is, $T(D_T \cap M) \subset M$, $T(D_T \cap M^{\perp}) \subset M^{\perp}$, and $P(D_T) \subset D_T$] if and only if $P(D_T) \subset D_T$ and $PT \subset TP$. If $D_T = H$, then T is completely reduced if and only if $PT = TP$.

8.2.9. If T is a symmetric operator in H and its spectrum consists only of real values, then prove T is either self-adjoint or essentially self-adjoint.

8.2.10. Let T be a symmetric operator in T. Prove the following:

(i) All points in the point spectrum and the continuous spectrum are real whereas the residual spectrum may contain nonreal values.

(ii) If nonreal $\lambda = a + bi$ is in the residual spectrum so is $c + di$, where d has the same sign as b.

8.2.11. Let T be a symmetric operator in H. Prove the following:

(i) the eigenvectors corresponding to distinct eigenvalues are orthogonal;

(ii) the closure $\bar{M}_\lambda$ of $M_\lambda = \{x \in H: Tx = \lambda x\}$ completely reduces T.

8.2.12. Prove that if T is a self-adjoint operator in H and P is the normalized spectral measure corresponding to T via $T = \int \lambda\, dP$, then for each $M \in B(R)$, $P(M)H$ and $(I - P)(M)H$ completely reduce T.

8.2.13. Let T be a closed linear operator in Hilbert space H with $\bar{D}_T = H$ (e.g., a self-adjoint operator in H). Prove the following:

(i) $\lambda \in \varrho(T)$ if and only if $\bar{\lambda} \in \varrho(T^*)$.

(ii) If $\lambda \in C\sigma(T)$ then $\bar{\lambda} \in C\sigma(T^*)$.

(iii) If $\lambda \in R\sigma(T)$ then $\bar{\lambda} \in P\sigma(T^*)$.

(iv) If $\lambda \in P\sigma(T)$ then $\bar{\lambda} \in P\sigma(T^*) \cup R\sigma(T^*)$.

8.2.14. Let T be an unbounded self-adjoint operator in H. Prove the following:

(i) $\lambda \in \varrho(T)$ if $R(T - \lambda I) = H$.

(ii) $\lambda \in C\sigma(T)$ if $R(T - \lambda I) \neq H$ but $\overline{R(T - \lambda I)} = H$.

(iii) $\lambda \in P\sigma(T)$ if $\overline{R(T - \lambda I)} \neq H$.

(iv) $R\sigma(T) = \varnothing$.

Appendix

C. Invariant and Hyperinvariant Subspaces

One of the best-known unsolved problems in functional analysis is the invariant subspace problem: does every bounded linear operator on a separable, infinite-dimensional, complex Hilbert space[†] have a nontrivial closed invariant subspace? The subject of invariant subspaces is very broad and many special cases of this problem have been solved. In this appendix we will restrict ourselves to two main theorems—one for compact operators due to Lomonosov and one for normal operators which is a natural application of the spectral theorems for normal operators. Both of these theorems not only give affirmative partial answers to the invariant subspace problem but also give somewhat stronger conclusions. The reader who wishes to delve more deeply into this problem may consult some of the references listed at the close of the appendix.

The following definition makes precise the concepts with which we deal in this section. From this point H will represent a complex Hilbert space and T a bounded linear operator on H. T is termed *nonscalar* if it is not a scalar multiple of the identity operator.

Definition C.1. A closed linear subspace M of H is said to be *invariant* under the operator T if $T(M) \subset M$. M is *reducing* for T if $T(M) \subset M$ and $T(M^{\perp}) \subset M^{\perp}$. M is *nontrivial* if $M \neq \{0\}$ and $M \neq H$. ∎

Note that when we speak of an invariant or reducing subspace we assume the subspace is closed.

[†] Recently C. J. Read (*Bull. Math. Soc. London* **16**, 337–401 (1984)) has exhibited a bounded linear operator, on a separable infinite-dimensional Banach space, which has no nontrivial closed invariant subspace.

The following remarks are easy to verify.

Remark C.1. (i) M is invariant for T if and only if $M^{\perp}$ is invariant for T^*.

(ii) M is reducing for T if and only if M is invariant under both T and T^*.

The basic motivation for the study of invariant subspaces stems from interest in knowing the structure of operators. If M is invariant under T, then T can be written with respect to the decomposition $M \oplus M^{\perp}$ as an operator matrix

$$\begin{pmatrix} T_{11} & T_{12} \\ 0 & T_{22} \end{pmatrix},$$

where $T_{ij} = P_i T P_j$ for $i, j = 1, 2$; P_1 is the projection onto M; and P_2 is the projection onto $M^{\perp}$. T_{12} will be the zero operator if and only if in addition $T(M^{\perp}) \subset M^{\perp}$, in which case T can be represented as a "diagonal" matrix of operators. Further knowledge of T may result if T_{11} and T_{12} can be similarly further reduced or if properties of T can be deduced from knowledge of the operators T_{ij}, such as knowledge of the spectrum.

If $\dim H < \infty$, H can be decomposed into a direct sum of invariant subspaces on each of which T acts in a simple manner (Jordan canonical form†). Likewise, a self-adjoint operator can be written as a direct sum of restrictions to invariant subspaces by means of the spectral theorem (see Problem 8.2.12). Interest in the invariant subspace problem stems partly from a desire to so decompose general operators on infinite-dimensional spaces.

Before we get into the main theorems of this section, let us explain the reason for some of the restrictions—infinite dimensional, complex, separable, closed, complete, and bounded—listed with the invariant subspace problem. We do this with the following remarks and three subsequent propositions.

Remarks

C.2. Nonzero linear operators (necessarily continuous) on finite-dimensional ($\dim > 1$) complex vector spaces always have nontrivial invariant subspaces—namely, the eigenspace corresponding to a nonzero eigenvalue, whose existence is guarenteed by the Fundamental Theorem of Algebra. This method of generating invariant subspaces does not work in the infinite-dimensional situation, however, since some operators have no

† See, for example, K. Hoffman, and R. Kunze, *Linear Algebra*, Prentice-Hall, Englewood Cliffs, New Jersey (1971).

nonzero eigenvalues. One may consider for example the shift operator T on $l_2((x_1, x_2, \ldots) \to (0, x_1, x_2, \ldots))$ or the operator V on $L_2[0, 1]$ given by $(Vf)(x) = \int_0^x f(t)\, dt$.

C.3. Bounded linear operators on real Hilbert spaces may not have any nontrivial invariant subspaces as a rotation in the plane exemplifies.

C.4. A nonzero linear operator A (bounded or unbounded) on a nonseparable normed linear space X always has nontrivial (closed) invariant subspaces, namely, the closed linear span of $\{A^n x \mid n = 0, 1, 2, \ldots\}$ for some $x \neq 0$ in X, which is separable and, therefore, proper.

The next result is due to Schaefer [35].

Proposition C.1. Nonzero linear operators on real or complex infinite-dimensional vector spaces always have proper (not necessarily closed) invariant subspaces. ∎

Proof. Let A be a nonzero linear operator on vector space V. It suffices to prove that an infinite-dimensional subspace W with basis $\{x_0, Ax_0, A^2x_0, \ldots\}$ has a proper invariant subspace. Define the linear function L on W by

$$L\Big(\sum_{i=0}^{n} \alpha_i A^i x_0\Big) = \sum_{i=0}^{n} \alpha_i.$$

Let S be the subspace $\{w \in W \mid L(w) = 0\}$. Then $\{0\} \subsetneqq S \subsetneqq W \subseteq V$ and S is invariant under A since $L(Aw) = L\omega$ for $w \in S$. ∎

In contrast, the following result is true if we insist on closed subspaces.

Proposition C.2. A bounded linear operator on a complex pre-Hilbert space may not have a nontrivial (closed) invariant subspace. ∎

Proof. Let V be the pre-Hilbert space of all polynomials on $[0, 1]$ with complex coefficients and with inner product $(p \mid q) = \int_0^1 p\bar{q}\, dt$. Let $M\colon V \to V$ be given by $(Mp)(t) = tp(t)$. Suppose S is a closed subspace of V which is invariant under M and has a nonzero element p. Let $\varepsilon > 0$ and let $q \in V$. If K is the closed finite set $\{x \in [0, 1] \mid p(x) = 0\}$, let U be an open set containing K with measure less than ε. By Urysohn's Lemma (Lemma 1.1) there exists a continuous function f such that $|f| \leq 1, f \equiv 1$ on K, and $f \equiv 0$ on $[0, 1] - U$. Applying the Stone–Weierstrass Theorem to $C_1([0, 1] - U)$ there exists a polynomial q_ε such that

$$\frac{\| q - (p+f)q_\varepsilon \|_\infty}{N+1} \leq \left\| \frac{q - (p+f)q_\varepsilon}{p+f} \right\|_\infty = \left\| \frac{q}{p+f} - q_\varepsilon \right\|_\infty < \frac{\varepsilon}{n+1},$$

where $N \geq \| p \|_\infty$ and $\| \quad \|_\infty$ is the sup norm over [0, 1]. This means

$$\begin{aligned}
\| q - pq_\varepsilon \|_2 &\leq \| q - (p+f)q_\varepsilon \|_2 + \| fq_\varepsilon \|_2 \\
&\leq \left[\int_U | q - (p+f)q_\varepsilon |^2 \, dm \right]^{1/2} \\
&\quad + \left[\int_{U^c} | q - (p+f)q_\varepsilon |^2 \, dm \right]^{1/2} + \left(\int_U | fq_\varepsilon |^2 \, dm \right)^{1/2} \\
&\leq \| q - (p+f)q_\varepsilon \|_\infty m(U) + \varepsilon + \| fq_\varepsilon \|_\infty m(U) \\
&\leq \varepsilon(\| q - (p+f)q_\varepsilon \|_\infty + \| fq_\varepsilon \|_\infty + 1) .
\end{aligned}$$

Consequently $q \in S$ since S is closed and $pq_\varepsilon \in S$ as S is invariant. This means $S = V$. ∎

The next proposition is due to Shields [37].

Proposition C.3. There is a linear transformation (not necessarily bounded) on a Hilbert space which has no nontrivial (closed) invariant subspaces. ∎

Proof. Let H be a (separable) infinite-dimensional Hilbert space. H then has c proper closed infinite-dimensional subspaces. Letting ω_c be the first ordinal number with cardinality c, there exists a one-to-one correspondence $\alpha \leftrightarrow M_\alpha$ between predecessors α of ω_c and the proper closed infinite-dimensional subspaces of H.

For each $\alpha < \omega_c$, let S_α be the statement: for each $\gamma \leq \alpha$, there exists a pair (f_γ, g_γ) with $f_\gamma \in M_\gamma$ and $g_\gamma \notin M_\gamma$ and $\{f_\gamma, g_\gamma \mid \gamma \leq \alpha\}$ is linearly independent. Clearly S_1 is valid. Let us assume S_β is valid for all $\beta < \alpha$. Let $V_\alpha = \operatorname{span}\{f_\beta, g_\beta \mid \beta < \alpha\}$. Since the algebraic dimension of V_α is less than c and the algebraic dimension of M_α is not less than c (see Problem 6.1.7), there is a vector $f_\alpha \in M_\alpha - V_\alpha$. The vector subspace V_α is not all of H and therefore contains no nonempty open set. Consequently $M_\alpha \cup V_\alpha \subsetneqq H$ and there exists a vector $g_\alpha \in H - (M_\alpha \cup V_\alpha)$. Consequently S_α is true. By transfinite induction S_α is true for all $\alpha < \omega_c$ so that there is a linearly independent set $\mathscr{S} = \{f_\alpha, g_\alpha \mid \alpha < \omega_c\}$ such that $f_\alpha \in M_\alpha$ and $g_\alpha \notin M_\alpha$ for each α.

Extend $\mathscr{S}$ to an algebraic basis $\mathscr{S} \cup \mathscr{E}$ of H. If T is defined on $\mathscr{S}$ by setting $T(f_\alpha) = g_\alpha$ and $T(g_\alpha) = f_{\alpha+1}$, then no matter how T is defined on $\mathscr{E}$, T will have no infinite-dimensional (closed) invariant subspaces because $Tf_\alpha = g_\alpha \notin M_\alpha$ for all $\alpha < \omega_c$. If $\mathscr{E}$ is finite, say $\{h_1, h_2, \ldots, h_n\}$, define

$Th_i = h_{i+1}$ for $i = 1, 2, \ldots, n-1$ and $Th_n = f_1$. If $\mathscr{C}$ is infinite, $\mathscr{C}$ can be well ordered and each element of $\mathscr{C}$ can be mapped by T to its successor. In either case when T is extended by linearity to H, T has no finite-dimensional invariant subspaces either since every linear combination of elements in $\mathscr{S} \cup \mathscr{C}$ is mapped onto a linear combination involving at least one new basis element. ∎

Even though the invariant subspace question is still unanswered, the analogous question involving reducing subspaces is settled. The following example gives an example of a bounded linear operator on a separable Hilbert space that has invariant subspaces (see [30, p. 45]) but no reducing subspaces.

Example C.1. The unilateral shift T on l_2 has no nontrivial reducing subspaces. T is given by $T(x_1, x_2, \ldots) = (0, x_1, x_2, \ldots)$ or by the rule $T(\sum_{i=1}^{\infty} x_i e_i) = \sum_{i=1}^{\infty} x_i e_{i+1}$. (The jth component of e_i is δ_{ij}.) Let $x \in l_2$, $x \neq 0$, $x = \sum_{i=1}^{\infty} x_i e_i$. Let n_0 be the smallest i such that $x_i \neq 0$. Let $y = T^{n_0+1}[(T^*)^{n_0+1}(x)]$. (Recall $T^* = S$ where $S(x_1, x_2, \ldots) = (x_2, x_3, \ldots)$ by Problem 7.4.1.) Thus $y_k = x_k$ for $k \neq n_0$ but $y_{n0} = 0$. Let M be the smallest subspace containing x that is invariant for T and T^*. Then $y \in M$ and so $x - y = (0, 0, 0, \ldots, 0, x_{n_0}, 0, 0, \ldots) \in M$. By applying T and T^* repeatedly to e_{n_0}, M is seen to contain all the basis vectors e_n for $n \geq 1$. Hence $M = H$. ∎

The following proposition gives some essential facts for the study of invariant and reducing subspaces.

Proposition C.4. Suppose $T \in L(H, H)$ and P is the projection onto the closed subspace M of H. Then (i) M is invariant under T if and only if $TP = PTP$.

(ii) M is reducing for T if and only if $TP = PT$. ∎

Proof. (i) If M is invariant under T and $h \in M$ then $TPh \in T(M) \subset M$ and so $PTPh = TPh$. Conversely, if $TP = PTP$ and if $h \in M$, then $Ph = h$ so that $Th = P(Th)$. Consequently, $Th \in M$ and M is invariant under T.

(ii) By definition, M is reducing for T if and only if M and $M^{\perp}$ are invariant under T. By (i) M is reducing if and only if

$$TP = PTP \quad \text{and} \quad T(I - P) = (I - P)T(I - P),$$

which is equivalent to

$$TP = PTP \quad \text{and} \quad 0 = -PT + PTP,$$

which is equivalent to $TP = PT$. ∎

A concept related to the notion of invariance is the concept of hyperinvariance.

Definition C.2. A closed subspace M of H is said to be *hyperinvariant* under the operator T if M is invariant under every operator S for which $ST = TS$. ∎

The question of whether every bounded linear operator on a Hilbert space has a hyperinvariant subspace is also unsolved. This question seems to be as difficult for nonseparable spaces as for separable.

The next theorem shows that not only do nonzero compact operators have invariant subspaces (a result discovered independently by J. von Neumann (unpublished) and N. Aronszajn and K. T. Smith [2]) but also hyperinvariant subspaces. This remarkable result was proved by V. Lomonosov [21] by using Schauder's fixed point theorem to produce invariant subspaces. The proof presented here of Lomonosov's result is due to H. M. Hilden and does not require any fixed point theorem. It can also be found in [23]. ∎

Theorem C.1 (*Lomonosov*).† Every nonzero compact operator K on a complex Banach space X has a nontrivial hyperinvariant subspace.

Proof. We can assume $\sigma(K) = \{0\}$. Otherwise X is finite dimensional [if $0 \notin \sigma(K)$] or K has a nonzero eigenvalue (see Theorem 6.25). In either case K has a hyperinvariant subspace, namely, $\{x \mid Kx = \lambda x\}$ for some nonzero eigenvalue λ. Therefore assuming $\sigma(K) = \{0\}$, the spectral radius formula (Proposition 6.27) implies that $\| (cK)^n \| \to 0$ as $n \to \infty$ for every complex number c.

We can also assume that $\| K \| = 1$. Let $x_0 \in X$ with $\| Kx_0 \| > 1$. Let $\overline{S_1(x_0)} = \{x \mid \| x - x_0 \| \leq 1\}$. Since $1 < \| Kx_0 \| \leq \| x_0 \|$, $0 \in S_1$ and $0 \notin \overline{K(\bar{S}_1)}$, a compact set.

† The result stated here is actually a special case of a more general result: If T and K are commuting operators on X with K compact and nonzero and T nonscalar, then T has a hyperinvariant subspace.

For each $y \in X$ define M_y to be the linear subspace given by

$$M_y = \{A(u): A \in L(X, X) \text{ and } AK = KA\}.$$

M_y is invariant under all operators which commute with K. Hence its closure is a hyperinvariant subspace under K. Since $M_y \neq \{0\}$ unless $y = 0$, it remains only to show that for some $y \neq 0$, $\bar{M}_y \neq H$.

Suppose on the contrary that for each $y \neq 0$, M_y is dense in X. Then for each $y \neq 0$, there is an operator A such that $AK = AK$ and $\| Ay - x_0 \| < 1$. Let $O_A = \{y: \| Ay - x_0 \| < 1\}$. O_A is open since if $y \in O_A$ with $A \neq 0$ and $\| x - y \| < (1 - \| Ay - x_0 \|)/\| A \|$ then $\| Ax - x_0 \| \leq \| Ax - Ay \| + \| Ay - x_0 \| < 1$. Consequently

$$\overline{K(\bar{S}_1)} \subset \bigcup_{A \in L(X,X)} O_A,$$

and by compactness there is a finite set $\{A_1, A_2, \ldots, A_n\}$ of operators that commute with K such that

$$K(\bar{S}_1) \subset \bigcup_{i=1}^{n} O_{A_i}.$$

Since $K(x_0) \in K(S_1(x_0))$, $K(x_0) \in O_{A_{i_1}}$ for some i_1. This means $A_{i_1}(K(x_0)) \in S_1(x_0)$. Then $KA_{i_1}K(x_0) \in K(S_1(x_0))$. This means for some i_2, $KA_{i_1}K(x_0) \in O_{A_{i_2}}$ or $A_{i_2}KA_{i_1}K(x_0) \in S_1(x_0)$. Continuing in this manner m times, we get

$$A_{i_m}KA_{i_{m-1}}K \cdots A_{i_1}K(x_0) \in S_1(x_0)$$

for some $i_1, i_2, \ldots, i_m$ from $\{1, 2, \ldots, n\}$. Let $c = \max\{\| A_i \|: i = 1, \ldots, n\}$. Since each A_i commutes with K,

$$(c^{-1}A_{i_m})(c^{-1}A_{i_{m-1}}) \cdots (c^{-1}A_{i_1})(cK)^m(x_0) \in S_1(x_0).$$

However $\| c^{-1}A_{i_j} \| \leq 1$ and $\| (cK)^m \| \to 0$ as $m \to \infty$, so that the sequence

$$((c^{-1}A_{i_m})(c^{-1}A_{i_{m-1}}) \cdots (c^{-1}A_{i_1})(cK)^m(x_0))_{m \in N}$$

converges to 0. But $0 \notin S_1(x_0)$, a closed set. This contradiction means M_y is not dense in X for some $y \neq 0$ and $\bar{M}_y$ is a nontrivial invariant subspace under K. ∎

Our final goal is to prove that every bounded normal operator on a Hilbert space H has a nontrivial invariant subspace. Actually this is a consequence of the Multiplication Operator Form of the Spectral Theorem for Normal Operators (Corollary 8.2) since subspaces of L_2 functions that vanish on a fixed set of positive measure are invariant for multiplication operators on L_2. We will, however, prove much more—that every nonscalar bounded normal operator has a nontrivial hyperinvariant subspace. To prove this we will use the Spectral Measure Version of the Spectral Theorem for Normal Operators (Theorem 8.7).

First we prove a preliminary proposition and lemma.

Proposition C.5. Suppose T is a bounded normal operator on the separable Hilbert space H and $T = \int \lambda \chi_{\sigma(T)}(\lambda)\, dP$ as in Theorem 8.7 (that is, $(Th \mid h) = \int \lambda \chi_{\sigma(T)}\, d\mu_h$ for each $h \in H$, where $\mu_h(M) = (P(M)h \mid h)$ for all M in $B(\mathbb{R}^2)$ and P is a normalized compact spectral measure on $B(\mathbb{R}^2)$). Then $P(M)H$ is reducing for T and $\sigma(T_{|P(M)H}) \subset \bar{M}$ for each M in $B(\mathbb{R}^2)$. ∎

Proof. $P(M) = \int \chi_M\, dP$ for each M in $B(\mathbb{R}^2)$ by Remark 8.2(iv). Therefore by Remark 8.2(v), $P(M)$ commutes with T and consequently $P(M)H$ is reducing for T by Proposition C.4.

To prove $\sigma(T_{|P(M)H}) \subset \bar{M}$, by Corollary 8.2 we can assume that T is a linear operator on $L_2(X, \mu)$ for some X and μ and that $T(f(x)) = G(x)f(x)$ for some bounded complex function G on X. Then $T_{|P(M)H}(f) = G\chi_{G^{-1}(M)}f$ (see proof of Theorem 8.7) so that $\sigma(T_{|P(M)H})$ is the essential range of G restricted to $G^{-1}(M)$ (see Example 6.17), which is contained in $\bar{M}$. ∎

Lemma C.1. If P is a spectral measure on $B(\mathbb{R}^2)$, the Borel subsets of $\mathbb{R}^2$, then for each M in $B(\mathbb{R}^2)$, for each $\varepsilon > 0$, and for each h in H, there exists a compact set $K \subset M$ such that $\| (P(M) - P(K))h \| < \varepsilon$. ∎

Proof. The measure on $B(\mathbb{R}^2)$ given by $\mu_k(M) = (P(M)h \mid h)$ is regular by Theorem 5.2. Therefore there exists a compact set $K \subset M$ such that $0 \leq \mu_h(M) - \mu_h(K) < \varepsilon$. Therefore

$$\| (P(M) - P(K))h \| = ([P(M) - P(K)]h \mid h) < \varepsilon. \qquad ∎$$

The following theorem will enable us to prove that every nonscalar normal operator has a nontrivial hyperinvariant subspace. It was proved by B. Fuglede† in 1950.

† A commutativity theorem for normal operators, *Proc. Nat. Acad. Sci. U.S.A.* **36**, 35–40 (1950).

Theorem C.2 (*Fuglede*). If T is a bounded normal operator on the separable Hilbert space H, $T = \int \lambda \chi_{\sigma(T)}\, dP$, and S is any bounded linear operator that commutes with T, then S commutes with $P(M)$ for every Borel set M [equivalently, $P(M)H$ is reducing for S for every Borel set M].

Proof. To show $P(M)H$ is reducing for S, it suffices to show that both $P(M)H$ and $P(M^c)H$ are invariant under S. For this it suffices to prove $P(M)H$ is invariant under S or equivalently that $P(M^c)SP(M) = 0$. Using the preceding, it suffices to prove that for compact subsets K_1 and K_2 of M^c and M, respectively, $P(K_1)SP(K_2) = 0$.

Let K_1 and K_2 be two such compact sets. Since $ST = TS$ and $P(M)H$ is reducing for T for each M in $B(\mathbb{R}^2)$, we have

$$\begin{aligned}
[P(K_1)TP(K_1)][P(K_1)SP(K_2)] &= P(K_1)TP(K_1)SP(K_2) \\
&= P(K_1)T[I - P(K_1^c)]SP(K_2) \\
&= P(K_1)TSP(K_2) - P(K_1)TP(K_1^c)SP(K_2) \\
&= P(K_1)TSP(K_2) = P(K_1)STP(K_2) \\
&= [P(K_1)SP(K_2)][P(K_2)TP(K_2)].
\end{aligned}$$

Since $\sigma(P(K_1)SP(K_1)) \cap \sigma(P(K_2)SP(K_2)) \subset K_1 \cap K_2 = \varnothing$ by Proposition C.5, we have by Problem 7.4.32 that $P(K_1)SP(K_2) = 0$. ∎

Corollary C.1. Every nonscalar normal operator has a nontrivial hyperinvariant subspace. ∎

Proof. If T is a nonscalar normal operator then $\sigma(T)$ contains at least two distinct points p_1 and p_2. [Otherwise, if we write T as $T(f) = Gf$ as in Corollary 8.2, and the essential range of G (the spectrum of T) contains only one point, then T is a multiple of the identity.] Let P be the compact normalized spectral measure corresponding to T and let

$$D = \{c \in \mathbb{R}^2 : |c - p_1| < \tfrac{1}{2}|p_1 - p_2|\}.$$

Then $\sigma(T_{|P(D)})$ and $\sigma(T_{|P(D^c)})$ are proper subsets of $\sigma(T)$ by Proposition C.5. Since $P(D)$ commutes with every operator that commutes with T, $P(D)H$ is hyperinvariant under T. ∎

The following corollary is actually true for any bounded linear operator T on the Hilbert space H (see [30, p. 32]). The proof, however, depends on developing a functional calculus for general operators as T. Nevertheless, with Fuglede's Theorem we can prove the result for normal operators.

Corollary C.2. If T is a bounded normal operator with disconnected spectrum, then T has a complementary pair of nontrivial hyperinvariant subspaces. ∎

Proof. If $\sigma(T)$ is disconnected, then since $\sigma(T)$ is closed there exist disjoint nonempty closed subsets C_1 and C_2 such that $\sigma(T) = C_1 \cup C_2$. Letting P be the normalized compact spectral measure corresponding to T we have $P([\sigma(T)^c]) = 0$ (see proof of Theorem 8.7), $I = P(\sigma(T)) = P(C_1) + P(C_2)$, and $H = P(C_1)H \oplus P(C_2)H$. Using Proposition C.5, $P(C_1)H$ and $P(C_2)H$ are reducing under T and $\sigma(T_{|P(C_i)H}) \subset C_i$ for $i = 1, 2$. Consequently, neither $P(C_1)H$ nor $P(C_2)H$ is a trivial subspace, and accordingly thus are a nontrivial complementary pair of hyperinvariant subspaces under T. ∎

There are many other results giving partial answers to the invariant subspace problem. For example, in [7] the following result is proved:

> If $T \in L(H, H)$ is *subnormal* (that is, there exists a Hilbert space $H_1 \supset H$ and a normal operator $S \in L(H_1, H_1)$ such that $T(x) = S(x)$ for all $x \in H$) then T has an invariant subspace.

Likewise, in reference [38] the following result appears:

> If $T \in L(H, H)$ is *hyponormal* (that is, $(T^*T - TT^* \geq 0)$ and if $\partial D \subset \sigma(T) \subset \bar{D}$, where D is the open unit disk in $\mathbb{R}^2$, then T has an invariant subspace.

In regard to generalizing Lomonosov's Theorem, Kim, Pearcy, and Shields [18] proved the following result:

> If T is a nonscalar operator on a Banach space and if $\dim(TK - KT)(X) \leq 1$ for some nonzero compact operator K, then T has a hyperinvariant subspace.

The interested reader who wished to pursue further study into the subspace problem should begin by consulting references [27], [28], [30], [38], [34], and other references cited in these works.

Bibliography

1. Akhiezer, N. I., and Glazman, I. M., *Theory of Linear Operators in Hilbert Space*, Frederick Ungar, New York (1961, 1963).
2. Aronszajn, N., and Smith, K. T., Invariant subspaces of completely continuous operators, *Ann. Math.* **60**, 345–350 (1954).
3. Bachman, G., and Narici, L., *Functional Analysis*, Academic, New York (1964).
4. Banach, S., *Théorie des Opérations Linéaires*, Monografje Matematyczne, Warsaw (1932).
5. Berberian, S. K., *Introduction to Hilbert Space*, Oxford University Press, New York (1961).
6. Berberian, S. K., *Notes on Spectral Theory*, Van Nostrand, Princeton, New Jersey (1966).
7. Brown, A. L., and Page, A., *Elements of Functional Analysis*, Van Nostrand Reinhold Co., London (1970).
8. Carleson, L., On the convergence and growth of partial sums of Fourier series, *Acta Math.* **116**, 135–157 (1966).
9. Davie, A. M., The Banach approximation problem, *J. Approx. Theory* **13**, 392–394 (1975).
10. Davie, A. M., The approximation problem for Banach spaces, *Bull. London Math. Soc.* **5**, 261–266 (1973).
11. Dunford, N., and Schwartz, J., *Linear Operators. Part I: General Theory*, Wiley (Interscience), New York (1958).
12. Dunford, N., and Schwartz, J., *Linear Operators. Part II*; Wiley (Interscience), New York (1964).
13. Fano, G., *Mathematical Methods of Quantum Mechanics*, McGraw-Hill, New York (1971).
14. Halmos, P. R., *A Hilbert Space Problem Book*, Van Nostrand, Princeton, New Jersey (1967).
15. Halmos, P. R., *Introduction to Hilbert Space and the Theory of Spectral Multiplicity*, Chelsea Publishing Co., New York (1951).
16. Helmberg, G., *Introduction to Spectral Theory in Hilbert Space*, North-Holland, Amsterdam (1969).

17. Hunt, R. A., On the convergence of Fourier series, pp. 235–255 in *Orthogonal Expansions and Their Continuous Analogs*, Southern Illinois University Press, Carbondale, Illinois (1968).
18. Kim, H. W., Pearcy, C., and Shields, A. L., Rank one commutators and hyperinvariant subspaces, *Mich. Math. J.* **22**, 193–194 (1975).
19. Lindenstrauss, J., and Tzafriri, L., *Classical Banach Spaces I: Sequence Space*, Springer-Verlag, Berlin (1977).
20. Liusternik, L. A., and Sobolev, V. J., *Elements of Functional Analysis*, Fredrick Ungar, New York (1961).
21. Lomonosov, V., On invariant subspaces of families of operators commuting with a completely continuous operator, *Funktsion. Anal. Prilozh.* **7**, 55–56 (1973). (Russian)
22. Lorch, E. R., On certain implications which characterize Hilbert space, *Ann. Math.* **49**, 523–532 (1948).
23. Michaels, A. J., Hilden's simple proof of Lomonosov's Invariant Subspace Theorem, *Adv. Math.* **25**, 56–58 (1977).
24. Naimark, M. A., *Normed Rings* (translated from Russian), P. Noordhoof, Groningen, The Netherlands (1964).
25. Nelson, E., *Topics in Dynamics I: Flows*, Princeton University Press, Princeton, New Jersey (1969).
26. Naylor, A. W., and Sell, G. R., *Linear Operator Theory in Engineering and Science*, Holt, Rinehart and Winston, New York (1971).
27. Pearcy, C., *Some Recent Developments in Operator Theory*, CBMS Regional Conference Series in Mathematics No. 36, American Mathematical Society, Providence, Rhode Island (1978).
28. Pearcy, C., and Shields, A. L., *A Survey of the Lomonosov Technique in the Theory of Invariant Subspaces* (Topics in Operator Theory), American Mathematical Society Surveys No. 13, Providence, Rhode Island (1974), pp. 219–229.
29. Prugovečki, E., *Quantum Mechanics in Hilbert Space*, Academic, New York (1971).
30. Radjavi, H., and Rosenthal, P., *Invariant Subspaces*, Springer Verlag, New York (1973).
31. Reed, M., and Simon, B., *Functional Analysis*, Academic, New York (1972).
32. Riesz, F., and Sz-Nagy, B., *Functional Analysis*, Fredrick Ungar, New York (1955).
33. Roman, P., *Some Modern Mathematics for Physicists and Other Outsiders*, Vol. 2, Pergamon, New York (1975).
34. Sarason, D., *Invariant Subspaces*, Mathematical Survey No. 13, American Mathematical Society, Providence, Rhode Island (1974).
35. Schaefer, H. H., Eine Bemerkung zur Existenz invarianter Teilräume linearer Abbildungen, *Math. Z.* **82** (1963).
36. Shields, A. L., *A Survey of Some Results on Invariant Subspaces in Operator Theory* (Linear Spaces and Approximation), International Series of Numerical Mathematics 40 (1977), pp. 641-657.
37. Shields, A. L., A note on invariant subspaces, *Mich. Math. J.* **17**, 231–233 (1970).
38. Stampfli, J. G., Recent developments on the invariant subspace problem (Operator Theory and Functional Analysis), *Res. Notes Math.* (Edited by I. Erdelyi) **38**, 1–7 (1979).
39. Stone, M. H., *Linear Transformations in Hilbert Space*, American Mathematical Society, Providence, Rhode Island (1964).
40. Sucheston, L., Banach limits, *Am. Math. Mon.* **74**(3), 308–311 (1967).

41. Taylor, A. E., *Introduction to Functional Analysis,* John Wiley and Sons, New York (1958).
42. Vulikh, B. Z., *Introduction to Functional Analysis for Scientists and Technologists,* Pergamon, Oxford (1963).
43. Whitley, R., An elementary proof of the Eberlein–Šmulian theorem, *Math. Ann.* **172**, 116–118 (1967).
44. Wilansky, A., *Functional Analysis,* Blaisdell, New York (1964).

Definition, Theorem, Proposition, Lemma, and Corollary Index

Propositions

Lemmas

Corollaries

Symbol and Notation Index

Miscellaneous notation

Subject Index

Errata for Part A

Page 190

In line 5, change "Suppose $a > b$" to "Suppose $a^{p-1} > b$." In line 10, change "The case $a \leq b$" to "The case $a^{p-1} \leq b$ (equivalently, $b^{q-1} \geq a$)."

Page 198, Problem 3.5.20

Change $\| f \|_1$ to $\| g \|_1$.

Page 230, Problem 4.2.19(iii)

Replace text beginning with "Show that ..." by

"Show that

$$\begin{aligned} g(t) &= 1, \qquad \text{if } \frac{1}{2^n+1} \leq t \leq \frac{1}{2^n}, \qquad 0 \leq n < \infty, \\ &= 0, \qquad \text{if } \frac{1}{2^{n+1}-1} \leq t \leq \frac{1}{2^n+2}, \qquad n > 1, \end{aligned}$$

and linear on the rest of (0, 1] is an example of such a function."

Page 258, Problem 4.3.12

In this problem, the symbols μ and ν are mixed up. The problem should read:

"... where $\mu_1 \ll \nu$, ..., and $\mu_3 \perp \nu$ and ... [Hint: By Theorem 4.7, $\mu = \mu_0 + \mu_1$, $\mu_1 \ll \nu$, and $\mu_0 \perp \nu$ ...]"